Exploring th(

By To

Exploring the Mindful Way

by
Tom Butler
With enabling support from Lisa Butler
We are Association TransCommunication Directors

First Edition, 2018

Websites
EthericStudies.org
ATransC.org

Paperback ISBN 978-0-9727493-7-4
ePub ISBN 978-0-9727493-6-7

Cover design by Tom and Lisa Butler
Art by Anne Wipf

Published by AA-EVP Publishing, atransc.org

Printed by Lightning Source, ingramcontent.com/publishers/print
ePub Distribution Catalogued by IngramSpark, ingramspark.com

Use License

© 2018 Tom and Alisa Butler

This book is intended to be used as an educational source. I reserve my rights under the copyright laws of the United States. Unless the Creative Commons rules noted below are followed, no part of this book may be used or reproduced in any manner whatsoever, including Internet usage, without my written permission, except in the case of brief quotations embodied in critical articles and reviews.

However, I offer limited use of this work under the Creative Commons Attribution-Noncommercial-Share Alike 3.0 Unported License. © 2018 by Tom and Lisa Butler.
See creativecommons.org/licenses/by-nc-sa/3.0/

With the Exception of material specifically marked with ©, you may:

Share: Copy and redistribute the material in any medium or format.

Adapt: Remix, transform and build upon the material with proper attribution.

Under no conditions, may the entire book, sections or chapters be reproduced without written permission from the author.

About the Cover

Title: I Ching 53 - Chien (Development - Gradual Progress
annewipf.deviantart.com/art/I-Ching-53-Chien-Development-Gradual-Progress-643401709

From the artist: "This picture belongs to a series of 64 landscapes inspired by

Artist: Anne Wipf

the 64 hexagrams of the I Ching (Yi King), the Chinese oracle book. I hope you'll like my interpretation.

"Words which inspired me: Tree on a mountain, bride, wild geese, shore, cliff, dry plateau, cloudy heights."

I thank Anne for allowing the use of her work for this book cover.

Back Cover

Immortal Self-Centric Mindfulness

The most important understanding seekers of spiritual maturity must come to is the difference between lucidity and hyperlucidity. Lucidity is the degree to which we are able to clearly sense information from our mostly unconscious mind. Hyperlucidity is a term used in the Implicit Cosmology for a complex of behaviors motivated by the belief we are lucid when we are actually only sensing what we have been taught to expect.

The second most important understanding is that lucidity is the seeker's objective, but that it is achieved in small steps. The only real conscious influence we have on our mostly unconscious mind is the expression of intention. This means that we must learn to consciously examine what we think is true. Mind changes only slowly, and so, the seeker's objective is to habitually express the intention to align perception with the actual nature of reality.

In the first book, *Your Immortal Self,* the process of consciously seeking greater lucidity is referred to as the Mindful Way. Many people practice mindfulness simply to improve personal wellbeing. A few step onto the Mindful Way to seek greater understanding of their immortal nature and the nature of the reality they inhabit. Even fewer remain as wayshowers for those who seek greater lucidity.

The fact of our immortality is explained in *Your Immortal Self*. This book, *Exploring the Mindful Way,* includes twenty-one essays explaining some of the more important concepts encountered on the Mindful Way. While you will benefit from first reading *Your Immortal Self,* there are sufficient explanations in this book to make it a stand-alone text.

Will you be a wayshower?

Acknowledgments

Admittedly, this book is all about reality according to Tom Butler. Should you disagree with what I have said, your disagreement is with me. However, much of what I think is true is anchored on what I have learned in the study of transcommunication with Lisa, so if you like the book, thank her.

Lisa has also been the motive force for our work in this field. This is important, because I think there are contributions in this book and *Your Immortal Self* (1) that will eventually further general understanding of our nature and the nature of reality. Had she not persisted in the study of EVP, I might have focused my interests on easier subjects such as inventing perpetual motion.

As I said in *Your Immortal Self,* the paranormalist community provided the background influence which gave me reason to compose this work in the first place. Every time someone made a faith-based comment as if it were fact, I hurried back to my computer to worked on this book.

Content

Exploring the Mindful Way i
Exploring the Mindful Way ii
Use License iii
About the Cover iv
Back Cover v
Acknowledgments vi
Content vii
Introduction to This Book 1
    Test Your Teacher 3
    Important Influences 4
    Mechanics 6
    Possible Errors 7
    References and Alternative Sources 8
Essay 1 Conditional Free Will 9
    About This Essay 9
    Introduction 9
    Anatomy of a Life Field 12
    Deterministic Influences 22
    Finding Free Will 32
    A Talisman 35
Essay 2 The Mindful Way 37
    About This Essay 37
    Purpose 40
    Mindfulness 40
    Teachers 40
    What We Do Now Matters 42
    Worldview 43
    Personal Reality, Local Reality and the Greater Reality 44
    Suspended Judgment 45
    Self-Determination 46
    The Mindful Way 46
Essay 3 Prime Imperative 49
    About This Essay 49
    Abstract 51

Introduction 51
Point of View for This Essay 51
Our Etheric Nature 54
Natural Law 57
Organizing Principles 58
Understanding 59
Purpose as Prime Imperative 60

Essay 4 Immortal Self-Centric Perspective 63
About This Essay 63
Introduction 64
Perspective 65
Our Body is a Complete Organism 65
Our Body as Avatar 66
Survival of Body Mind 68
Personal Style and Astrology 69
Balance 71
Irrational Behavior 71
Degrading Avatar Relationship 72
May I Introduce Myself? 74

Essay 5 Ethics as a Personal Code for Mindfulness 75
About This Essay 75
Introduction 77
Morality Versus Ethics 78
First Ethical Consideration 80
Ethical Treatment of Human Research Subject 81
A Useful Code of Ethics 81
Ethical Conduct is a Lifelong Learning Experience 82

Essay 6 Paranormalist Community 83
About This Essay 83
The Paranormalist Community 85
A Divided Community 86
A Community Divided Cannot Stand 88
Cultivating a Common Culture 90

Essay 7 Clarity of Communication 95
About This Essay 95
Introduction 96
People Have a Style of Learning. 97

Selective Understanding 100
Selective Attention 101
A mismatch of agenda impairs communication. 102
Worldview changes in small increments 105
Discussion 107

Essay 8 How We Think 109
About This Essay 109
Introduction 111
Terms 112
How We Think 117
First Sight Theory 119
Implications of Unconscious Preprocessing of Thought 121
Lucidity 123
Hyperlucidity 125
Discussion 126

Essay 9 Consensus Building in the Paranormalist Community 129
About This Essay 129
Introduction 131
A Fractured Community 134
Importance of Sharing Ideas 141
Spiritual Anatomy 143
Understanding is Relative 147
Cooperative Community 151
Justified 157
Belief without understanding is Faith 158
Summary 160

Essay 10 Skeptic 163
About This Essay 163
Abstract 164
Introduction 165
Skepticism and Scientism 166
Tells of a Skeptic 166
Comparing the View of Science with the View of Skeptics 167
Organized Harm to Society 168
A Case Study: Government Acting on Skeptical Views 169
A Case Study: Skeptical Control of the Media 170
Healthy Skepticism 171

Skeptical About Skeptics.. 173
Essay 11 Pseudoscience...175
    About This Essay.. 175
    Introduction... 177
    Fact of Paranormal Phenomena is the Issue 178
    The Scientific Method... 179
    Inappropriate Science... 180
    Pseudoscience 181
    Alternative Terms for Pseudoscience.. 190
    Scientism 191
    Relative Scientism.. 191
    Community Response... 191
Essay 12 Concerns with Wikipedia...193
    About This Essay.. 193
    General 194
    Treatment of Subjects 195
    Wikipedia Editing Rules 196
    Who Can Edit Articles 196
    The Skeptical Community.. 197
    Personal Attacks 197
    Why This Is Important 198
    What can be Done?.. 199
    Navigation Guide for Wikipedia 202
    Conclusion 204
Essay 13 Arrogance of Scientific Authority..205
    About This Essay.. 205
    Background.. 209
    Science in the Paranormalist Community 211
    Human Research Subject... 213
    Libel and Slander 214
    Opinion About Arrogance of Scientific Authority............................... 215
    From My Experience.. 217
    Update 220
Essay 14 Open Letter to Paranormalists..223
    About This Essay.. 223
    Introduction... 224
    About This Letter 225

The Paranormalist Community 225
Theories of Reality 226
Experiencing Phenomena 229
What You Need to Know About Science 236
Qualified to Practice Science 238
Pseudoscience 242
Science and the Paranormalist Community 244
Blind People and an Elephant 251
Concluding Comments 251
Comments from the Media 251

Essay 15 Let's Talk About God 253
About This Essay 253
Introduction 255
Building a Cosmology 256
Implicit Cosmology 259
As Above, So Below 260
Unconscious Perception 262
Personal Reality 262
Self-Organizing Reality 263
God and Gods 265

Essay 16 What is it Like on the Other Side 267
About This Essay 267
Introduction 268
Old Models of Reality 270
Transition 272
Next Venue for Learning 281
Moving On 282

Essay 17 The Hermes Concepts 285
About This Essay 285
Introduction 286
Who Was Hermes? 287
Paraphrasing the Emerald Tablet 294
The Foundation Concepts Associated with Hermes 295
The Three Aspects of a Teacher 297
Important Metaphysical Concepts Attributed to Hermes 298
The Seven Organizing Principles of the Kybalion 299
My Introduction to the Hermetic Concepts 303

Resetting the Old Concepts 306
Essay 18 The Razor's Edge **309**
    About This Essay 309
    Introduction 310
    Origin of the Upanishads 311
    Katha Upanishad 311
    Universal Message 315
Essay 19 Progression, Teaching and Community **319**
    About This Essay 319
    Spirituality 319
    Cooperation 320
    Friends as Teachers 321
    Selfless People 321
    A Vision About Collectives 322
    First a Student, Then a Teacher 322
    The Nature of Understanding 323
    The Process of Gaining Understanding 325
    We Exist to Learn 326
    The Prime Imperative as a Spiritual Obligation 327
    The Transition Experience 327
    Self-Realization 328
Essay 20 Law of Silence **331**
    About This Essay 331
    Introduction 332
    Suspended Judgment 333
    It Takes a Community 336
    Practicing the Law of Silence 340
    Community 341
Essay 21 Informed Regret **343**
    About This Essay 343
    Introduction 344
    As it Happened for Me 345
    My Learned Point of View 350
References and Alternative Sources **357**
Index **369**

Introduction to This Book

Your Immortal Self: Exploring The Mindful Way (1) is my second version of a guidebook to the other side. The first was *Handbook of Metaphysics*. (2) Both were written in response to an urge to gather and present information which might help all of us on our journey to spiritual maturity.

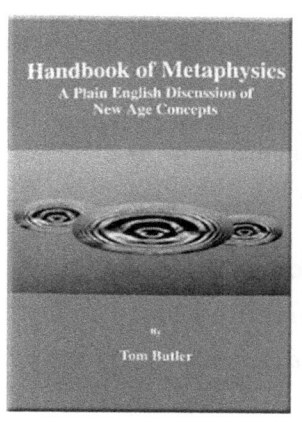

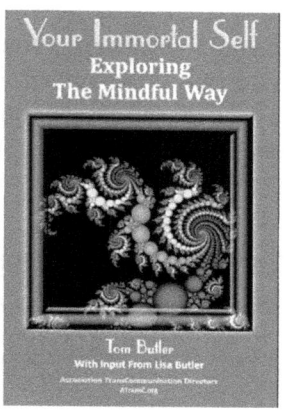

Published 1994   Published 2016

Handbook of Metaphysics was my first book. It was intended to provide the kind of foundation information I felt people needed to understand their paranormal experiences. It was written before the Internet, and as it turned out, I did not have the necessary experience to realize the publisher was not giving me the kind of guidance needed by new authors. Consequently, there are structural errors I would not make today.

The *Handbook* is useful in a basic way, but it is out of date as compared to *Your Immortal Self*. Even though a few copies of the book are still available from the publisher (no profit to me), I do not recommend that you take time to read it.

As a conservative paranormalist, I am probably one of the most pragmatic people you will meet. This translates into an almost obsessive desire to focus on the more objective aspects of these phenomena. This should be important to you because I am asking you to accept the possibility of survival as a fact, and not as a belief. You need to trust me well enough to at least tentatively accept the validity of the concepts I present. Doing so will help open your mind to the more abstract concepts.

If survival is a fact, the implications of survival are actionable. By that, I mean that understanding our spiritual nature should suggest actions we can take to better align our thinking to be more in agreement with our immortal self.

Your Immortal Self begins with an explanation why the Survival Hypothesis needs to be considered when evaluating evidence related to all things paranormal. That argument is in Discourse 1: *Trans-Survival Hypothesis*. Discourse 2: *Introduction to the Implicit Cosmology* provides a detailed disclosure of the implications of survival. It is followed by many discourses that explain important elements of the cosmology, such as Discourse 2: *Organizing Principles of Formation,* Discourse 6: *Etheric Fields* and Discourse 7: *Life Fields*.

Essay 2: *The Mindful Way* in this book is based on Discourse 11.

Essay 4: *Immortal Self-Centric Perspective* is based on Discourse 8.

Essay 19: *Progression, Teaching and Community* is based on Discourse 12.

Your Immortal Self, Section II: *Community* is concerned with the paranormalist community in which we must study phenomena related to our psychic ability and survival.

For the same reason, this book, *Exploring the Mindful Way*, includes essays written in an immortal self-centric context about paranormalist community issues. Community is the habitat which we must experience to gain understanding. As such, it is important that we have at least a passing acquaintance with its nature.

Your Immortal Self, Section III: *Transcommunication* is concerned with some of the phenomena related to communication across the veil. I highly recommend Section III: Transcommunication if you wish to learn how to work with survival-related phenomena. To do so, you will need to read *Your Immortal Self*, as *Exploring the Mindful Way* has little specifically about transcommunication. It is useful, though, to understand the nature of your immortal self to underpin your study of transcommunication.

This entire book is written from the perspective of the Trans-Survival Hypothesis with close attention to the immortal self-centric perspective. Each essay is complete in itself. I have attempted to follow the way of a teacher by arranging them as concepts, experiences and effect.

Test Your Teacher

An important concept in the ancient wisdom schools is the idea that seekers most ask to be taught. However, having found a likely teacher, seekers are expected to test their teacher before agreeing to be a student. In the ancient wisdom schools, seekers often study under the same teacher for many years. The relationship should not be taken lightly.

There is also the question of the foundation concepts taught by the school. What is the point of view taught in the school? For instance, in contemporary terms, does the teacher follow the Normalist school of thought in thinking there is no such thing as paranormal phenomena? In those terms, I am a Dualist and accept the evidence that we are immortal self. See: "A Divided Community" (Page 86) in Essay 6: *Paranormalist Community* (Page 83).

When reading an article about anything important to you, it is always good to examine the author's credentials. This is especially true for paranormalist subjects because there are so many people with a view of the nature of reality that is simply not supported by the evidence.

Sadly, many people in our community are happy to assert opinions under cloak of their unrelated academic authority, apparently without caring about the ethical implications of not telling their readers what they actually think is true. For instance, the parapsychological journals routinely include research reports written from the perspective of anomalistic psychology, but rather than disclosing they are attempts to debunk phenomena, the articles appear to be honest explorations of how people experience the phenomena. There is even a recent book about EVP that includes material strongly reminiscent of the anti-survival writing of anomalistic psychology.

> Anomalistic psychology is the study of unusual human experiences with the intention to prove they have a non-paranormal explanation. (58)

Beyond advanced training in their specific subject, doctorates are supposed to be trained in critical thinking, technical writing, research design and advanced use of libraries. As a four-year engineering graduate, I was trained in the same, but did not have some of the advanced technical courses, nor did I undergo the rigorous coaching for theses writing. Even though my engineering degree is probably more technically rigorous than required for an advanced degree in psychology or philosophy, you should expect a Ph.D.

in one of those fields to be better trained in critical thinking, a most important skill for paranormalists.

If you do not have a college degree, it is reasonable for me to expect that you are not as well trained in critical thinking, research design or practical report writing. Certainly, I know my skills fall short of many Ph.Ds. I have encountered.

This is not to say a person without a college degree is inferior, it is just a fact of life that all of us must consider. The same must be said about the subject area in which a person is academically trained. My degree in electronics included considerable training in physics and some in chemistry. I am close to a master's degree in math, but there is no way I can or should claim academic authority held by a Ph.D. in physics or chemistry.

By the same token, I have fifty or so years studying various aspects of metaphysics. Of course, I read books ... hundreds, it seems! In 1992, I turned that reading for personal improvement into research for the *Handbook of Metaphysics*. (2)

Interspersed with my reading, hands-on practice and writing has been numerous total emersion courses such as those offered by The Monroe Institute, (3) The Silva Method (4) and Delphi University (5). It has been my belief that the best way to understand a system of thought is to immerse myself in that system for a time. That has resulted in multi-year experiences in systems such as BOTA, (6) Eckankar (7) and even ordination in Spiritualism. (8)

Lisa and I have been the Directors of the ATransC since 2000. As directors, we have been directly involved in research, study, production of phenomena and sitting with practitioners, including hosting sessions with some of the most active physical mediums of our time.

The entirety of my experience has been from the critical eye of an engineer. Lisa and I consider ourselves amongst the most pragmatic people you will meet in this community. While this has turned out to be a handicap, considering that popularity is part of leadership, it should give you assurance that we are going to base our comments on the most objective view available.

Important Influences

Many of the ideas explored in *Your Immortal Self* came from some of the situations I often encounter in day-to-day living. For instance, the skeptic's complaint that we have no theory explaining survival phenomena particularly informed my efforts with the Implicit Cosmology. It was from my days as a Wikipedia editor that I learned the depth of the skeptic's faith in orthodox science.

Of course, I have been influenced by parapsychological research. As you would see in Essay 13: *Arrogance of Scientific Authority* (Page 205) and Essay 14: *Open Letter to Paranormalists* (Page 223) in the book, my study of that research and interaction with parapsychologists has also informed my dissatisfaction with the way transcommunication practitioners have been treated.

Some theories have had an important influence on my point of view. You will see that I am particularly impressed with Rupert Sheldrake's work on morphogenetic fields and James Carpenter's work on perception. The study of psi phenomena by parapsychologists has established an important foundation for discussing survival. But don't be too impressed. Psi-to-survival phenomena is a little like the way gravity is to rocketry. Rocketry is all about thrust and payload, of which gravity is only one factor, albeit a major one. All mental functioning involves psi functioning, but the study we are interested in is the relationship between conscious self and mostly unconscious mind. As James Carpenter proposes in First Sight Theory, the real study is in how the mind process information.

Existing metaphysics and ancient wisdoms have shaped New Age thought. Of course, religions have an influence, but they were first influenced by the same ancient wisdoms and philosophy, which was eventually corrupted in response to the need for social engineering.

The principles governing our nature believed to be taught by Hermes, probably 6,000 years ago in Egypt, are essentially the same as those we seek to understand today. The Tarot is a relatively modern interpretation of those ancient wisdoms. And, in fact, combined with contemporary understanding, the Major Arcana of the Tarot represent an important tool for personal progression.

More contemporary developments are leading to the rediscovery of the ancient wisdoms; however, it is the more contemporary view I present in my writing. With that said, I find it useful to point out how our current understanding tends to verify some of the concepts of ancient wisdom.

Transcommunication phenomena have been the most important influence on my writing. They are also the most objective. Electronic Voice Phenomena (EVP) provides a means of testing trans-etheric influences. Survival is a trans-etheric experience. Thinking is a trans-etheric process. All of these are psi functioning, but it is the transformation of psi influence to physical objectivity that provides important hints about our immortal nature.

The better part of my life has been in pursuit of understanding about my own nature. This has given me insight into what needs to be done to increase lucidity, to have a sense of the variety of phenomena, to effectively product phenomena and to recognize the confusion of false positives.

It is difficult to assess the degree to which I have gained lucidity. For me, clear sensing of the etheric is more an emergent knowing then voices or pictures. After so many years of gradual development, it is probably no longer possible for me to know what it is like not to have this clarity. As a person trained to solve problems in my work as an engineer, I have always begun with a mental exercise. This means it is also difficult to distinguish the knowing of simple problem solving and the knowing that comes from my friends on the other side. Is there really a difference?

Much of my writing begins with that sense of knowing. While I am not so vain as to think I am a channel for discarnate geniuses, I am also not so arrogant as to claim I did this on my own. All of us work as a community. *Your Immortal Self* and this book are designed to help you make that a conscious collaboration.

Mechanics

Each essay in this book was originally written as my effort to understand an aspect of the Mindful Way. Writing an essay about a subject I am trying to understand helps me focus. I speak of the need for participation in a cooperative community for personal progression. In a way, writing an essay serves as my cooperative community. (Perhaps with a little help from my friends on the other side.)

The result of this approach to essay writing is that each easy is a standalone document. My theme is always *"How can this subject be understood in terms of the Trans-Survival Hypothesis?"* As such, most essays share more or less the same foundation concepts. That means, when they are compiled into a book, the supporting concepts are often repeated.

I am saying all of this to explain that you will find parts of each essay that seem familiar. When you do, resist skimming past those parts. Each explanation of the concepts is tailored to show how they apply to the topic. In this way, I hope to help you understand the concepts you need for progression by showing how they apply to real-life subjects.

Some of the essays, such as Essay 2: *The Mindful Way* (Page 37), are taken directly from *Your Immortal Self*, but typically with a little updating. You will find many of these essays on EthericStudies.org but those may be out of date. If you notice the year the essay was originally written, you can guess that my understanding has evolved. The version for this book has been evolved and some integration with the other essays has been done to clarify communication.

Sidebars

> I use indented paragraphs like this with a left-side bar to indicate an interjection or comment that breaks the flow of the text, but which is intended to add clarity. These replace the more usual footnotes.

Possible Errors

You will find mechanical, grammatical, and possibly, logical discontinuities in this book. We spent nearly five $5,000 having *Your Immortal Self* proofread and copyedited. Even so, since it was first published, I have found enough mistakes to feel the need to reissue the book. Yet, it is likely more errors remain.

Proceeds from sales of *Your Immortal Self* are being used to pay for the initial cost of publishing, including proofing. At the rate the book is selling, it will probably never pay for those costs. Yes, I am writing to a very small audience, but the cost simply does not warrant the benefit of a slightly improved product, so I am the proofreader. I was once told by an English teacher that, based on my entrance exams, I should not be in college. Perhaps he had a point.

The logical review of my work is even more difficult. Academically trained metaphysicians will probably not care for my self-taught approach. From my experience with parapsychologists, few have the necessary background to logically review the Implicit Cosmology. Possibly, this is one of those *"Write it and help will come"* situations. Because of this lack of peer vetting, it is wise for you to focus more on the usefulness of the message than on the metaphysics.

As I sometimes do on the website, I am asking for input from you, should you see problems in structure that need to be fixed. The website contact tool is available for your suggestions, comments and questions. Depending on interest, the Idea Exchange will be available, as well. You will see in the essays that I consider collaboration essential for our progressions.

References and Alternative Sources

The more you know about the subject, the better prepared you will be to apply them to your daily living. It is up to you to do the work. That is why so many alternative sources are listed in the References. I do not necessarily endorse the supporting material. The references are provided to give you access to further background, or in some cases, more detailed explanations. Consider the References a study guide.

The book is available as an eBook and a paperback. I have not included as many internal links in the eBook version as I did in *Your Immortal Self*, but still, the eBook version is designed to make study a little easier. You may also want to keep the *Glossary of Terms* (9) open on your computer. It is at ethericstudies.org/glossary-of-terms/ under the **Glossary** Tab. Alternatively, a more current version is at the back of *Your Immortal Self*.

Because of the standalone nature of each essay, you need not have read *Your Immortal Self*. When I would normally refer to *Your Immortal Self*, I have usually been able to refer to another essay for the more detailed explanation I intend. With that said, this is a companion book to *Your Immortal Self*. I do recommend that you read both.

> It is my practice to include sidebars containing information that should be helpful but break the flow of the argument. Sidebars do not work well in small-format books or eBooks, so such comments are set aside with a different color and offset like this.

Essay 1
Conditional Free Will
2017

About This Essay
Whether or not we have free will has always been a concern in my work with metaphysics. A cornerstone of the cosmologies I work with is the idea that we are creators and that we create by imposing our intended order. But, what if what we express is not what we intend?

The question became increasingly present in my daily activities as I began the initial layout of this book. The way I respond to such perturbations in my life is to write an essay.

It is during the process of composing my thoughts to write, that I am able to access the scraps of memory related to the subject at hand. At my age, there are a lot of scraps and the ruminating can take days.

The way I put it, my study of the Hermetic Wisdoms has taught me to learn everything I can about everything. This is not so much *a study* as it is *mental taking note for future reference*.

Evidence for the Trans-Survival Hypothesis (10) is evidence of our etheric anatomy and that is evidence of the concept of intended order. As you will see in this essay, our ability to express intended order is impaired by environmental influences which include physical, biological, social and spiritual principles.

Essay 2: *The Mindful Way* (Page 37) was going to be the first in this book, but by the time I finished *Conditional Free Will*, it seemed best to make that essay first. Conditional Free Will essay explains the best reason ever for you to step onto the Mindful Way. Also, the detailed explanation of your spiritual anatomy will help you understand the rest of the book.

Introduction
The assumption that we have free will is part of the foundation on which we build our sense of self. Free will is usually characterized as the ability to decide for ourselves what to do next or how to react to information. Other ways of saying free will include self-determination and freedom of choice.

While most of us assume we have free will, it turns out that many philosophers and scientists think we might not. Here is a brief overview of contending theories (a more detailed discussion of these influences is provided under *Deterministic Influences* (Page 22):

Determinism

In philosophy, the idea that the operation of reality is deterministic means our choices are possibly predetermined. Determinism is something of an umbrella term for a number of deterministic influences. (11) In terms of human behavior, it means that our present is determined by our past. Deterministic influences include our genes, prior experiences, social dynamics and cultural influences. In physical processes, it means that a process is bound by natural principles.

Nature-Nurture

Is a person's temperament predominantly the product of social, environmental influences such as growing up in an academically inclined family versus one more focused on sports (Nurture)? Or, is a person's temperament something that is set at birth; perhaps carried in our genes (Nature)? Also see *Spiritual Instincts* (Page 26).

Nature's Habit

The blueprint for the way biological organisms are formed, a process known as morphogenesis, has considerable momentum. That is, organisms change over time but do not abruptly change. An instance of a species formed today is virtually the same as that born yesterday. Rupert Sheldrake referred to this blueprint as *Nature's Habit*.

Human instincts are a behavior version of Nature's Habit which tends to determine how an organism will behave. Just as with morphogenetic momentum, behavior also has momentum so that our human body has essentially the same guiding instincts as all humans. These instincts tend to dominate our behavior at birth and are only moderated by reason as we gain in rational maturity.

God's Will

A widely held view is the religious one in which our fate is thought to be predestined as God's will. This would seem to argue that whatever happens to us is not our fault but is our fate. This appears to be the ultimate surrender of self-determination, in which our only responsibility is to be a righteous believer.

Predetermined by Agreement

This seems like a New Age theory, but it is actually very similar to God's will, in that our actions are possibly predetermined by trans-etheric influences beyond our control once we enter into a lifetime. The idea is that, before we were born, we entered into an agreement with one or more other personalities to facilitate specific life experiences. This might be an agreement to be a mate, but there is no reason it cannot be an agreement to kill a person. In this view, people killed in a war would have agreed prior to being born to experience a violent death.

This follows the argument that we must have a specific kind of experience to gain a specific kind of understanding. If this concept is true, we may be both the benefactor of the agreement and the supporting actor. This would suggest that we have free will prior to agreeing on an action but are deterministically guided for the life of the contract.

Some argue that we might decide not to participate in the agreement when it is time to fulfill the contract. If so, that would be an expression of free will, but a violation of a prior agreement. There is also the likelihood that we would not have the presence of mind to consciously decide our behavior.

Spiritual Instincts

If we accept the evidence of our immortality, then God's will and pre-lifetime agreements are not out of the question. As I will explain below, we may have entered into this lifetime, this venue for learning, to gain understanding about some aspect of reality. Our free will is how we decide to respond to instincts we inherit from our local source. I refer to them as spiritual instincts to distinguish them from our human's instincts.

It seems pretty clear that our free will is limited to some extent by deterministic influences. If I jump off of a cliff, the natural principle of gravity assures I cannot change my mind. Being born a man predicts different actions in life than being born a woman. Human instincts dominate our behavior, especially if we do not learn to manage them. There is little doubt that some of us are very different than others in our family, but the influence of family and culture determines who we are if we do not consciously act to make it otherwise.

Anatomy of a Life Field

This essay is written from the point of view that we are immortal personalities temporarily entangled with a human for this lifetime—a person. That is the essence of the Trans-Survival Hypothesis (10) which I explain in detail in *Your Immortal Self*. (1) The model describing our life field is included in the Implicit Cosmology, (12) which is based on implications of survival, current understanding and theory derived from our work with transcommunication. For the explanations offered in this essay to make sense, it is important that you are familiar with that model. As such, it is briefly explained here.

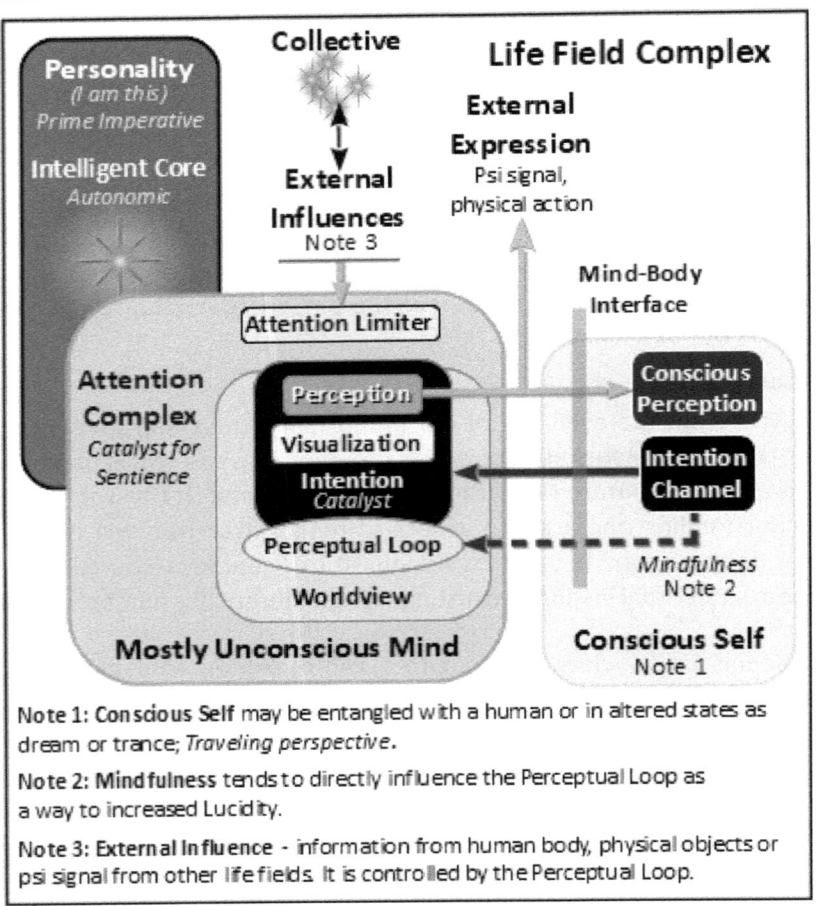

Life Field Complex

- Personality (I am this) — Prime Imperative
- Intelligent Core — Autonomic
- Collective
- External Influences (Note 3)
- Attention Limiter
- Attention Complex — Catalyst for Sentience
- Perception
- Visualization
- Intention — Catalyst
- Perceptual Loop
- Worldview
- Mostly Unconscious Mind
- External Expression — Psi signal, physical action
- Mind-Body Interface
- Conscious Perception
- Intention Channel
- Mindfulness (Note 2)
- Conscious Self (Note 1)

Note 1: Conscious Self may be entangled with a human or in altered states as dream or trance; *Traveling perspective*.

Note 2: Mindfulness tends to directly influence the Perceptual Loop as a way to increased Lucidity.

Note 3: External Influence - information from human body, physical objects or psi signal from other life fields. It is controlled by the Perceptual Loop.

> Please note the meaning of *implication*. As I use it, **implication** means **the consequences of what has been stated must be true for the original statement to be true**. If we are immortal self, the

implications that must be considered include the idea that we are not our human body, we are a person for a reason and we are likely influenced in some way by our human. If we do not accept those consequences, we cannot rationally accept the idea that we are immortal.

A *field* is defined here as *a set of elements with related characteristics which are bound into a system by a common influence.* In this model, life fields are the basic building blocks of reality. (13) A hypothetical Source life field is the top field in a hierarchy of nested fields. (14) Thus, the Source life field is the body of reality.

This follows the same model proposed by Rupert Sheldrake in the Hypothesis of Formative Causation (15) In that, the morphic field organizing a human organism is proposed as the top field for all of the sub-fields representing the various components of the body. In effect, our life field is a sub-life field of Source and exist in Source's field of influence as an aspect of Source.

A Source life field and our relationship with it is not important to a discussion about free will, except to establish a boundary for this model. The important elements of this boundary include:

- Our real home is described here as the etheric, which appears to be a form of conceptual space.
- Reality consists of life field and the expressions of life fields.
- All is thought. Expression follows the **Creative Process: Attention on an imagined outcome to produce an intended order.**
- Fields of influence are the conceptual equivalent of physical objects. As such, the Source life field is reality.
- Everything in the Source life field (everything in reality) is within the Source scope of influence.
- Everywhere is here. Parapsychologists refer to this as nonlocality; however, it is more correct to say that everywhere is local.

To understand the limits of our free will, it is necessary to understand how our mind processes information. That means understanding the anatomy of our life field. But before I explain, it is necessary to clarify that I am not a psychologist, nor am I trained in any of the mind sciences. The model I use has been developed from black box analysis based on known and hypothetical input and output signals of an imaginary container representing our life field. Engineers use the same approach for designing electronic circuits. The result is a set of functional areas inside the container which will respond to known inputs to produce known outputs. Done right, the resulting functional areas of the model can be used to predict previously unnoticed signals.

People who are well informed about current thought concerning how our mind works will probably not recognize or accept this model. I suppose one of the reasons for this is that few people actually trained in the subject are willing to include survival or transcommunication assumptions in its design. Just be clear that this is a useful tool for understanding the concepts I wish to discuss and not one that is likely to show up in academic literature.

The Life Field Complex Diagram (above) represents the model which has resulted from black box analysis of our etheric anatomy. It is based on the assumption that the Trans-Survival Hypothesis is mostly correct, (10) that the Hypothesis of Formative Causation is essentially correct (15) and First Sight Theory is a reasonable model for our thought processes. (16) Supporting information includes the way we think transcommunication works and current parapsychological research results concerning psi functioning.

As shown in the diagram, the major functional areas of our life field are:

Intelligent Core

Personality, which is our immortal aspect and source of our purpose. This is our *I am this*. It is this aspect with which we seek to become more consciously integrated.

Attention Complex

Our mostly unconscious mind which represents the memory, perception and expression processes.

Conscious Self

Our conscious perspective as *I think I am this*. Conscious self is a *traveling perspective* which, during a lifetime, seems to rest in the head of our human avatar. However, when sleeping or in an altered state, our perspective is disassociated from our body and free to roam etheric space. This freedom is limited by the Principle of Perceptual Agreement, which itself, is governed by Worldview. (17)

In this model, processes such as expression and perception are referenced in a very fundamental form. It is up to you to extrapolate how they apply to any one circumstance. The essays I write are intended to explain ideas that seem to be important for following the Mindful Way, such as Essay 17: *The Hermes Concepts* (Page 285), Essay 3: *Prime Imperative* (Page 49), Essay 9: *Consensus Building in the Paranormalist Community* (Page 129) and Essay 5: *Ethics as a Personal Code for Mindfulness* (Page 75).

Environmental Signals

Our mind is an etheric thing. Our body is physical. Well, at least we assign physicality to all of the information that comes to us via its senses. Given this physical-etheric difference, there must be some kind of conversion of information from physical to etheric. It seems reasonable to speculate that our human brain is where that conversion occurs. The effect is that all of our body senses must be converted to etheric signals. In turn, all of our physical expressions such as speech or motion are converted by the brain from etheric to physical commands.

Environmental information consists of signals from your body and psi signals from the etheric. Here, *signal* is used in the sense of *expressed information*, more as a gestalt-like thoughtform than a stream of information. James Carpenter posited that everything in reality expresses a psi signal and that all of our expressions are accompanied by a psi or psychokinetic signal impressed into the etheric. (17) (18) His *Personalness Corollary* [#1] and *Weighting and Signing Corollary* [#6] describe how we tend to pay attention to or ignore information, depending on its importance and our interest.

> As you read this explanation, keep in mind that what you think of as who you are is not your body. You are not in your head. You are a nonphysical personality entangled with your body in an avatar relationship for this lifetime.
>
> That means you are aware of the book or computer screen you are reading as information sensed by your body's eyes and converted into a psi signal which you, as an etheric personality, are able to access.
>
> Put a different way, your body is continuously sending sensed information as bioelectrical impulses passed from your five senses to your brain. We think it is in the brain that the information those signals carry is converted into nonphysical psi signals which can be used by your nonphysical mind to develop a picture of the physical world that you visualize as being inhabited by your body.

As I will explain below, how those signals are presented to your conscious awareness is decided by your worldview functional area which represents what you have been taught. Thus, you create the world you think your body is in based on what you have been taught.

Catalyst for Sentience

Our mind is in etheric space where concepts are things in a similar way that objects are things in the physical. To model mind, it is necessary to identify influences, functions and states as concepts. The most important state is *attention* because, without it, the mind functions would be dormant. Think of the attention state as a precondition for sentience. Functionally, it does little more than act as a sort of integrating catalytic influence which we would characterize as the life force. I include it here for completeness to bound the perception-expression functions.

Intention is in the model as two forms. It is an influence which we consciously express toward mind. It is also a catalyst for the perception and expression processes. It represents the intention to perceive or express.

The catalyst concept may seem to add unnecessary complexity, but not including it would leave a huge hole in the model. Once you have digested the model, you will see that these elements might point toward important realizations.

Perception

Notice that the Perception Functional Area in the Attention Complex is closely associated with the Visualization Functional Area and that both are in the Intention Functional Area. They are part of the Perceptual Loop discussed below.

As well as I can tell, all of our thoughts, actions and perceptions occur in response to an environmental signal; either current as in seeing or smelling something, or historical as in considering an old insult. When we are triggered to react, the intention to do so is little more than an impulse. That impulse to react initiates a process that begins with the Visualization Function. The Visualization Function creates a characterization of the environmental signal based on Worldview and submits the result to the Perceptual Loop.

Visualization is probably a gestalt-like thoughtform form of characterization, rather than a single picture. The environmental signal might be from the body's five senses or a psi signal from either side of the veil. Physical objects exist as they are perceived by a person (conscious self-avatar). The visualization precedes the objective experience.

The characterization formed by the Visualization Functional Area is compared with the contents of Worldview to determine if it is familiar. This might be a *many-tries* process as the characterization is adjusted to closer agree with Worldview. If sufficient agreement is found, the characterization is submitted to the Perception Functional Area. An action in response to the environmental signal is generated as a psi signal (perhaps psychokinetic) and possibly as a signal to the brain to change the body in some way. A signal is also sent to conscious self to produce conscious awareness of the signal. Again, this is awareness of what has come out of the Perceptual Loop and not necessarily a true representation of the actual signal.

Attention Limiter

External Influences
⇩
Attention Limiter

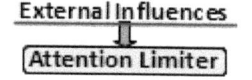

The Attention Limiter acts as a filter to screen out environmental information that is of no interest to us. Environmental signals include information from our etheric personality (core intelligence), other personalities such as those in our collective, loved ones and friends on the other side. Signals from our body's physical senses are also filtered by the Attention Limiter. Our control of this filter is limited to the extent we are able to control the Perceptual Loop to control the contents of Worldview (lucidity). The *Reject* outcome of the Perceptual Loop discussed below, is a signal to the Attention Limiter to ignore such information in the future.

In science fiction, a common issue is how telepaths are able to function if they are bombarded with telepathic signals from everyone. This functional area answers that worry. In principle, we are able to sense virtually every signal in reality, but our previously expressed threshold of interest protects us. The challenge is in setting that threshold. Our human instincts have it set to detect threats, food and opportunities for mating. Do our spiritual instincts have it set to detect opportunities for greater understanding?

Worldview

Worldview
Personal Reality

Worldview is like a database which is populated by our human's instincts and what we have been taught by our family, teachers, religions, media and experiences. It presumably includes a degree of understanding inherited from our collective and a sense of specific purpose inherited from our core intelligence (personality), which is described here as spiritual instinct. Worldview is the most influential aspect of our life field. It is *Nature's Habit* for our life filed as described in the Hypothesis of Formative Causation (15) and the yardstick by which environmental information is measured.

Worldview has considerable momentum, in that once a decision is made, once information is integrated into Worldview, it is very difficult to change. Expect that it must be changed in small increments. Also, expect that everything we consciously perceive is decided by how it is characterized based on Worldview.

Perceptual Loop and Worldview

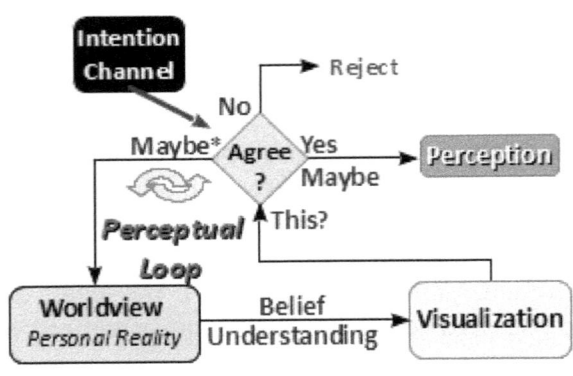

We know that our mostly unconscious mind has a mechanism to accept or ignore some information, decide what will be allowed into our worldview, what will be presented to our conscious mind and how that presented information is characterized. We are pretty sure this is also true of what we express into the environment and the signals sent to our body. This mechanism is modeled in the Implicit Cosmology as the Perceptual Loop. Information that passes through the Attention Limiter next enters the Perceptual Loop.

The Perceptual Loop process begins with an attempt to characterize the information (visualize) based on what is in Worldview. The information comes as a gestalt-like thoughtform and must be characterized based on familiar information. An important point to keep in mind while modeling our mind is that we create the present based on the past.

The initial characterization is then compared with the contents of Worldview. The visualized form is **not yet** submitted to conscious self as perception or sent into the environment as expression.

The process of visualization and then comparison with Worldview is probably a very rapid, iterative one resulting in many tries. Each try would produce a slight modification of the original input so that after many tries, the visualized version may have drifted quite a bit from the intended input.

> This process is a little like when a repair person needs a specialty tool that is not present, and so looks in the toolbox for something that will work in its place. The Perceptual Loop produces a visualization of reality that is already in memory and that matches at least some of the characteristics found in the incoming gestalt thoughtform.
>
> That information is visualized and held up to Worldview in an *"Is this what you mean"* manner. It is the *"Yes, that will have to do"* result that is sent to conscious awareness. Thus, the more we know, the more tools we have in the toolbox, the better our perception might agree with actual reality.

There are probably many extenuating considerations involved in the comparison. For instance, was the original input accompanied by a sense of urgency? Did the signal come from a friend? Was it an ordinary signal from the body? The three primary states expected to come from the Perceptual Loop are:

Reject

If no agreement is found between what is in worldview (familiar) and what is visualized, the environmental information is simply rejected as if it never passed the Attention Limiter. That *Reject* outcome is probably fed back to the Attention Limiter as a modification of the filter.

Conditional Accept

A second possible outcome of the Perceptual Loop *Agree* decision is *"Yes."* However, the incoming information is probably in the form of a thoughtform which is not a format we are able to consciously experience. The perceptual process produces a version of that information based on more familiar symbols represented in Worldview. Thus. if there is some amount of agreement between sensed information and Worldview, a characterization of the information is offered to conscious awareness.

This is an important point. What we become aware of is not the raw information we sensed. It is a characterization of it that is in sufficient agreement with memory to be considered familiar. We consciously experience a version of the information, a result described in mental mediumship as a colored message. This argues that any information access is potentially colored by cultural contamination. This is especially true of transcommunication.

This colored result is especially influenced by expectation. We are more apt to experience what we expect. Conversely, we are less apt to experience information if we have previously expressed disbelief in the subject.

The Perceptual Loop is used for expression as well. What we intend to express into the environment, say speaking to a friend or a command to the body to pick something up, is colored by Worldview. In terms of the psi signals we send to the environment, our worldview is doing the speaking. All we do is express the intention. If we intend to be forgiving of an offending person, the signal that person receives may well be a *drop-dead* message instead, if that is how we previously informed our worldview. Even if our spoken message is loving, the psi signal we send to the person's Attention Complex may be that *drop-dead* message if we feel that way but have forced a kinder spoken response.

Ambiguous Accept

The sensed information may be ambiguous, meaning that it seems familiar but is not specifically defined in Worldview. In this case, it has the potential of being integrated into Worldview and submitted to conscious perception as a modified version of familiar information, but now *updated* with the new influence.

It is this output of the Perceptual Loop on which we have conscious influence. By consciously intending to see things as they are, consciously questioning perception and avoiding making an *Accept-Reject* (believe or not) decision, we are able to encourage the Perceptual Loop to be more accepting of sensed information as it is, rather than as it compares to Worldview. This is the enabling concept of the Mindful Way as I speak of it in my writing, and which I explain in great detail in *Your Immortal Self: Exploring the Mindful Way*. (1)

Be aware that training the perception-expression processes to respond as you intend is a small change-at-a-time process that should be developed as a lifestyle.

Expression

When we decide to speak, act or even when we think about something, we initiate a perceptual process to visualize what we want to say or do. The difference between a fantasy and an expression is the intention to make it so. As it is modeled here, **the creative process is attention on an imagined outcome to produce an intended order.** (19) Expression then, is a process consisting of intention initiating the visualization of what is intended, perceiving that visualized outcome and then intending to express it into reality.

This process is moderated throughout by the Perceptual Loop just as if it were sensed information. As such, expression is the outward influence of perception. We tend to think, do, speak and feel based on Worldview.

This is central to the idea that we create our reality, and we do so, based on what we know, which is contained in Worldview. The idea is that we cannot express what we cannot visualize. And the visualization process is based on what we have been taught.

Lucidity

To be lucid means to have clear perception of our usually unconscious mind. More to the point, it means being able to consciously sense environmental information as it was intended and not as it is colored by the perceptual processes.

It is arguable that none of us are completely lucid, so we speak in terms of degrees of lucidity. As a practical matter, the average person has virtually zero lucidity, and always senses the world as he or she has been taught. In that regard, the average person's personal reality is essentially the same as all the others in the community. The assumption here is that the community has a local sense of actual reality which is usually not helpful for a person seeking spiritual maturity.

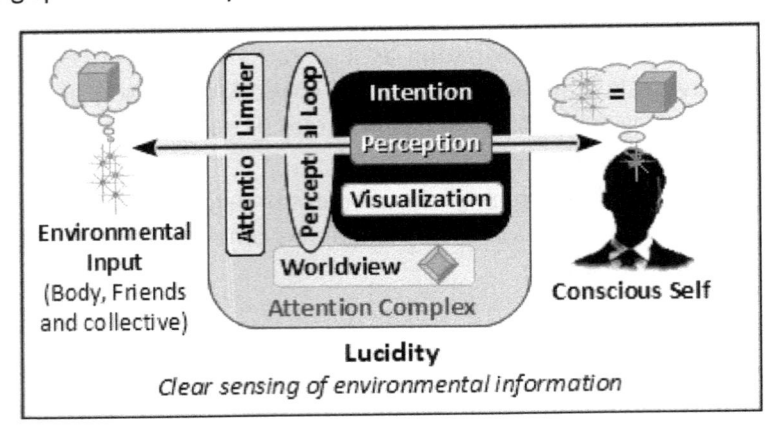

Figure: Attention Complex showing Environmental Input (Body, Friends and collective) connected via Attention Limiter and Perceptual Loop to Intention, Perception, Visualization, Worldview, and Conscious Self. Lucidity — Clear sensing of environmental information.

Our personal reality is defined by our worldview. The process of aligning our personal reality with the actual nature of reality begins when we realize that there is a difference and consciously seek to change. The intention to change is expressed to the Perceptual Loop as increased curiosity, consciously turning attention toward things that seem real, examining consequences of beliefs and questioning every action.

Deterministic Influences

Determinism is an argument that must be considered in three parts. For physical systems, it assumes that naturally occurring principles in nature determine the behavior of physical processes. For the behavior of biological systems, it assumes that behavior is inherited by way of a genetic predisposition. Assuming the Survival Hypothesis is allowed, the third part is a set of influences emanating from our etheric (nonphysical) aspect.

Physical Principles

For the purpose of this essay, the rules governing operation of the physical world can be generalized by saying that the behavior of everything is constrained in some way by rules which are thought to apply to all of the physical universe. For instance, an important fundamental relationship in electrical circuits is known by Ohm's Law, which states that voltage (V) is equal to the to the current I (I) times the resistance (R): $V = I \times R$. This relationship should be equally valid on the other side of the galaxy.

I use the Mandelbrot Set to demonstrate how a simple equation can represent an infinitely complex aspect of reality. The set consists of all the numbers from -2.0 to +1.0 on the real axis and -1.5 to +1.5 on the imaginary axis. Using the simple equation:

$$Z_{n+1} = Z_n^2 + C, \text{ where } Z_0 = C$$

the plot shown below is produced by calculating each point a predetermined number of times, using the result as the beginning value of C for the next cycle. If the result approaches infinity, the point is assigned black on the plot. Otherwise, it is typically assigned a color or intensity based on the size of the resulting value. There is a more detailed explanation in *Your Immortal Self*, and an earlier version in "The Cosmology of Imaginary Space" Discourse on EthericStudies.org. (20)

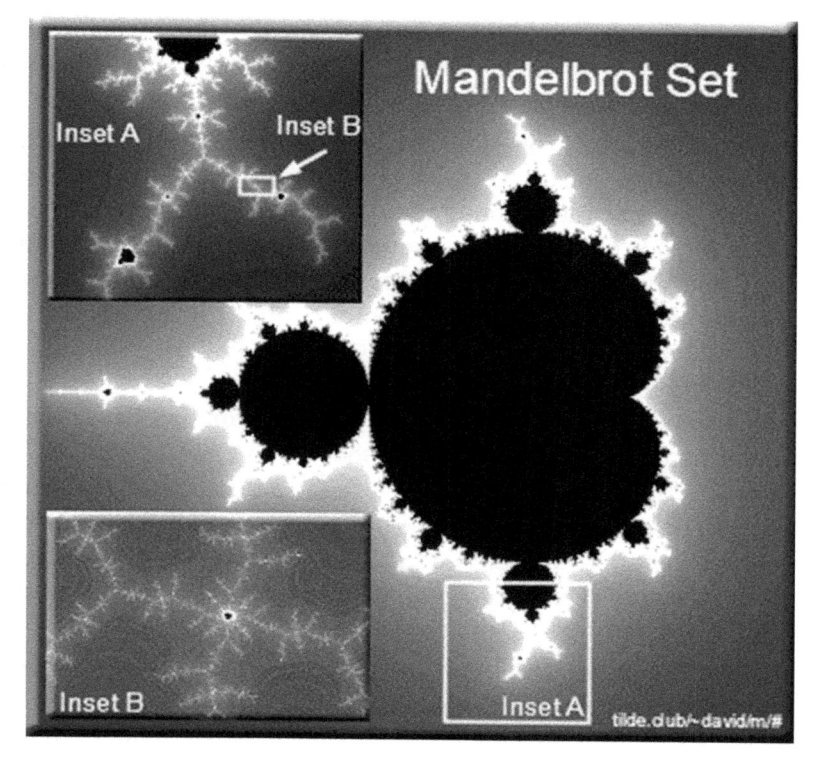

The top figure, sometimes known as The Apple Man, is a fractal, meaning that it is repeated many times in the plot, but at different scales; by selecting beginning points of increasingly small values (for instance Real Number: -1.165, Imaginary Number: -0.288).

Inset A in the *Navigating Within the Mandelbrot Set* Diagram (below), is an enlargement of the area marked *Inset A* at the bottom of the larger figure. Inset B shows part of Inset A for which the plot has been further *telescoped*. In it, you can just begin to make out a miniature version of the Apple Man fractal. There is an infinite number of such fractals that become visible as the formula is used as a *telescope* to calculate points with ever small coordinates.

The message is that the complexity of the physical world is organized by relatively simple rules. The same can be said of the greater reality, but with a different sort of rules.

Genetic Predisposition

This is an area for which I have little training. The idea is that each of our cells contains 23 chromosomes, each containing hundreds or thousands of genes, which in turn, contain our DNA. All the characteristics of the body are dictated by the DNA within the genes. In the deterministic models, it is argued that our personality is also coded into our genes.

The research is pretty clear that the blueprint for our physical body is in our genes. It is also logical to think that much of our temperament is influenced by our body, its abilities and differences in appearance from the norm. However, it is not so definite that our temperament is determined by our genes, and evolution does not seem to completely explain how the characteristics of a species change over time.

Nature's Habit

Rupert Sheldrake developed the Hypothesis of Formative Causation as a possible explanation for how morphogenesis is controlled. (15) Morphogenesis is the process in which a cell is differentiated into another cell. Remembering that an organism begins as a single cell, all included cells have the same genes. How a cell knows to divide into a skin cell or bone cell, for instance, is one of the mysteries of nature that has yet to be reasonably well modeled using mainstream science.

Formative Causation holds that morphogenesis is managed by way of fields which represent each element of an organism. A human body has a field, as does its skin, bones and cells, each kind with its attendant field. The field is defined by a set of instructions which orchestrates the formation and activity of its part of the organism. This is a nested hierarchy of fields model.

The instructions are based on what Sheldrake referred to as Nature's Habit. In other words, the attendant morphic fields (aka morphogenetic fields) cause their part of the organism to form and function by way of morphic resonance based on how that part has always been formed. This theory does allow for gradual changes based on successful, creative solution to environmental challenges which may be inherited by the morphogenetic memory of the species.

In the Implicit Cosmology, each instance of a species is expected to be a complete life filed as a member of a collective representing that species. If the Hypothesis of Formative Causation is correct, the collective would share a single intelligent core containing the body image (memory and Nature's Habit). That is, if there are a billion instances of a variety of dog, then their collective consists of a billion top dog fields integrated by one intelligent core.

The idea of a shared intelligent core is proposed to account for the mystery of what guides morphogenesis, to provide a means of transmitting evolutionary changes to all the species and to agree with the model used in the Implicit Cosmology to explain survival phenomena.

All members of a species biologically share the same genetic code, but we do see variations in temperament amongst members of a species. As I understand it, the Hypothesis of Formative Causation's provision for inheritance of instincts does not account for differences in temperament.

Differences in temperament remains an open question. The Implicit Cosmology was designed to model a person as a life field (who we really are) entangled with another life field (our human avatar). There is nothing in the model forbidding it to be applied to a different animal, such as a dog or bird. In that case, the difference in temperament in animals can be explained using the Implicit Cosmology.

In the concept of transmigration (related to reincarnation), through many life cycles, a soul is born into increasingly complex avatars until it finally joins with a human. The idea that we might have once been a plant, bug or dog does not set well with most of us. But, most of us who accept survival, also accept that we will eventually find ourselves in a different venue for learning … not necessarily on earth, and therefore, not necessarily as a human. That is little different than transmigrating from a bug to a human. I do not know, and the subject is a little outside of this cosmology; however, be aware that our animal friends may well have been us at one time or could be next time around.

Human Instincts

The human brain can be considered in two parts: forebrain and brain stem. According to "Brain Structures and Their Functions," the brain stem: *"is responsible for basic vital life functions such as breathing, heartbeat and blood pressure. Scientists say that this is the 'simplest' part of human brains because animals' entire brains, such as reptiles (who appear early on the evolutionary scale) resemble our brain stem."* (21)

The forebrain supports rational thought, so an assumption in the Implicit Cosmology is that the brain stem has evolved to support survival of the human body and the forebrain has evolved to support the entangled personality. In this view, it is reasonable to expect that, like other animals, humans would get along pretty well without an entangled personality.

Human instincts are a behavioral version of Nature's Habit. They become part of our worldview and dominate our behavior at birth. We spend the majority of our life trying to manage them, our success in which is sometimes known as rational maturity.

A brief Internet survey did not produce a useful list of instincts. It is pretty clear that they all relate to the urge to perpetuate the species. The challenge is to distinguish between our human's instincts and our spiritual instincts. For instance, curiosity is sometimes included in lists for human instincts. Curiosity about the habitat is useful for survival, but it also contributes to gaining understanding through experience.

Mimicking play is an important trait for survival and procreation, as it teaches adult skills such as child care, hunting and protecting the family. We see that a playful attitude is common amongst our trans-communicators, so play may be a fundamental instinct.

Spiritual Instincts

The assumptions for my comments about spiritual instincts are that we are immortal personality entangled with a human avatar as a person. Also, that we enter into a lifetime for the purpose of gaining understanding about the nature of reality as it is expressed in this venue. Further, that each of us is a member of a collective of other personalities with which we share our understanding, and in turn benefit from the collective understanding. (you and I are not necessarily in the same collective) These assumptions are detailed and justified in *Your Immortal Self*. (1)

We may be in this lifetime for secondary purposes, such as helping a member of our collective have a specific experience. But in every experience rests an opportunity to gain understanding. I posit here that each of us has an instinct to gain understanding through experience that came with us into this lifetime and will follow us into whatever new venue we find ourselves in after this lifetime. See Essay 3: *Prime Imperative* (Page 49).

When we are mindfully seeking to understand the nature of our experiences, we are *going with the flow* of our instincts. Life is easier in a spiritual sense. This is the path toward free will ... at least freer will; certainly self-determination.

When we remain steeped in our human instincts, we are resisting our urge to gain understanding. Spiritually, life is harder. This is the path dictated by circumstance and our human's survival instincts.

The idea that we have instincts that are concerned with more than simple human existence is addressed in the *Katha Upanishad* [1-III-4 through 1-III-8]. In it, the teacher explains to the seeker the importance of discernment in life's experiences with an eye toward understanding may avoid the necessity of future lifetimes. (22) See Essay 18: *The Razor's Edge* (Page 309).

Natural Law

The *physical science view* of determinism is that principles are a natural result of fundamental forces such as gravity, atomic-level influence and constants such as the natural rate of decay or the rate at which a field loses strength as one moves away from the source. Biological processes also evolve out of these fundamental principles. That is the physical view held by mainstream science.

The *metaphysical view* of principles governing the operation of reality is known by many as the Principles of Natural Law. Taking the metaphysical point of view makes sense if you accept the premise that reality is entirely thought, and we create the physical with our mind. Before you discount this mental point of view, remember that it is an overly simple way of saying that we are immortal beings and that the physical is an aspect of the greater reality, which we and other immortal beings express according to our collective consciousness. This is the natural consequence of our survival beyond this lifetime.

One of the most important rules I follow in metaphysics is that magic is not allowed. I make a distinction between the personality of which I am an aspect created to experience the physical, and personalities that hold the physical venue for learning in their imagination. The idea is that many personalities use the physical as a school, but that there are others in charge of running the school. Every indication is that these *school entities* did not imagine the physical as it is today but did so as an evolution of trial and error. Reality is very efficient, and everywhere we look, complex evolves from simple. If that is true, the best way to create the physical I can think of is to begin with a handful of simple principles and let them do the creative work.

Natural Law can be thought of as riding above the physical principles. They are concerned with the relationship of a person (personality entangled with a human) with the etheric. Think of them as something of a roadmap for the operation of a person. The degree to which a person understands and is able to live in accordance with the principles is a measure of the person's progression. The principles are thought to be everywhere the same. (23)

Historical View

My first attempt to develop a cosmology included the concept that Source expressed the experiential aspect of itself from which we evolved, and a formative aspect representing a parallel nested hierarchy of personalities. (2) I was influenced by the idea of nature spirits apparently involved in formation, care and evolution of the various aspects of reality. The concept is so deeply embedded in New Age thought that it is difficult to ignore.

The Implicit Cosmology does not specifically call for a hierarchy of formative aspects of Source. It does include the idea that the various venues, such as the physical, are imagined and maintained by probably many personalities. However, I also argue that the life field concept is the fundamental building block and organizing principles are the formative rules for the creative process.

What differentiates a personality into a venue builder as opposed to an experiencer remains a detail to be worked out. It is clear that we are all creators, and it is our spiritual urges that guide us in our experiences. Perhaps we will spend a while holding a venue in our mind. Perhaps we are doing that now. Certainly, we create venues for our *little me* to try out ideas. We are both experiencers and the formative aspects represented by nature spirits in lore, as our actions are constrained by organizing principles.

Spiritualism

The National Spiritualist Association of Churches (NSAC) defines **Natural Law** as an *"ascertained working sequence or constant order among the phenomena of nature."* (8) That *constant order* is considered the expression of Infinite Intelligence (non-anthropomorphic god, Source, Nature). The

NSAC has a statement of understanding known as the *Declaration of Principles*. (24) Principles 1, 2, 3 and 7 are relevant to this discussion:

1. *We believe in Infinite Intelligence.*
2. *We believe that the phenomena of Nature, both physical and spiritual, are the expression of Infinite Intelligence.*
3. *We affirm that a correct understanding of such expression and living in accordance therewith, constitute true religion.*
7. *We affirm the moral responsibility of individuals and that we make our own happiness or unhappiness as we obey or disobey Nature's physical and spiritual laws.*

Hermetic Teaching

Natural Law is also an important part of the Hermetic systems of thought. The more commonly cited principles are from *The Divine Pymander of Hermes Mercurius Trismegistus*. (25) Hermes was thought to have lived in Egypt 6,000 years ago. Many believe he represents the source of important concepts concerning the operation of reality. From *The Kybalion*: (26)

1. **The Principle of Mentalism.**

   The all is mind; The universe is mental.

2. **The Principle of Correspondence.**

   As above, so below; as below, so above.

3. **The Principle of Vibration.**

   Nothing rests; everything moves; everything vibrates.

4. **The Principle of Polarity.**

   Everything is dual; everything has poles; everything has its pair of opposites; like and unlike are the same; opposites are identical in nature, but different in degree; extremes meet; all truths are but half-truths; all paradoxes may be reconciled.

5. **The Principle of Rhythm.**

   Everything flows, out and in; everything has its tides; all things rise and fall; the pendulum-swing manifests in everything; the measure of the swing to the right is the measure of the swing to the left; rhythm compensates.

6. **The Principle of Cause and Effect.**

   Every Cause has its effect; every effect has its cause; everything happens according to Law; chance is but a name for Law not recognized; there are many planes of causation, but nothing escapes the Law.

7. **The Principle of Gender.**

   Gender is in everything; everything has its masculine and feminine principles; Gender manifests on all planes.

I discuss these in more detail in Essay 17: *The Hermes Concepts* (Page 285).

Implicit Cosmology

While developing the Implicit Cosmology (12) as it is implied by the Trans-Survival Hypothesis, I found certain concepts were involved throughout. It was evident that they are foundation concepts on which reality is formed. Thus, I defined thirty-eight Organizing Principles. (27) As it turns out, only a few resembled the usual Natural Laws which evolve out of the Hermetic teaching. (28) The principles are in three degrees of granularity so that concepts like *Collective* and *Field* are associated under *Reality*, concepts such as *Attraction* and *Life Field* are associated with *Formation* and concepts such as *Attention* and *Transition* are related to *Personality*.

Do not be overly concerned with the names of these. A different model might describe different terms because of different granularity or perspective. The underlying principles would be the same for every version of a reality model even though the nomenclature might be different.

Organizing Principles are active agents, rather than immutable laws. In principle, each life field is a creating intelligence and the principles regulate the creative process. (19) They might be able to be superseded or ignored but they are always present and do exert an influence.

That influence becomes more important when attention is turned toward an applicable characteristic. For instance, the *Field* Organizing Principle simply represents a fundamental characteristic of how reality self-organizes. Things exist as fields. However, life field as a formative agent does not come into play as a concept until purpose is considered.

The influence of Organizing Principles is always present and becomes a factor as we turn our attention toward a visualized field with the intention to express it into the environment. That is just a specific way of saying that a concept becomes a factor when a person intends to apply it in some way. Creation of intended order will be much more difficult if the visualization and intention are inconsistent with organizing principles governing a concept.

The idea of immutable laws is a traditional part of systems of thought that seek to incorporate the Hermetic Principles. As I model Organizing Principles, they are expressions of Source's understanding of itself. The model also holds that we are aspects of Source that exist to satisfy its curiosity about itself. If this is true, as we gain understanding about the nature of reality, and return that to Source, presumably, Source will learn a more complete self-image of itself. That would potentially result in changes in the underlying principles. As such, if Source is still learning, then Organizing Principles are evolving.

Here are a few examples from "Discourse 3: Organizing Principles" taken from *Your Immortal Self*: (27) (Early versions of these are at ethericstudies.org/organizing-principles/.)

In the category of **Reality**

Hierarchy: A hierarchical relationship exists between Source, aspects of Source and subsequent expressions of those aspects.

Prime Imperative: Aspect personalities inherit purpose from their source.

In the category of **Formation**

Aspectation: The influence of intention on an imagined result expresses aspects of reality which are a subset of personality's personal reality

Perceptual Agreement: Personality must be in perceptual agreement with the aspect of reality with which it will associate.

In the category of **Personality**

Personal Reality: Perception of reality defines personal reality.

Self-Determination: Personality's behavior is limited only by the Organizing Principles.

Worldview: Worldview is a learned response moderated by understanding.

Retro Familiar Storytelling

This is something of an undocumented concept. As a *Sunday Society Meeting Medium*, I must always be alert for new ways my mind will fool me. I define the *retro familiar* concept as **Modification of an initial memory or perception based on secondary feedback from the environment.**

In mediumship, this is seen when the medium elaborates on the message based on the sitter's feedback. For instance, the medium might say *"I see a big tree."* The sitter might respond with *"I like to read under a big tree."* In Retro Familiar Storytelling, the feedback from the sitter might produce a response from the medium as *"Yes, I see that. You like the warmth from the sun."* This is coloring based on information that was not in the originally sensed message but that seems like it was in retrospect.

Any modification of how a person senses memory, a psi signal or visualization based on feedback from the environment should be considered a form of storytelling that possibly colors the original sense beyond its initial meaning. This effect is probably best seen as a sloppy mental habit that can be managed through mindfulness.

The *retro-familiar* response is a form of hyperlucidity. (29) It should be considered a problem for free will because the response provides positive feedback to the perceptual processes, possibly indicating an erroneous belief is correct.

> To be clear, this concept does not indicate that the person is faking in the sense of *"Yes, I meant to say that."* Because current awareness is being presented to conscious self in real time, the Perceptual Loop can too easily link the sitter's feedback with the medium's sense of the message. This may also be related to how false memories are formed.

Finding Free Will

As a practical matter, our freedom to make informed decisions about our life is a function of how well we understand the processes which limit free will, and the success we have in managing those influences. In practice, we are pretty much on automatic (no free will) until we know to take control. We then have increasing free will as we gain understanding. Here, I use the Organizing Principle of **Understanding** from the Implicit Cosmology which is defined as **Perception of reality as it is and not as it is believed to be, with emphasis on underlying principles.** (12) (27)

There is something of a threshold of acquired understanding about our personal nature, beyond which we begin to recognize the need to deliberately seek greater understanding. It is crossing that threshold which I describe as stepping onto the Mindful Way. Until we have taken that most important step, our free will remains more illusion than fact.

The Tower Key 16 of the Tarot as illustrated by Dr. Paul Foster Case (30)
Meaning: Breaking up of old mental structures to make way for greater clarity; Culmination of understanding leads to realization and possibility of further understanding; Relates to the Dark Night of Soul and the Dawn that follows. (176)

Paradoxically, the less free will we have, the less aware we are of our lack of free will. Consequently, it is unlikely a person will cross that mindfulness threshold without some form of outside influence. That is typically in the form of a personal crisis brought on by a growing realization that accepted truths are not so true after all. This is a concept that has been understood by spiritual teachers since the time of Hermes. Consider *The Tower*, Key 16 of the Tarot. I use the Case deck as taught by the Builders of the Adytum:

To this Key is attributed the stage of spiritual unfoldment called Awakening, because it represents the flash of clear vision which reveals to the searcher the true nature of his being which has previously been hidden from him because of the bondage of his consciousness. (30)

Personal Responsibility

Of course, each of us must be responsible for our actions. Well, at least we must be accountable, else our society could not be based on the rule of law. But, is it possible to have personal responsibility for our actions if we do not have conscious control over our perception?

Perhaps the highest expression of personal responsibility is the sacrifice of ourselves for a perceived higher purpose. I read somewhere that our instinct for survival is hierarchical. Many of us will fight to survive, but readily sacrifice ourselves for our family. However, we will sacrifice ourselves for our country at the expense of our family. The United States depends on an all-volunteer military, which means we depend on citizens willing to sacrifice their life for the good of our country.

This paradox is pointed out by David Ropeik in his article *The Greatest Threat of All: Human Instincts Overwhelm Reason*: (31)

> *You woke up each day last year and went about your business as any human does, compelled by deep and ancient instincts to do the things necessary to get yourself safely to bed at night.*

And later:

> *We are compelled from the deepest level of our genes and survival instincts to taking more from the system than it can provide and put back in more waste than it can handle, and no amount of human brain power outwit the natural instincts that are driving us 150 miles an hour toward a cliff.*

And last:

> *Dangerous, because the belief that our intellect can provide the tools and enlightened leadership that will ride to the rescue, arrogantly denies the inescapable truth that we are still mostly instinctive animals, each of us compelled by deep subconscious urges to do what we can as individuals to survive today; and the day after that, and everybody else, are just not as much of a concern.*

Ropeik was addressing how being controlled by instincts allow us to ignore greater, less obvious threats to our survival. This is the problem of our human's instinctive response to cultural influences (Worldview) versus our conscious self's mindful examination of our actions.

Personal responsibility cannot be executed without examination of our every action from the perspective of understanding of Natural Law. Without that understanding, and realization that our mostly unconscious mind only lets us be aware of what we have previously believed to be true, our free will is an illusion.

Taking Control of Free Will

The assumption of this essay is that we entered into this lifetime with a purpose, but because entanglement with our human is so complete, most of us have lost sight of that purpose. Probably for all of us, the dominance of our human's instincts has overshadowed the fact that our body is a faithful servant and not our actual self.

A common theme in New Age literature is that transcendent spiritual teachers deliberately enter into a lifetime to help those of us who are still in the physical. They do so knowing the risk that they might lose control of their spiritual maturity by succumbing to the belief that they are their body. The physical is a compelling temptress to whom we willingly surrender our self-determination.

Consider what has been explained in this essay, seeking to take conscious control of your thought processes is perhaps the most important step you can take to gain spiritual maturity. Essay 9: *Consensus Building in the Paranormalist Community* (Page 129), includes quite a lot about taking control of the thought process. Also, understanding the *Life Field Complex* model discussed above gives you the necessary tools for relating what you have learned to other situations.

The most important thing to remember is that taking control is a deliberate, lifelong process. Done right, it will become a way of life that moves your current body-centric perspective to an etheric, immortal self-centric perspective.

A Talisman

A few years ago, the phrase, *"Just because you can, doesn't mean you should"* became stuck in my mind. From time-to-time, I worry it like a dog worries a rock. It came to me when a jacked-up pickup came past me way too fast, way too noisy and so high that, in a collision, the average compact would roll under the bumper.

The truck may have been legal, but it was not compatible with civil society. In a word, the owner was antisocial; thumbing his nose at the right thing to do. The truck was a rolling example of "Just because you can, doesn't mean you should."

As I wrote this essay, I tried to think of a touchstone or talisman of sorts that would help us know when we are not expressing free will. I think the answer is this phrase.

Ask yourself, "Should I do this?" This is not a question if it is the right thing to do because that is a moral question entirely dependent on social norms. It is an ethical question that is dependent on spiritual instincts. Is it something that a mindful person would do?

References and Alternative Sources
Listed at the end of the book beginning on Page 357.

Essay 2
The Mindful Way
2014

About This Essay
You may know mindfulness as a form of meditative stress reduction. It has evolved from Hinduism, and later from Buddhism. It is taught as a way of living in the now by being aware of our feelings, senses and the world around us as calm participants, rather than hassled victims of daily living.

The first mention I can find of mindfulness is in the *Katha Upanishad*, Verse III, Line 8. This 3,000 to 4,000 years-old text is discussed in Essay 18: *The Razor's Edge* (Page 309). I have included the most pertinent lines here: (22)

> 1-III-3. Know the Self to be the master of the chariot, and the body to be the chariot. Know the intellect to be the charioteer, and the mind to be the reins.
>
> 1-III-4. The senses they speak of as the horses; the objects within their view, the way. When the Self is yoked with the mind and the senses, the wise call It the enjoyer.
>
> 1-III-5. But whoso is devoid of discrimination and is possessed of a mind ever uncollected - his senses are uncontrollable like the vicious horses of a driver.
>
> 1-III-6. But whoso is discriminative and possessed of a mind ever collected - his senses are controllable like the good horses of a driver.
>
> 1-III-7. But whoso is devoid of a discriminating intellect, possessed of an unrestrained mind [unmindful*] and is ever impure, does not attain that goal, but goes to samsara.
>
> 1-III-8. But whoso is possessed of a discriminating intellect and a restrained mind [mindful*], and is ever pure, attains that goal from which he is not born again.
>
> 1-III-9. But the man who has a discriminating intellect as his driver, and a controlled-mind as the reins, reaches the end of the path - that supreme state of Vishnu
>
> > *I have added [unmindful] and [mindful], as that is the terminology used in some of the other translations.

In terms of the Implicit Cosmology, these lines say that we are in an avatar relationship with our body. Becoming aware of this presents a path to spiritual maturity. Without that realization, we will not gain that most important realization.

The *Katha Upanishad* preceded Hinduism, which adapted the concepts to the Yoga Tradition as mindfulness. However, one of the relevant root terms from the Buddhist line of development means *remembering*, as in "Remember the Buddha." This is a reminder that ancient wisdoms can take on different meaning through translation from language to language.

Today, most versions of mindfulness are reformatted ancient wisdom directed toward techniques for improving quality of life. In fact, that is exactly how it is intended here. The value I wish to add is to explain these concepts from an immortal self point of view, rather than the usual point of view that we are our body.

From the perspective of immortal self, the way we experience our world as explained by current science and lessons learned from the study of transcommunication, we know the study of mindfulness needs to include a focus on how to manage perception. This begins with understanding the role Worldview plays in forming our personal reality. My assumption is that aligning Worldview with the actual nature of reality may well improve quality of life. That alignment is an important part of the Mindful Way.

A common reaction people have to my writing about mindfulness is irritation that I am trying to make mindfulness something it is not. This complaint usually comes without the person reading the essay or knowing the history of mindfulness. The term emerged into my way of thinking as I looked for a way to describe how our study of EVP had evolved into the study of transcommunication, and how that evolved into a search for understanding about how we, as experiencers, are affected by these phenomena.

The *Spring 2014 ATransC NewsJournal* (32) was our last issue. The first version of *The Mindful Way* essay was included as my effort to explain the *so what* of our study. I have since dedicated myself to finding other ways of explaining the main point we have learned from transcommunication. That is, what we do now matters for the rest of our existence. My challenge is to find a way to convey to you the urgency I feel about the need for us to understand the implications of our existence.

From my experience as a life-long seeker, mindfulness is the most important tool available for us to manage perception. Our perception, what we consciously experience, is a function of Worldview, and that is really just a database of memory and instincts. Since the only conscious influence we have on Worldview is the expression of intention, mindfulness is just a way of saying that we learn to be aware of the intention we are expressing to our mostly unconscious mind. Improving awareness can be generalized as improving lucidity.

Last Issue #129:

Spring 2014

ATransC NewsJournal

For us who are in a lifetime, stepping onto the Mindful Way is the rational response to the realization that we are immortal. It is for that reason this essay was included in *Your Immortal Self*. The other essays from that book which are included here are Essay 4: *Immortal Self-Centric Perspective* (Page 64) and Essay 19: *Progression, Teaching and Community* (Page 319).

To be clear, while I am suggesting that you adopt a mindfulness approach to seeking spiritual maturity, I am not licensed or qualified to advise you to do so as a therapy or remedy. Any path that takes you in that direction can be expected to provide the side benefit of a more agreeable life, but my intention here is to shine a light on a way for you to gain spiritual maturity.

Purpose

The phenomena of transcommunication appear to have a purpose beyond the reassurance it offers to loved ones. After examining mediumistic messages from the other side and *revelations* brought by past teachers, it is easy to imagine that our etheric communicators are trying to teach us about the reality of our immortality by showing us they exist. This essay is written as an exploration of the idea that the EVP messages in our recorders, or the paranormal images we find in our photographs, are a new way of telling us that we are part of a larger community. Perhaps it is up to us to understand what that means.

Mindfulness

The terms *mindfulness* and *mindful living* have become catchphrases for right living, but not in a pretentious way or in an attempt to tell us what to do. People speak of mindfulness almost in a reverent tone, as if the concept relates more to God than to daily living. Always, it is used to offer guidance in how to improve our life, how to be all that we can be.

Discussions about the phenomena of transcommunication are usually about technique and quality of examples. Who is talking may be discussed, especially if the information seems to come from a loved one, but the question of continuous life seldom comes up. While in fact, considered from the perspective of our immortal self, transcommunication may actually be all about our immortality. If this is true, then learning to live mindfully may be the most important ability we can learn.

> It is noteworthy that, in a typical Spiritualist meeting, mediumship is demonstrated as a spirit greeting, rather than a comprehensive message. The spirit greetings are demonstrated to show the truth of our continuous life.

Teachers

With proper controls, Instrumental TransCommunication (ITC) can be a rich source of information about the other side. For instance, we have learned from EVP that we should expect a life review during our transition. We know they can see us, and we know our communicators sometimes *get together* with friends on the other side. We also know that there are changes in their ability or need to communicate so that some do not *report in* for years and some make contact right away, but after a while, seem to *move on*.

Mediumistically acquired information, sometimes referred to as channeled material, must be considered with reservation because we know cultural influences can color messages. Even so, consistency amongst communicators seems to add credibility to some messages. (33)

The fabled Hermes of ancient Egypt continues to be an important teacher of mindfulness. The only document credited to Hermes that seems reliable is *The Emerald Tablet*. (28) In Line 2, he speaks of *The One Thing*, which represents the expression of Source as the organizing principles involved in *The Great Work* of the Hermetic tradition. (34)

> 2. And as all things are from only One Thing, by will of the one God, so all things have their origin in this One Power, by adaptation to their individual purposes.

The One Power is the Creative Process (19) by which reality is adapted to satisfy imagined purpose. The Creative Process is attention on an imagined outcome to produce an intended order.

The Great Work is all about the path followed by seekers to gain understanding. The lessons involved in this are virtually the same as those brought by many more contemporary teachers. The message is that a person benefits from learning to live in accordance with the true nature of reality.

Jesus is another important wayshower. A review of teachings attributed to Jesus, as found in Aramaic-to-English translations, shows that he taught that our *I Am* presence exists in the greater reality and that our transition out of this lifetime is toward our *I Am* presence: *"Where that I Am really is, there you already are, and you can be, consciously"* (from Luke 24:38-49). The *"...and you can be, consciously"* part is a direct reference to lucidity. See the "Lucidity" (Page 123) section in Essay 8: *How We Think* (Page 109).

Jesus also taught the unity of humankind, that one person's actions reflect on all people. (35) A transcript of Hans Bender's words as conveyed by Kai Mügge during a séance can be read on the ATransC website (atransc.org/hans-benders-message/). (36) To paraphrase, Bender explained that we are not alone and that how we view the other side has a lot to do with how we experienced it during our transition. He said that what we are doing here affects the other side and that we can project negativity into the greater reality which can cause problems for others.

Jane Roberts' Seth material (37) appears to be a reliable source of information about the other side. (33) Three important *instructions* from Seth are:

- People create their own reality.
- People exist in more than one aspect of reality at once.
- The only wrong act is to violate oneself or other life.

The common message from all of these sources is:

- Who we really are, our *I am this* personality, always exists in the greater reality.
- We are able to connect with our etheric aspect through *right thinking*.
- How we think now, affects us and others now and beyond this lifetime.
- It is for us to learn to live in accordance with the true nature of reality.

This understanding is not based on one person or one organization teaching religious doctrine. Think of it as the handbook for *right living* given to us by our friends on the other side.

What We Do Now Matters

With close examination of ITC messages, a pattern begins to emerge that tells us much about the person. While the messages appear to be paranormal, it has been noted by many researchers that different practitioners are apt to record rather different kinds of messages from the same situation.

To illustrate, Lisa and another person went into a dark room of a reportedly haunted building and recorded for EVP. Lisa is a pragmatic, levelheaded witness and recorded EVP containing useful information. The other person delighted in being scared and expected scary EVP, and in fact, she recorded scary EVP. In both cases, the messages were clearly paranormal, but their character tended to agree with the practitioner's temperament and Worldview.

As it turns out, it appears people's expectations are projected onto their experiences. This has been noted in what has become known as the Sheep-Goat Effect. In that, people who are more psi-sensitive (psychic) tend to have more paranormal experiences. In his book, *First Sight: ESP and Parapsychology in Everyday Life*, (38) James Carpenter explained a hypothesis, based on evidence currently being presented in parapsychology, which holds that people are always informed about the world via their natural psychic sensing. Further, he argues that people are constantly psychokinetically influencing their world.

What all of this means is that we also see with our inner senses (first sight) and always have some influence on our world with our intention, which is based on what we think is true.

Worldview

Engineers design models for systems they are trying to understand. One way to develop a model is to figuratively put the subject in an imaginary *black box* with the known inputs and outputs clearly defined. The trick is then to think of what would have to happen inside of the box in response to the inputs to produce the outputs. Not knowing for sure what is inside the box, engineers usually solve the problem by theorizing a model with functional areas inside the box.

> I refer to a model as a theoretical set of functional areas defined to produce the effect proposed by a hypothesis. As in black box analysis, the model need not be factually correct, but it must functionally produce the intended results. In terms of the etheric, a model is a thought exercise.

Researchers have found that people imagine what they are experiencing, and the information for that imagining comes from the Worldview database. If the incoming information agrees with the database, then it will actually be experienced by the person. If it does not match the database, then it will either be changed to agree with the database and experienced in that changed form or outright rejected. Refer to the *Functional Areas for Perception and Expression Diagram* (Page 44).

The way we express ourselves involves the same processes. Something causes us to react, and however that initial stimulus is translated by Worldview, an imagined reaction is developed. At that point, it is just a fantasy, but if we intend to act, then, what is visualized is expressed in some way. The rest of the story is that, with that intention to act, we begin to psychokinetically influence the world. (16)

Using this model, it becomes evident that Worldview plays an important part in our lives. By all indications, we are born with a more-or-less empty Worldview database populated only with our human's instincts. It appears reasonable to argue that we do begin with a degree of understanding so that one might say that a child is *an old soul* if born with more than average understanding about the world. It also seems reasonable to say that the average person's worldview is full of what has been taught by teachers, parents, clergy and the media. Much of that is simply local custom or popular wisdom.

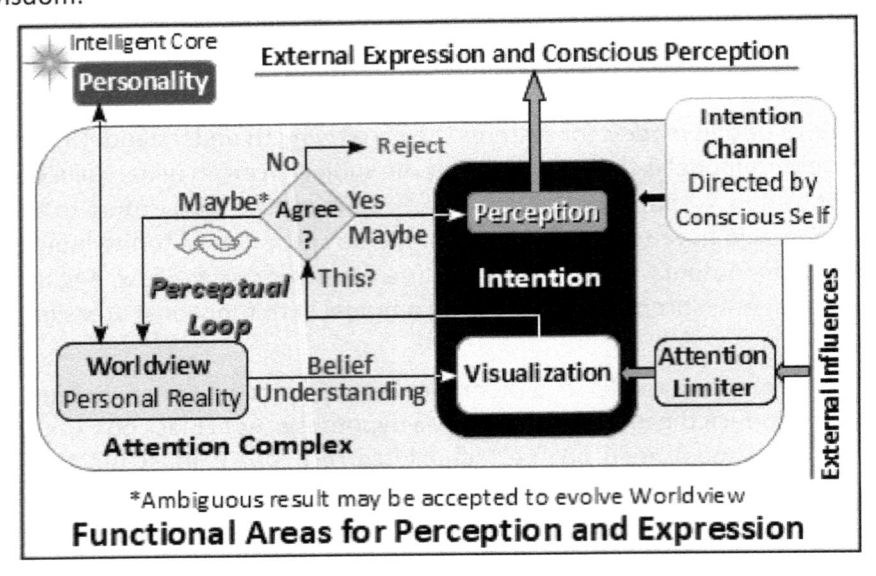

```
┌─────────────────────────────────────────────────────────────┐
│  Intelligent Core    External Expression and Conscious Perception │
│   Personality                                                │
│                                          Intention           │
│              No ──► Reject               Channel             │
│       Maybe* Agree Yes                   Directed by         │
│         ◊    ?   ► Perception          Conscious Self        │
│      Perceptual  Maybe                                       │
│       Loop   ↑This?    Intention                    External │
│                                                     Influences│
│    Worldview      Belief    Visualization  Attention         │
│   Personal Reality Understanding            Limiter          │
│              Attention Complex                               │
│                                                              │
│   *Ambiguous result may be accepted to evolve Worldview      │
│        Functional Areas for Perception and Expression        │
└─────────────────────────────────────────────────────────────┘
```

Personal Reality, Local Reality and the Greater Reality

Of course, there is only one actual reality, but there are differences in the way people experience that one reality. This is all about the individual person, so it is important to understand that each of us has a local reality which is that part of the greater reality which we are aware of, and more importantly, to which we pay attention. Our hometown is part of our local reality, but there are likely parts of it we are actively aware of and other parts that only provide background for the sense of *town*. Our neighbor will have a slightly different local reality and someone living in another country will hardly be aware of most of what we think of as real.

The greater reality just is. It does not have the capacity to be positive or negative. The same can be said of local reality: it just is. How we perceive our local reality is rather different. For instance, where we live just is, but it has characteristics such as good, bad, warm or uninviting, depending on how we think of it. Our personal reality is how we perceive our local reality; what we think of it. Right or wrong, as far as we are concerned, our personal reality is the real reality and that is determined by our worldview—what we have been taught but biased by whatever understanding we have achieved.

In the Mindful Way, we learn to examine our worldview to see if what we believe is true makes sense. The idea is to align personal reality with local reality; the true nature of reality and not what we have been taught to think is true.

Suspended Judgment

Rethinking what we believe to be true may seem paradoxical. If we believe something to be true, how can we tell if we should change our mind or even examine the belief? In practical application, the Mindful Way is a life-long process, a path to be followed one-step at a time, so how does one begin in the middle of a lifetime?

An effective way to begin mindful living is to make a conscious decision to have an open mind. We do this by taking conscious control of the process our mind uses to consider new information. The *Functional Areas for Perception and Expression Diagram* (Page 44) represents a model for how a person experiences information from the environment.

We visualize what we are experiencing in a very fast, mostly subconscious reaction to information from our environment. This visualization is based on what we have been taught, which is in our worldview. If the incoming information agrees with what we expect, say a friend on the phone or the door opening when we turn the handle, then it will be experienced. If it does not agree with what we visualize, it may not be noticed, as if we are blind to it.

An important characteristic of this comparison between what we expect and what we encounter is that a close agreement will likely result in perception of the information as well as feedback that can modify Worldview with an ambiguous *maybe*. In other words, we learn. As what we learn begins to consistently agree with reality, it becomes understanding. While we are told that Worldview shapes our first after-death experiences, it appears that it is this understanding that persists beyond this lifetime.

The idea is to learn to monitor the decision that comes out of that comparison. The idea of suspended judgment is that we seek to just experience and not decide if we accept it or not. People tend to automatically reject things they do not understand. With suspended judgment, the decision to accept or reject is not made without allowing time to consider the experience in the context of more information.

Self-Determination

We have to decide ... everything. If not what we experience, then we must at least decide how to react. Self-determination also means that we create our world. Again, not necessarily the brick and mortar places and things we live in, for we live in a collectively visualized venue for learning. For sure, we decide how to react to these things.

Two people might have essentially the same experience, but each will remember it in a different way. A person who is in the habit of thinking things always go wrong will likely remember it as a bad experience; however, a person who is generally optimistic about life is likely to remember it as a good experience or at least as a learning experience. It is all about attitude and that is a learned thing.

Here too, suspended judgment can help. Whatever we think the world is like, we can learn to consciously intercept that *"Oh, it's awful"* response with either a *"wait and see"* or an *"it has a good side"* response. You may be thinking that this is idealistic, but it works. Once it becomes a habit to intercept those internal decisions, there is more room for alternative explanations for what we experience. An *awful* reaction tends to stop further consideration of alternative explanations.

We are always psychically interacting with your environment. How we think of incoming information also has a lot to do with how that information continues to develop. It is likely that a positive or at least neutral response will encourage a more beneficial effect in our environment.

> It is helpful to know when interacting with others if they are fundamentally afraid of the world. From what we learn about a person through normal interaction, we can ask, *"Do they think they live in a friendly world or a scary world?"* Self-determination is always colored by how we feel about the fundamental nature of our world.
>
> Fear of our world can become a basic part of our decision making without our realizing. For instance, owning a gun is a fear reaction. Belief in original sin is a fear reaction. Our human is very afraid of the unknown, and that fear will color our thoughts about situations in which there are many unknowns, such as the dark and social commitments. Aggressiveness is often a fear reaction.

The Mindful Way

This is an abbreviated discussion about the Mindful Way. The main message is that what we do now will follow us for the rest of our existence—here and hereafter. The more our personal reality agrees with the actual nature of reality, the more progress we will make in our evolution toward a spiritually mature personality; understanding begets understanding.

The key is to stop and think before we react. To paraphrase Jane Roberts' Seth, perhaps the only sin is to impose our will on others. We must learn to stop and think about how our actions affect others. We are citizens of our community, the world ... and the greater reality. We psychically interact with it so that our feelings about another person in some way affect that person.

The only right we have is to decide what we think of our world and how we will react to what we decide. We are the only judge as to how well we are doing and that is not based on what we have been taught but on understanding we have gathered during our existence.

In an ideal world, people would just naturally be mindful of how they are doing as citizens. Laws to enforce behavior considered common decency today would be unnecessary because people would be mindful of how their actions might affect others. Of course, we do not live in an ideal world, but that is the point. We are also a society of people whose personal reality is very different than the actual nature of reality. The ideal of the Mindful Way is to evolve a society of people who understand they are part of a community.

References and Alternative Sources
Listed at the end of the book beginning on Page 357.

Intentional Blank Page

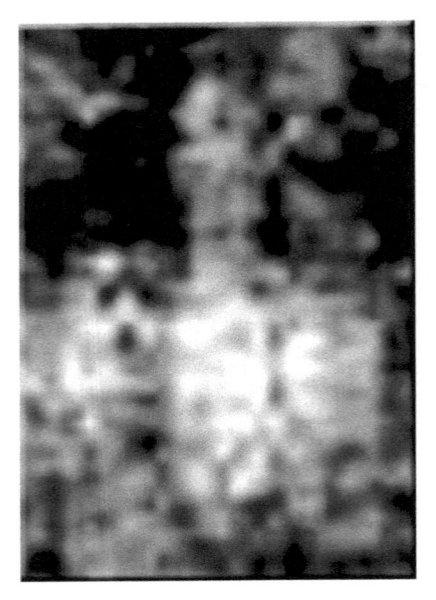

Video-loop ITC Image collected by Tom and Lisa Butler.

Possibly a man holding a small white dog, as if posing for a portrait. In the color version, it appears the man is wearing a dress uniform. His face is distorted.

Essay 3
Prime Imperative
2014

About This Essay

In speaking of our spiritual aspect, we depend on many concepts that may be true as a matter of popular wisdom, but that might not actually be true. For example, I speak of our spiritual progression as if it is a factual characteristic of who we are. But, I am assuming we are expected to progress in the sense of increasing spiritual maturity. The idea comes as a consequence of the assumed efficiency of nature and our assumed immortality, which in turn comes as a consequence of the assumption that we will survive beyond this physical lifetime for a reason. But is it true? From whence come our spiritual instincts?

Many of the essays I write begin as an effort to address a nagging sense that I am making assumptions for which I have not established a reasonably rational foundation. The idea that we have spiritual instincts is such an assumption. This essay is my effort to establish that rational foundation.

The etheric aspect of reality is, by definition, nonphysical. Anything nonphysical is conceptual, and still today, necessarily theoretical. Consequently, any rational argument I make about the etheric is based on theories proven by theories. This is not an acceptable, logical argument. However, it can be argued that each theory inherits validity if all of the related theories are considered ... and are at least marginally supported by the evidence.

In fact, much of what we know about our physical world is based on credibility inherited from a few objective observations. The resulting principles work very well.

The proof for survival I presented in *Your Immortal Self* (1) is based on consideration of many forms of trans-etheric phenomena. For instance, Electronic Voice Phenomena (EVP) are well-established as actual phenomena. (39) It is reasonably well established through forensic-quality analysis (40) and abundant anecdotal evidence that EVP are a form of trans-etheric influence. It is not as well established that they are initiated by discarnate personalities, but the discarnate origin of EVP inherits validity from other theories, especially those supporting mindfulness and content analysis of the Seth Material. (33)

In this essay, I have attempted to develop a logical argument to say that we exist for the purpose of gaining understanding about the nature of reality. I also argue that we have a spiritual instinct or urge to gain understanding that tends to influence our every action.

The ideas of purpose and spiritual instincts are foundation concepts in the New Age and Spiritualist communities. Perhaps one of the most influential books for me was Robert Ardrey's 1961 *African Genesis*. (41) I did not realize it at the time, but looking back, I can see that Ardrey became something of a role model for me. His process involved watching animals in their natural habitat, and so, it was from him that I learned about the naturalist approach to science.

His observations of how animals behaved in their natural habitat were interesting, but it quickly became clear that they translate into human instincts. For example, I learned from him about the way birds establish a pecking order. This behavior extends to other animals in the wild, and certainly to humans.

Ardrey caused quite a controversy when he claimed that animals are natural killers. Probably the only line I remember from the original *Star Trek* television series was Captain Kirk saying, *"Yes, we are killers. But we can choose not to kill today."* In that one line, is acknowledgment that we have human instincts which drive us, but we also have a rational aspect which is able to moderate those instincts. That, with the conscious decision to moderate our behavior, is a good explanation of mindfulness.

The work I have done in the study of transcommunication and survival has been largely that of a naturalist observing the phenomena in its natural habitat. As it should be for any good naturalist, these observations have evolved into a theory about their nature.

Another lesson learned from Ardrey is his concept of Territorial Imperative. He proposed that humans have evolved an instinct to take and hold territory. The important idea I have taken from this is that others also think humans are driven to possess. Again, it is not the territory part I am getting at. It is that our human has survival instincts which govern its every action. It also has an overriding urge to be the top human in all circumstances to further the objective of those instincts.

From the perspective of this essay, the territorial imperative can be thought of as a pinnacle instinct. Our human's foundation instincts are concerned with survival of the species, and most particularly, survival of our human's gene pool. Territory represents a master urge intended to enable realization of those foundation instincts.

Our immortal self's pinnacle instinct can be seen as the urge to gain understanding about our local reality. The foundation spiritual instincts it enables include teaching, learning, cooperation and testing. The stronger the community, the more able we are to fulfill the objective of our spiritual instincts.

Abstract

The foundation assumption of this essay is that a person is an etheric personality entangled with a human body for a lifetime. And, that a person enters into a lifetime to gain understanding about the nature of reality through daily living experiences. It is argued here that part of gaining maturity is learning to manage the influence of human instincts while responding to the urge inherited from our etheric nature to turn toward particular experiences. Understanding is seen as the objective. Curiosity is seen as the perspective. Mindfulness is seen as a way of life.

Introduction

If we are more than our body, if we are spirit beings only temporarily entangled with a human body during a lifetime as a person, it seems reasonable to think we are a person for a purpose. It is also reasonable to think this purpose relates to who we are as immortal personalities, rather than a purpose related to our human and just this lifetime.

Our first purpose is to live a good life, wherever it takes us and whatever we do for a living. It is through life experiences that we have opportunity to gain understanding. In support of our first purpose, our second purpose is to turn toward those experiences that challenge us so that we might gain understanding.

Point of View for This Essay

The point of view taken in this essay is that our real home is in the greater reality, which I refer to here as the etheric. The supporting concepts are concerned with our collective, but in order to have a beginning and end to who we are, I say that reality is the body of Source. In Spiritualism, this is Infinite Intelligence.

> This is also the omnipresence aspect of the biblical God but not the omniscience or omnipotence aspects.

We are an aspect of Source. Well, we are more likely an aspect of a personality that is many rounds of aspectation removed from Source. To be consistent with the way we imagine things, each aspect likely produced many aspects of itself. Thus, probably the best way to look at our relationship with Source is to think of Source's life field as a nested hierarchy of life fields.

Think of those nested groups as collectives. The collective we are part of is formed of many aspects of a single local source. I think that, someday, we will once again become one with our local source, just as it will eventually become one with its local source. Think of a local source as the nexus personality for a collective. This is not a personality to be worshiped. It is to be respected as the source of our purpose for being. I will get to that purpose in a moment.

A useful way of thinking of a nexus personality and its collective is to remember the mental process you follow to decide about something. Suppose you are thinking about going shopping. The first step is to visualize what you are shopping for and where. You might imagine yourself being there and looking at the merchandise; perhaps trying on something. Perhaps you will realize the price and forget the who thing.

When we imagine a place in our mind, in which we imagine ourselves, we create a venue for learning and an aspect of ourselves to experience the venue. We give that *little me* a purpose to experience (shop) and the self-determination to thereby gain understanding (cost). The imagined venue is an aspect or subset of our sense of reality and our *little me* is a subset of who we are as we imagine ourselves.

Now here is the important part. The imagined venue is in our life field. It will forever be part of our reality as a memory and we will remember that *little me* experience forever. All of the experience will have become part of our sense of reality. All of this will be in the form of understanding which will ever so slightly modify our understanding of reality.

Our venue for learning is the physical universe. It appears to be an imagined aspect of reality that is shared by many nexus personalities, so that you may well be in a different collective than your neighbor. Certainly, your nexus personality will have a little different reason for expressing you into a lifetime. All of us have the shared purpose of gaining understanding, but probably for different aspects of this venue or from different perspectives.

We are influenced by the instincts of our human body. But we begin with different circumstances and with a different degree of prior understanding inherited from our nexus personality. So, while we share the need to assure survival of the human species, we differ in how we approach life and how well equipped we are to find meaning in daily experiences.

Cooperative Collective

I once had a waking vision of the face of a clock. It was suspended in the air, face-up, but tilted toward me a little so that I could see that there were many black specks scurrying about on the white surface. My impression was that they were little stick people like those I might draw in a hurry. The space between three and four o'clock was an open hole and a few stick people had apparently fallen through. The people on the face of the clock were somehow helping the people who had fallen through the hole. As the hour hand made a complete circuit, the ones in the hole came to the surface and a few of the others jumped into the hole.

My sense was that the stick people were all part of a collective of personalities, a soul group if you want, and they were doing all they could to help their fellows who had entered into a lifetime, symbolized by falling through the hole. I knew that they were helping one another to progress by gaining understanding, and that none of them would be able to move on until all had made sufficient progress. The hour hand represented a lifetime.

Probably not all of them, but many of my helpers, friends, and guides, both on the other side and in the physical, are part of my collective. I am never very far from them, and just as I am a student, in turn, I am teacher, for we must all move as one.

Immortal Self

Your first reaction about what I have said so far might be that you do not want to return to your nexus personality if it means you will disappear like your *little mes* more or less disappear in your memory. I have thought about this a lot. As I understand the metaphysics, a good way to think of your relationship with your nexus personality and your collective ... and ultimately with Source ... is as a chorus.

Your nexus personality is the chorus. Individual personalities in your collective are the members of the chorus. During a lifetime, you step out before the chorus and sing a solo. You are backed by the harmonizing sounds of the other members. When you have finished your solo, you step back to join the chorus. At no time has your conscious self become anything less. Rather, your presence is made greater by the chorus.

There is nothing in the metaphysics I have studied indicating we will lose our sense of individuality.

Our Etheric Nature
To understand our purpose, we must first understand our nature.

Popular Wisdom
Religious leaders tend to consider perpetuating the species and assuring supremacy of their religion as sacred work of God. For them, survival is a matter of going to a heaven of one kind or another to receive their just rewards.

In mainstream science, the theory of evolution holds that organisms undergo random mutations. (42) Survival of the mutations depends on how useful they are to the survival of the species. There is no intention involved in the theory of evolution, just chance.

In a real sense, mainstream science is religious about what it accepts as truth. When it comes to agreement between mainstream and survival-based philosophy, experience has taught us that it is not a matter of having enough quality evidence, but rather, it is a matter of honest examination of existing evidence. So, the first idea we need to consider is that, what is taught by religions and mainstream science, is biased toward faith-based beliefs; faith in the Bible or faith in science.

Life Fields as the Basic Building Block of Reality
As shown in the diagram below, a field can be thought of as a number of elements bound together by a common influence acting as a nexus. In a life field, an intelligent core acts as the nexus for the field and is ultimately who we really are. I refer to this intelligent core as personality. From the perspective of conscious self, personality is normally an unconscious presence that maintains our form and purpose. It functions as our higher self or soul.

Conscious self is the experiencing aspect of a life field. Our conscious self is also who we think we are and represents our perspective of reality.

Between personality and conscious self is our mostly unconscious mind. Mind is like a computer that considers inputs from the environment and makes decisions about how to react depending on how that information compares to what is in the database called *Worldview*. As we enter into a lifetime, our worldview is populated with human instincts inherited from the human's body mind. Also, at birth, Worldview includes what can be thought of as spiritual instincts and a degree of understanding inherited from personality.

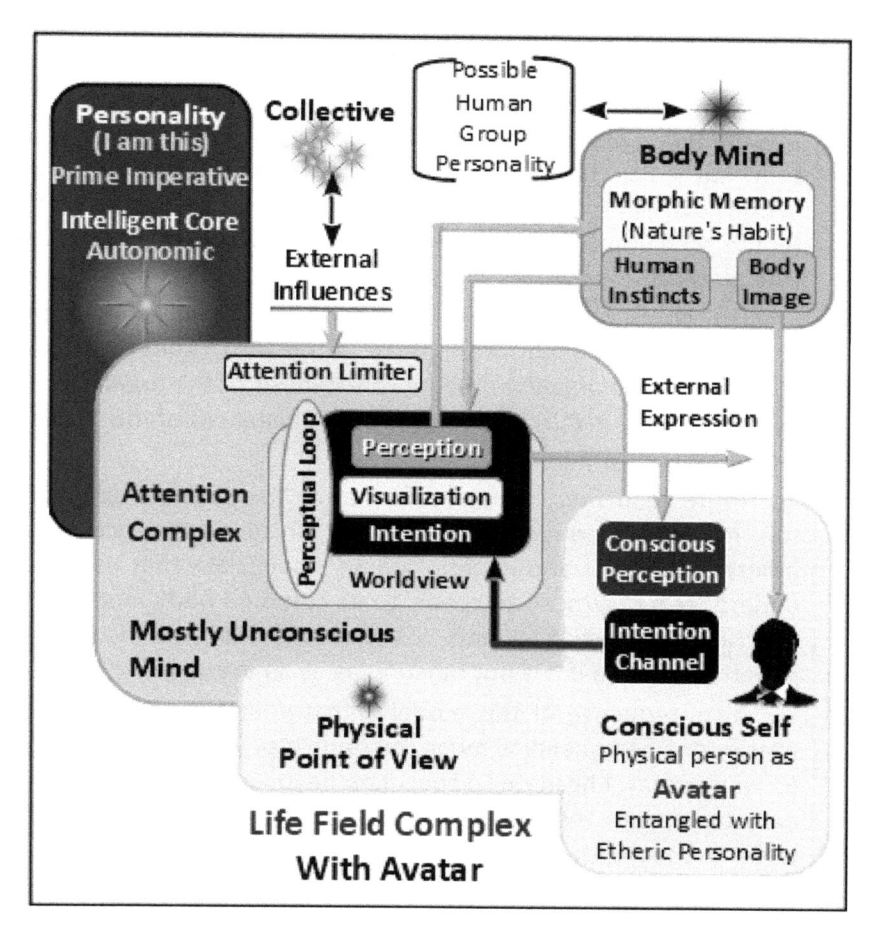

```
┌─────────────────────────────────────────────────────────────┐
│                            Possible                          │
│   Personality    Collective  Human                           │
│   (I am this)                Group         Body Mind         │
│  Prime Imperative           Personality                      │
│  Intelligent Core                        Morphic Memory      │
│     Autonomic                            (Nature's Habit)    │
│                     External             Human     Body      │
│                    Influences          Instincts   Image     │
│                                                              │
│               Attention Limiter          External            │
│                                          Expression          │
│                   Perception                                 │
│    Attention      Visualization                              │
│    Complex        Intention             Conscious            │
│                   Worldview             Perception           │
│                                                              │
│   Mostly Unconscious                    Intention            │
│   Mind                                  Channel              │
│                                                              │
│              Physical                Conscious Self          │
│           Point of View             Physical person as       │
│                                           Avatar             │
│         Life Field Complex          Entangled with           │
│           With Avatar               Etheric Personality      │
└─────────────────────────────────────────────────────────────┘
```

As we gain in maturity during this lifetime, we learn to manage our human's instincts, but in the process, our worldview is populated with cultural wisdom and what we learn from friends, schools, religions and the media. Most people live a life dictated by their human's instincts as they are moderated by cultural wisdom. Some, however, turn toward a more mindful way. It is in the nature of the Mindful Way that we see evidence that we have a purpose beyond simple survival of the species.

The consequences of duality and survival are that our human must be an independent life form. If it is, then it must have its own core intelligence. (15) The difference, I think is that our human is part of a collective with a shared Worldview. That is, our body's life field is like ours but with many conscious selves in the form of instances of that species, rather than the single conscious self we experience. The argument for this is a complex one, of which I have attempted to make sense in *Your Immortal Self*. (1)

> As it is modeled in the Implicit Cosmology (12), every instance of life is a life field with a Worldview function. Consequently, I speculate that each life fields representing a physical organism has a Worldview. But focus on the difference between an etheric life form (in the psi field) and aspects of reality to which we assign physicality (the physical).
>
> The physical organism is thought to be organized by a group consciousness thoughtform. Thus, a memory common to the physical species organizes the morphogenesis of the organism. That is, *Nature's Habit*, (15) is shared across all instances of the species as a common body memory.
>
> There seems not enough information to speculate about what this means in terms of a physical organism's sentience beyond inherited instincts and genetic coding. I speculate that our human undergoes a return, so to speak, to its common body memory, and that behavior of each instance of the species is entirely instinctual or genetically coded if it is not functioning as an avatar.
>
> A consequence of this model is that your pet may well be an avatar for a personality in much the same way your human functions as your avatar. I hesitate to speculate beyond that because it is too easy to get into the subject of Hinduism's transmigration of souls, which is beyond the scope of this essay.

The Mindful Way

Our mostly unconscious mind is our receiver of information. The best theory I have seen for this is that everything in reality produces a psi signal. (38) Our mind receives these psi signals but ignores all but those which are important to our body's wellbeing, and which are either directed to us or in which we have intended an interest.

The body's five senses must be translated into psi signals, presumably in the brain. A consequence of what our body senses being processed by mind is that it can as easily ignore what we physically sense as it can ignore information from other minds.

Mind has a set of functional areas which, in effect, asks Worldview if it recognizes the information. This question and answer may occur many times in sequence until mind either decides to ignore the information, modify it into a form which agrees with Worldview or modify Worldview to recognize the information in future encounters.

The result of this *"Do you recognize this?"* process is sent to conscious self, so that what we actually experience is a version of the original information based on Worldview. This tends to produce conscious perception that conforms with cultural expectation.

> I refer to this as cultural contamination because it is a hinderance to lucidity.

The only influence we have on mind is the expression of our intention. In mindfulness, we learn to intend to experience reality as it is, rather than how we are taught to believe it is. The idea is that we do not necessarily know the actual nature of reality because of cultural contamination. In effect, we live in a personal reality which is a version of actual reality. Through mindfulness, we evolve our perception toward the actual nature of reality by habitually questioning the implications of our perception.

Worldview has considerable momentum so that it tends to only change in small increments. First Sight Theory (38) teaches us that we can influence Worldview by intending certain behaviors. This tends to cause mind to turn toward one kind of information and away from others. By examining our every assumption with questions such as *"Does this make sense?"* or *"How will this affect others?"* while intending that we do no harm or intending that we see things as they are, we can begin the long process of aligning our personal reality with the actual nature of reality.

Natural Law

Natural Law is defined by the National Spiritualist Association of Churches (NSAC) as an *"Ascertained working sequence or constant order among the phenomena of nature."* They explain: *"Natural laws are simple statements* *of the orderly working of the universe and all that is in it. They represent the constant outward expression of what we can expect to happen in any given situation."* (8)

Infinite Intelligence is the Spiritualist equivalent of Source spoken of in this essay and the book, *Your Immortal Self*. It is understood that Natural Law is an aspect of Infinite Intelligence. As it is in the NSAC Declaration of Principles, (24) it is also understood that the goal of a Spiritualist is to learn to recognize and understand the principles and learn to live in accordance with their dictates. We have self-determination, but it is influenced by Natural Law. This philosophy is embodied in four of the nine principles of the Declaration of Principles:

1. We believe in *Infinite Intelligence*.
2. We believe that the phenomena of Nature, both physical and spiritual, are the expression of Infinite Intelligence.
3. We affirm that a correct understanding of such expression and living in accordance therewith, constitute true religion.

And

7. We affirm the moral responsibility of individuals and that we make our own happiness or unhappiness as we obey or disobey Nature's physical and spiritual laws.

Organizing Principles

The Implicit Cosmology, described in *Your Immortal Self*, has been developed from the implications of survival. It includes thirty-eight Organizing Principles. (27) There was no attempt to adhere to known natural laws which came to us by way of the Hermetic Teachings (28) and are from the human perspective. Organizing Principles became evident during the design of the Implicit Cosmology and are from an etheric personality perspective.

The concept that there are organizing principles is noted as one of the eight **Organizing Principles** related to formation of reality. It is defined as **Reality operates according to a body of Organizing Principles which are inherent from Source's creative expression.**

From the supporting text:

Source is modeled here in the sense of from whence it came rather than as a father god. Since reality is Source's life field, everything in reality is governed by the same rules of behavior governing Source and its expressions. Given the existence of Source and Organizing Principles, a logical argument could be composed to describe how the whole of reality might self-organize.

Curiosity

For the purpose of this discussion, the most important organizing principle is **Curiosity** defined as ***Curiosity is the source of attention.*** It is proposed in the Implicit Cosmology (12) that our curiosity is inherited from Source which is seen as self-aware and curious about its nature.

The *Anticipation Corollary* of First Sight Theory, (38) which essentially states that mind seeks to anticipate events, provides reasonably good support for this idea of a personality having a natural tendency to seek to understand its environment in order to anticipate changes.

In mindfulness, we express curiosity with the intention to understand how experiences align with what we think is true and what we actually experience. The key concepts here are *intended*, which is a conscious influence on mind, and *understand*, which is comprehension of the relationship between expression and perception.

Understanding

Thus far in this essay, I have argued that we are spiritual beings temporarily entangled with a human during this lifetime and that we are part of a collective of personalities who cooperate to gain understand which is intended to satisfy the curiosity of our nexus (top) personality.

If this argument is reasonably correct, then the reason for our existence is to seek understanding. Understanding in this sense is characterized as comprehending the principles of Natural Law or Organizing Principles. That is, our purpose is to come to understand the fundamental rules by which reality operates ... and their consequences.

In the ancient wisdom schools, students are often referred to as seekers. Seeking involves the process of intending to find understanding in experience. But more important, seekers look for experiences that will help them understand specific aspects of reality. You may be familiar with the idea of initiations. Seekers undergo initiations intended to establish that they have gained specific understanding.

During an initiation, seekers are challenged to answer questions about subjects that are unrelated to what they were taught. The expected answer requires sufficient understanding about what they were taught. In this way, understanding is seen as a universal knowing that is independent of the experiences which taught the understanding.

Understanding is Relative

Understanding is relative. By that, I mean that awareness of an underlying principle often leads to recognition that even more fundamental principles are involved. For instance, borrowing something from a friend brings the responsibility to assure its safe return. You might think the understanding to be gained is the nature of responsibility. It is, but as we proceed, it may become clear that it is also that borrowing obligates us to offer a favor in return, so obligation is a new understanding.

There are many possible concepts to understand that come from the initial decision to borrow something but underlying all of them is the important concept of cooperative communities. I define the **Cooperative Communities** Organizing Principle as *"An effort to express understanding is necessary for progression. Collectives are inherently cooperative communities. A person is attracted to communities of like-minded people cooperating to facilitate progression."*

The cooperative community concept applies to many kinds of interaction amongst people. Understanding one should enable you to apply it to others.

Purpose as Prime Imperative

Religion pretty much began with the message that our purpose is to follow the path of the Great Work, which is to understand *"The One Thing"* as it is addressed in the *Emerald Tablet*.

The *Emerald Tablet* (28) is widely believed to be one of the few surviving documents from the Hermetic Teaching thought to have been set forth some 6,000 years ago in Egypt. The text is very esoteric if you are not familiar with metaphysical concept. I have taken the liberty of paraphrasing it in Essay 17: *Hermes Concepts* (Page 285). (28) The paraphrase is included here. (Line numbers in parentheses are from Essay 17.)

Here, Hermes as talking to his students in conversational terms.

Lesson Name:

The Truly Great Work

    a. I can tell you as your teacher that your thoughts and your deeds are directly related so that your thoughts affect your expression, and your perception of that expression affects your thoughts. (Line 1)

b. Reality is both singular as Source and the expression of Source according to its intention. This expression of intention represents ordering principles which govern the adaptation of reality to individual purpose. The world you live in is an aspect of the greater reality as it is expressed by way of the Creative Process. (Line 2)

c. The Creative Process requires the visualization of the imagined purpose with the intention to make it so. (Line 3 and 4)

d. You, the person as an etheric personality entangled with a physical body, are the creator in this lesson. (Line 5)

e. And so, the creative influence produces all things in reality. The Creative Process finds expression through the informed intention of the person. (Line 6)

f. It is necessary to learn to distinguish between that which is part of actual reality as expression moderated by organizing principles and that which is perceived as real, but which is actually illusion. (Line 7)

g. Increased understanding of the actual nature of reality is contributed by the student to the collective of personalities in the greater reality, and thus merged, becomes available to the student as more profound understanding. (Line 8)

h. As such, you will find that understanding leads to clear sensing which enables a person to experience reality as it is, rather than as you have been taught. (Line 10 and 11)

i. Your increased understanding achieved through the Great Work may lead you to better living and increased stature in your community. (Line 9)

j. Thus, I have told you how the world has been created. But be mindful that these truths are not evident to those who have not stepped onto this path of learning. (Line 12)

k. As the teacher of this hidden way, I represent the three parts of a teacher. That is, I represent the understanding of the One Thing and The Great Work, I am an example of how you may integrate this understanding into daily life and in me you can see the possibilities of living this path. I am three times accomplished: as a teacher, role model and a citizen. (Line 13)

l. And now you understand the Truly Great Work. (Line 14)

It would be easy to translate the One Thing of the *Emerald Tablet* to mean god, but the Hermetic Teaching is about one god and many aspects of that god. Those aspects are the organizing principles which govern the operation of reality. Those principles are modeled as the expression of Source. In that sense, the One Thing is Source (God).

What I like to refer to as The Prime Imperative (23) is described in Line 6. Gaining understanding means learning to see reality as it is, rather than as we are taught.

References and Alternative Sources
Listed at the end of the book beginning on Page 357.

Essay 4
Immortal Self-Centric Perspective
2014

About This Essay

In his book *Far Journeys*, Robert Monroe described some of the lessons he had been taught while astral traveling. In one, he described two etheric tourists who came on a tour near the Earth Plane. Tourist AA decided he wanted to enter into a lifetime to see what it was like in the physical. Tourist BB was worried for him and advised against such folly. His fear was that AA would enter into a lifetime and soon forget who he was. (43)

AA did enter into a lifetime, and as BB feared, soon forget he was an etheric being only temporarily entangled with a human for a lifetime. As BB worried about AA and remained near to help as he could, AA entered into lifetime after lifetime, slipping ever further into the illusion of physical life.

As the story goes, there came a time in which AA began to remember his true nature and began the gradual process of regaining memory of his actual self. After many lifetimes more, he finally completely remembered his etheric nature and was able to escape the Earth Plane.

Ascended masters volunteering to enter into a lifetime in order to help humankind is an often-repeated concept. The idea is that even the spiritually advanced personality risks succumbing to the illusion of thinking they are their body and forget their spiritual nature.

Think of it. We are born into a body in which human instincts dominate. The brain is not fully formed, and therefore is not able to fully function as a transmitter-receiver for mind. Parents treat the person as if it is the body.

By the time the baby is sufficiently developed for the conscious self to begin exerting an influence, the experiencer aspect of the entangled personality (conscious self) is already convinced it is the body. The cycle only changes when the person finally achieves the presence of mind to question beliefs and its human's instinctual behavior. It is then that something like mindfulness becomes a viable tool to gain further understanding. The rest depends on the availability of teachers, a cooperative community and the person's determination to gain understanding.

This essay is inspired by my observations of the way people struggle to remain on the Mindful Way. Even after deciding to live mindfully, the illusion of being the body often biases lessons learned to cast everything in a body-centric perspective.

The intention of this essay is to convince you to stop and reconsider your perspective as part of your daily mindful behavior. When you ask yourself if what you are sensing is real, also ask yourself if you are sensing it from your human's perspective or your immortal self's perspective.

Learning to maintain an immortal self-centric perspective is an essential tool for progression. It is for that reason this was included in *Your Immortal Self* as Discourse 8. The other essays from that book are Essay 2: *The Mindful Way* (Page 37) and Essay 19: *Progression, Teaching and Community* (Page 319).

Introduction

This discourse is about our nonphysical nature: our conscious mind, memories, and that mostly unconscious part of our mind we sometimes meet in our dreams. It is for you to decide, but it has been my experience that our happiness and progression depend on our informed life choices. If the Implicit Cosmology (12) is reasonably correct, it must be understood that what we do in this lifetime will affect the rest of our eternity.

This subject is important because understanding the more universal nature of who we are will have an important influence on how we understand the implications of our life experiences.

Put a different way, we have been taught from birth that the limits of our word are what we can see, smell, taste, feel and hear. In fact, our physical senses only mark the limits of our human body's physical world.

Scientists who are willing to at least consider the nonphysical nature of mind tell us that even information from our five senses must be translated into the same psi form before their information can be considered. This means that our real sensory functions are better related to our core personality than to our physical body.

If this is true, then we have spent a good part of our lifetime totally misunderstanding what *real* actually means. That is a pretty shaky foundation on which to make life-changing decisions. The solution, of course, is to change our perspective through education.

Perspective

The perspective from which a question is considered has a lot to do with how it is answered. Our normal perspective is that of being our human body. We look at the world through our human's eyes from the perspective of within our human's head. This is natural. We have had that perspective since birth. However, if we are immortal personalities temporarily hosted by our physical body, the more correct perspective is to say *"This is my human, but my true self is not physical. I am not my body."* We might say that the usual perspective is *body-centric* but that the more correct perspective is *immortal self-centric*.

Here, *personality* is used to indicate the intelligent core of the complex of influences and functional areas that composes our whole self. In this view, personality is the immortal intelligent core while the human body is an avatar-host for personality's conscious self.

Our Body is a Complete Organism

As an organism that has evolved on this planet, our physical body is probably fully functional without our entanglement. It is proposed in the Hypothesis of Formative Causation (15) that the morphogenetic development of physical organisms, such as our body, is organized by a morphic field which supports the body image and the necessary rules for its development (known as morphogenesis).

Morphic fields are life fields but are described in terms of a physical organism. They include functional areas for Worldview (memory, Nature's Habit), perception, levels of consciousness and a means of expression. As such, the human body life field is essentially the same as our personality life field. By that, I mean that our worldview is directly comparable with Nature's Habit. The one major difference appears to be that our instinct is to gain understanding through experience and our body's instinct appears primarily to survive in the physical world.

The message is that, during our physical lifetime, we share Worldview and many of our perceptual and expressive abilities with our human. However, as shown in the *Life Field Complex with Avatar* Diagram (Page 67), the morphic field for our avatar also has a functional area representing *Nature's Habit*. The Body Image functional area is comparable to our personality's intelligent core.

Our Body as Avatar

Based on current understanding, our human body has a consciousness which is not completely suppressed during a lifetime. In fact, the relationship, is at least to some extent, a symbiotic one in which our daily choices are often greatly influenced by our human. A person gains in maturity by learning to cooperate with the body consciousness while remaining true to the ideals of balance and progression toward greater understanding.

It helps to understand how to distinguish between our body's instincts and our understanding. To do this, it is necessary for a person to recognize the difference between the inherited urge to understand and the body's instinctual urge to survive and perpetuate its kind.

It is also a challenge to distinguish between the beliefs that are taught by our local culture and the actual nature of reality as it is understood when those beliefs are superseded by understanding. This can best be accomplished when we are aware that we are not our body.

Avatar

In Hinduism, an avatar (from Sanskrit for *descent*) is characterized as a deliberate descent of a deity to Earth. The term can be translated into English as *incarnation*. **A person is an immortal self entangled with a human in an avatar relationship.**

In the avatar relationship, the personality remains associated with the etheric aspect of reality, but its perspective is the personality's conscious self, as it thinks it is the physical body. In trying to understand this relationship, the most important thing to remember is that we do not know much about how humans might behave if they did not have an entangled personality. While it is necessary to make a few assumptions for informed speculation, the underlying rule should be that humans are life forms which deserve respect and good care as hosts that enable our existence in the physical.

The relationship between etheric personality and the human avatar is shown in the *Life Field Complex with Avatar* Diagram (Page 67). Notice in the diagram that the Attention Complex (middle) is shared by the etheric personality (upper-left) and the complex representing the human body consciousness (upper-right). The actual human body is in the box marked Avatar (lower-right). This relationship is discussed below.

Cooperation Between Personality and Avatar

At the time of personality's entanglement with its human avatar, its Worldview is populated with a subset of understanding and the urge to gain further understanding. All of this is inherited from a source personality. As is shown in the *Life Field Complex with Avatar* Diagram (Page 67), the conscious self and its human avatar share Worldview. This shared memory is the main point of this discussion, because beginning at birth, the dominant conscious self must learn to manage the human influence, even as it learns to adapt to its local reality.

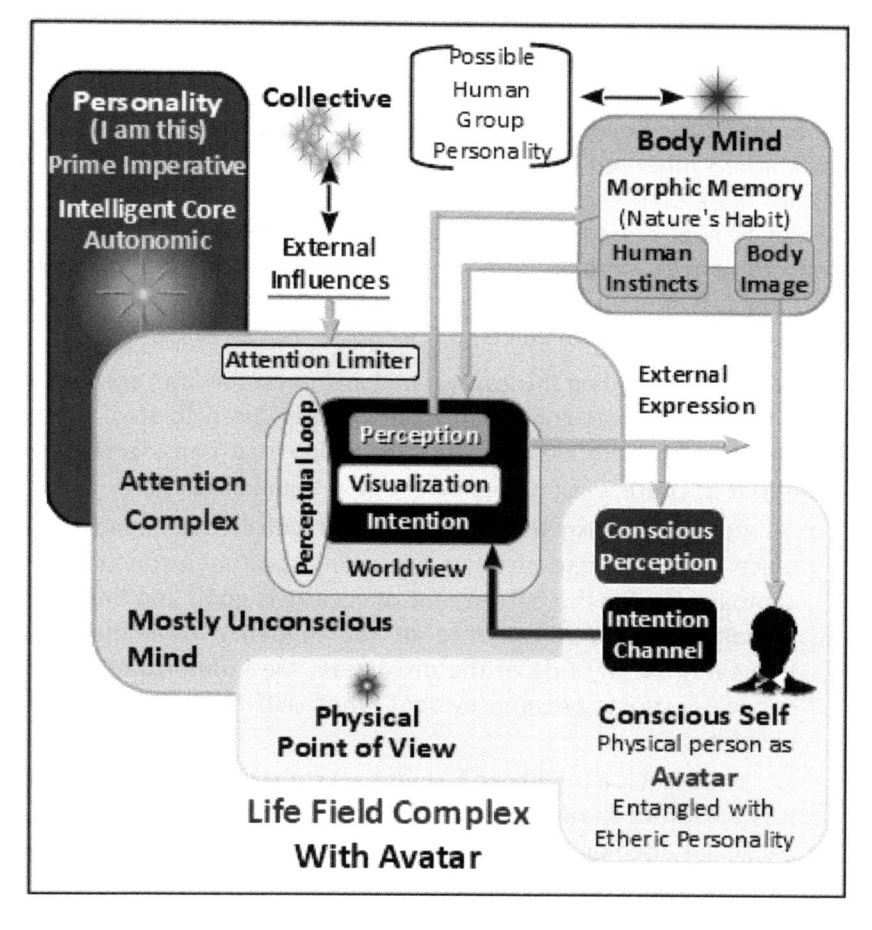

Life Field Complex With Avatar

An example of the human influence's persistence is demonstrated by the way we frequently hear in transcommunication that newly transitioned communicators still identify with their physical bodies. Some communicators even report something like a *getting well* period as they become accustomed to healthy mind and body. This suggests that people remain under the influence of the human body image well into transition.

Understanding the avatar relationship can help us live a productive life from the perspective of following inherited urges from personality (spiritual instincts) while managing our body's instincts. The challenge is in learning how to distinguish which are our body's issues and which are real to our personality. Essay 2: *The Mindful Way* (Page 37) addresses this.

Each avatar relationship begins with a blend of personality and human body traits. One way to make an informed guess as to what those traits are is to look for those that we were born with that influence our behavior today. If the avatar hypothesis is correct, then the human part of our worldview should be memory, beliefs and instincts related to survival. Understanding and inclinations inherited from our etheric personality should be related to increasing understanding or progression of personality. For instance, responding to peer pressure would support herd or tribal safety but it would also tend to restrict learning. In a similar way, compulsive behavior would seem to suggest human instincts while habitual curiosity would seem to support learning (spiritual instincts).

> While proofreading this essay, I had an interaction on Facebook with a fellow who was convinced Hillary Clinton tried to steal the 2015 election from Bernie Sanders. He clearly had considerable anger about it. There was no possibility of a rational discussion.
>
> Political activism that is not balanced with an open mind is a likely place to find human instincts overriding spiritual instincts. Actively campaigning for the betterment of society is good and important. I submit doing so is evidence of the influence of immortal self. However, for any side of the discussion, the influence of hindbrain survival instincts become evident when activism turns to belief in absolutes.
>
> Tribalism is a bright red flag when it comes to political ideologies. It is a form of herd behavior that manifests as dogged loyalty to a particular ideology. This behavior seems only possible if human instincts are in full control of the member of the tribe.

Survival of Body Mind

It is difficult to say that the human exists just as an avatar for etheric personalities, and then to say that all life forms have a personality. The collective model probably applies for all life forms, but is there a difference in the character of consciousness? We know that animals exhibit self-conscious behavior, clearly indicating that they are more than simply a collective mind, but is there a difference in purpose?

The physical organism body consciousness is modeled in the Implicit Cosmology as being part of a collective, as well. The difference is that it is more of a group consciousness as described in the Hypothesis of Formative Causation. (15) In that model, a morphic field has access to a worldview-like database that is a collection of *Nature's Habits* for that family of closely related organisms. For the human, it represents how the human body has evolved since its origin.

This is an area that needs much more consideration. It is important to know that there is nothing in the Trans-Survival Hypothesis that argues it only applies to people. How it might apply to a loving pet is beyond my speculation at this time; however, there is nothing in the cosmology that says other animals cannot be avatars for immortal self.

Personal Style and Astrology

Just as people are born left or right-handed, people are born with basic personality traits that tend to influence their behavior. Personal styles are cataloged and studied in psychology as a means of understanding human behavior. (44) They have also been adapted to teach salespeople how to relate to customers. In 1981, David Merrill and Roger Reid published a book reporting their study of corporate personnel interactions which became something of an industry standard. (45) They noted four main personal styles:

Analytical: Thinking, thorough, disciplined

Amiable: Supportive, patient, diplomatic

Driver: Independent, decisive, determined

Expressive: Good communicator, enthusiastic, imaginative

Each basic style is typically further divided so that a person might be seen as a Driver-Analytical or a Driver-Expressive. The point of these styles is that people likely begin dealing with a situation from the perspective of one of these styles. The question is whether or not this inclination is inherited from personality or from avatar.

The idea of astrology is that people's behavior is influenced by the astrological conditions at the time of their birth. A year is divided into twelve signs based on the ancient zodiac and each indicates a different set of personality traits. The *Personal Styles* Diagram (Page 70) shows a suggested relationship between astrological signs and the four personal styles.

It is important to note that personal styles and astrology are not being recommended here. They are used to demonstrate that we tend to display personality characteristics that are evident at a very young age and which tend to shape our lives.

Aries	Sagittarius	Aquarius	Libra
Driver Self-initiative	**Driver-Expressive** Where others fear to go	**Expressive-Driver** Visionary Speculative thinking	**Expressive** Cooperative relationships
Intuition	Intuitive thought		Thinking
Leo			Gemini
Driver-Amiable Steward and hope	**Driver-Analytical** Mission of discovery	**Expressive-Amiable** Minister to the heart	**Expressive-Analytical** Brings order
Intuitive feeling	Fire	Air	Logical thinking
Pisces	Water **Balance**	Earth	Taurus
Amiable-Driver Self-sacrifice	**Amiable-Expressive** The great cause	**Analytical-Driver** Methodical researcher	**Analytical-Expressive** Teacher
Empathetic			Empirical thinking
Cancer	Scorpio	Virgo	Capricorn
Amiable Brotherhood Of man	**Amiable-Analytical** Desire and intent	**Analytical-Amiable** Discrimination	**Analytical** Functional integrity
Feeling	Self-control	Sensory feeling	Empirical

Personal Styles with Astrological Houses
Styles and signs are ways of modeling observed human behavior. This is not an endorsement of either model.

There appears to be general agreement between the systems; however, the personal styles are more often described in terms of information acquisition or community, while the astrological signs tend to emphasize the same sort of characteristics usually associated with instincts. For instance, an

Aries, who is a person born between March 21 and April 20, is described as (amongst other qualities) impulsive, physical and driven. Aries is related to a Driver in the Personal Styles Diagram, and a Driver is described in terms like independent, decisive and determined. A Capricorn (December 23 to January 20) is described as (amongst other qualities) instinctive, over-reacting and moody. A Capricorn compares to an Analytical who is described in terms of thinking, thorough and disciplined.

Since there is so little known about this from the immortal self-centric view, it must be left for us to take the initiative to study and self-analyze. The interpretation of astrological signs and personal styles offered here is just an opinion, and your experience may be different. Again, these are examples indicating the kind of cues you would look for if you conduct a self-appraisal.

Balance

Notice the center circle in the *Personal Styles* Diagram (Page 70) labeled *Balance*. As you come to better understand yourself and your avatar, and therefore gain in maturity, you will find yourself converging on the middle way. One of the most important secret lessons taught by ancient wisdom schools is that balance is the middle way toward maturity. This does not mean a balanced person would never express extreme behavior, only that such a person would consciously do so, tactically for a purpose, and then return to the center without attachment for the outcome of his or her actions. See the Essay 2: *The Mindful Way* (Page 37).

Irrational Behavior

Do you have a pet, perhaps a dog or cat? If so, you may have noticed a lot of behavior that seems to be irrational. For instance, your pet might be fearful of an unexpected object in the backyard, or it might be unreasonably afraid of men, especially strangers. Sure, you would think it is just behaving like an animal. Animals often seem irrational. That is one of the distinguishing characteristics between humans and animals.

But what if you have irrational behaviors? Are you afraid of the dark? Do you go out of your way to avoid strangers? Is it difficult to communicate with people of the opposite sex? Are these behaviors rational? Do you have reasons for them which are more than just an excuse?

Everyone has fears that might be more exaggerated than circumstances would seem to require. The usual way of dealing with them is to talk the person out of the fear, either by appealing to logic or by showing that there is no need for fear, in effect, to wear out the fear response.

In the body-centric view, we naturally appeal to the rational mind, which is that part of us that is supposed to be logical, thinking and learning. However, in the immortal self-centric view, it can be seen that the human animal is the source of irrational fear. Because of the entangled avatar relationship, the human instincts are part of the shared Worldview, which is in turn, the governing factor of the Perceptual Loop of a life field's Attention Complex. As such, too often, the strongest and first response to an external influence is the human animal's fight or flight reaction. See the "Perceptual Loop and Worldview" (Page 18) in Essay 1: *Conditional Free Will* (Page 9).

It is natural that a young person will more often respond to circumstance from the human's instincts, but over time, it is expected that the person will more often have a rational response. That is the idea of maturity. Even so, people are plagued all of their lives with an unrecognized human influence. So long as they think with a body-centric point of view, it will be natural for them to seek to suppress the animal response, rather than understand the necessity to more directly manage the Perceptual Loop through mindfulness.

A better model for therapy might be to learn how to appeal to the animal in us, rather than the rational personality. We need a human whisperer more than we need a psychologist.

Degrading Avatar Relationship

There would be circumstances in which personality is forced to withdraw from the human avatar. Of course, physical death of the human is the usual reason, but personality might also withdraw if the body is physically healthy but is otherwise no longer able to support the expression of conscious self.

The inability of the brain to continue functioning as a physical-to-etheric transducer in cases of senility would be such a situation. This would not necessarily be a forced withdrawal, but there may be little reason for personality to continue the avatar relationship if personality's intention to gain understanding through experience in the physical can no longer be supported.

Personality's withdrawal from the human may be temporary, as is seen in the case of physical injury to the brain or coma. The reports of a persona with advanced senility who becomes temporarily lucid is another example that might be explained by temporary voluntary withdrawal.

Keep in mind that this model recognizes that the human is a complete life form. As is illustrated in the *Life Field Complex with Avatar* Diagram (Page 67), the Body Mind functional area includes human instincts which are filtered in the Attention Complex before presentation to conscious perception of the organism. In the event the personality disengages from the human, the human's perception would no longer be moderated by the expression of intention informed by personality.

A human that has experienced a lifetime entangled with personality would not be well equipped to continue functioning as a person without the rational influence of conscious self. At the same time, the Worldview function shared during entanglement would continue to be part of the human's perception and expression processes. The missing influence would be intention expressed by conscious self.

Extrapolating what I know about human behavior, and relating that to the Implicit Cosmology, I would expect to see the human-minus conscious self to physically appear the same, but to exhibit changes in behavior which should be predictable from the model.

Worldview is memory, and the Perceptual Loop in the Attention Complex is moderated by intention. Without rational intention informed by personality, the perception-expression process would tend to present stream-of-consciousness thoughts to the human's conscious state. This would look like waking expression of the kind of mostly random dream imagery we experience in light sleep. (46)

The human's instincts would no longer have the rational influence of conscious self, and would therefore, have greater influence on the perception-expression process. This would manifest as more animal-like responses and a tendency to fixate on or obsess about otherwise trivial things in the environment. This would likely include rapid, swings of emotion from passive happiness to aggressive behavior.

> Terminal lucidity is a term used to describe how people near the beginning of their transition, and who suffers from dementia, suddenly become coherent. They are able to say goodbye to their loved ones before their physical death. If the entangled personality has disengaged from the human, terminal lucidity may be explainable by personality's return for the transition. Other unusual terminal behavior might also be explained by examining the personality-avatar relationship.

May I Introduce Myself?

With these considerations, then, how should we think of ourselves? How do we think with an immortal self-centric perspective? Each of us will have different takes on this but there are general points that should be considered.

To begin, try thinking of yourself as two people. For instance, I am Tom Butler. *Butler* is the family name of my body, and it serves to give you a sense of its lineage. Perhaps we can call my body Mr. Butler. *Tom* is a good name for my personality in this lifetime, so a proper introduction might be *"Hi, I am Tom, and this is Mr. Butler."*

Mr. Butler was born in May, and according to astrology, should be stubborn, think habitat is very important and have a strong dependence on tactile sense. When compared to characteristics of other signs, this is pretty accurate. According to the avatar model, these characteristics should provide hints about the nature of Mr. Butler.

My astrological chart always seemed to indicate more what I need to overcome than what drives me, but I also fall into the Analytical-Expressive side of the personal styles chart, which is where I feel comfortable. Again, according to the avatar model, these characteristics should provide hints about Tom's nature.

References and Alternative Sources
Listed at the end of the book beginning on Page 357.

Essay 5
Ethics as a Personal Code for Mindfulness
2014

About This Essay

Right behavior has always been a focus for us as directors of the Association TransCommunication. We are also ordained Spiritualists, for which we are mindful of the need to represent Spiritualism in good light. Most importantly, we have accepted the evidence that we are immortal selves and what we do now will have an influence on the rest of our existence.

It is because of our immortality that we pay so much attention to Natural Law and the implications of our actions. Being in my seventies, many of the people I have known have made their transition. Some of my past relationships have not ended in the best light, yet I know that the fact they are possibly no longer in the physical does not mean I no longer need to worry about the consequences of my actions. There will come a time in which I will want to honestly examine my actions. The fewer such times I need to face, the better, so I am trying to learn to be nice.

Being nice may be a learned response, but most of the lessons we are taught in school about being nice are seriously contaminated by cultural norms. For instance, it was acceptable for me to light up a cigarette after dinner … still at the dinner table … even in a restaurant. Remember those days?

So, being nice is a relative attitude. Ethics are not. Even though people will try to explain ethics in terms of social norms, ethics are a bedrock concept which must be based on our immortality.

Seth is a good teacher for this. When I was submitting the *Handbook of Metaphysics* (2) manuscript to various publishers, one told me that he had 600 channeled books in his cue waiting to be reviewed. Anyone with a little time on their hands can write a philosophical dissertation and call it channeled. Most I have examined show they are fatally flawed by cultural contamination. The words attributed to Seth tend to rise above local norms.

I also like Jane Roberts' Seth Material (47) because at least one study indicates the information attributed to Seth does, indeed, come from a personality other than Jane or her husband. There is reason to accept its validity as originating from a discarnate personality. (33) Besides that, it is difficult to find metaphysical lessons that ring so true. We must consider our teachers wherever we find them.

For this essay, the most important advice we have received from Seth is that we should not violate others. I like to paraphrase that as *"I must not impose my will on others."* As a litmus test, I ask if my actions interfere with another person's self-determination. My bedrock ethics begins there. All else come for that litmus test.

Here are a few practical applications of the *Thou Shalt Not Violate* Clause:

- Loud noises can evoke all kinds of reactions in involuntary listeners. Making a person anxious or angry with loud music or aggressive actions is both a violation and can cause strong reactions. As a practical matter, the violator depends on the victim's respect for the rule of law to prevent an aggressive response. Alternatively, the victim may simply be too fearful of retaliation to speak up. Else, there might be mayhem.

- It is fine to select your friends and associates but check your reasons. If you avoid someone because of what you have been taught or because they are different, it is possible you are being prejudiced. Prejudicial actions are a form of imposing your will on others. Also, they potentially deprive us of opportunities for greater understanding.

- Behavior that is intimidating to others may be an imposition of your will on them. For instance, a man in typical ridding leathers and club logo on his back, passed us on a large, loud motorcycle. He had a pistol strapped to his leg. My first reaction was concern that the man was not safe to be around. His every action seemed designed to impress his will on people around him. Not to mention that he is likely the last sort of person we want to see packing a device only intended to kill.

- The Academic-Layperson Partition is a self-imposed separation by people with an advanced degree from people without an advanced degree. Again, it is fine to prefer to be around peers, but science is a service to humanity, to a great degree paid for by the public through grants and the funding of universities. Ignoring the obligation of science is a blatant violation of citizens' trust in the form of intellectual prejudice. The violation is in the form of the abdication of duty implied by title.

You can probably find similar examples from your experiences. The point is that we often unconsciously violate others by acting without examining how our action affects others. That is the opposite of mindfulness.

This essay began as my effort to understand whether or not I was out of line for thinking a supposed scientist had violated a research subject. Essay 13: *Arrogance of Scientific Authority* (Page 205) was the first writing to come of my study. As I considered writing Essay 14: *Open Letter to Paranormalists* (Page 223), it occurred to me that a code of ethics had important potential for our spiritual seeking.

As you read this short essay, imagine yourself as a Seth trying to explain the importance of seeking spiritual maturity to an interested but uninformed audience. How would you word the message? Use those words to help yourself see the importance of ethics to your spiritual maturity.

The Pledge

Introduction

The National Spiritualist Association of Churches (NSAC) Declaration of Principles functions as an outline for a personal ethics code based on the understanding that our *"... existence and personal identity continue after the change called death"* (Principle 4). When it comes to ethics in daily living, the personal responsibility described in Principle 7 sets the tone for the highest standard of right living. It reads: *"We affirm the moral responsibility of individuals and that we make our own happiness or unhappiness as we obey or disobey Nature's physical and spiritual laws."* (See nsac.org under the Spiritualism Tab for more on this) (8)

The simplest understanding to come from Principle 7 is first that we are responsible for our actions. And second, we benefit from understanding and living in accordance with the principles governing the operation of Nature. But how do we do the work to understand those organizing principles?

The Mindful Way represents techniques intended to help a person learn to be present in everyday life. The idea is to become aware of our actions by habitually questioning what we do and why. Mindfulness is a powerful way of coming to understand Natural Law and teaches our unconscious mind to replace belief in our worldview with beneficial understanding. It is our worldview that informs our mostly automatic responses to daily living. (48)

Morality Versus Ethics

Virtually all of the definitions of *morality* and *ethics* I have found are based on a body-centric perspective with no consideration of our immortality. To make my point, I have taken something of a backdoor approach by including reference to two different philosophical essays:

Morality

"The Definition of Morality" by Bernard and Joshua Gert provides an excellent discussion of morality and ethics. The authors suggest that one of two dominant perspectives on morality might be taken by theorist: (49)

1. *Descriptively to refer to certain codes of conduct put forward by a society or a group (such as a religion), or accepted by an individual for her own behavior, or*
2. *Normatively to refer to a code of conduct that, given specified conditions, would be put forward by all rational persons.*

The authors propose that: *"If one uses 'morality' in its descriptive sense, (Perspective 1) and therefore uses it to refer to codes of conduct actually put forward by distinct groups or societies, one will almost certainly deny that there is a universal morality that applies to all human beings."*

They explain that: *"Those who use 'morality' normatively* (Perspective 2) *hold that morality is (or would be) the code that meets the following condition: all rational persons, under certain specified conditions, would endorse it."* And later in the essay, "... *"virtually all hold that 'morality' refers to a code of conduct that applies to all who can understand it and can govern their behavior by it, ..."*

Ethics

"Aristotle's Ethics" by Richard Kraut includes this overview of Aristotle's point of view about ethics in the Preamble to the essay. In part: (50)

Aristotle follows Socrates and Plato in taking the virtues to be central to a well-lived life. Like Plato, he regards the ethical virtues (justice, courage, temperance and so on) as complex rational, emotional and social skills.... What we need, in order to live well, is a proper appreciation of the way in which such goods as friendship, pleasure, virtue, honor and wealth fit together as a whole. In order to apply that general understanding to particular cases, we must acquire, through proper upbringing and habits, the ability to see, on each occasion, which course of action is best supported by reasons.

Therefore, practical wisdom, as he conceives it, cannot be acquired solely by learning general rules. We must also acquire, through practice, those deliberative, emotional and social skills that enable us to put our general understanding of well-being into practice in ways that are suitable to each occasion.

Consider this about Aristotle's philosophy from the perspective of our immortality and the idea of a cooperative community.

Kraut concludes his essay with:

(Based on Aristotle's writing) ... *Human beings cannot achieve happiness, or even something that approximates happiness, unless they live in communities that foster good habits and provide the basic equipment of a well-lived life.*

The study of the human good has therefore led to two conclusions: The best life is not to be found in the practice of politics. But the wellbeing of whole communities depends on the willingness of some to lead a second-best life—a life devoted to the study and practice of the art of politics, and to the expression of those qualities of thought and passion that exhibit our rational self-mastery.

Both terms have very similar meaning, but in practical use by contemporary society, morality is most often defined in terms of what the organization expects of its members. This could be a religious moral code, a corporate code of conduct or an institutional one such as the expected behavior of college students. Even when a code includes references to ethics, the code virtually always requires compliance with social norms which are typically described in terms of morality.

The point of view suggested by the Trans-Survival Hypothesis (10) is that we are immortal self and must not be governed solely by cultural (local) norms. If we succumb to behaving according to what is socially right, rather than what is spiritually right, we effectively abdicate our responsibility to seek spiritual maturity.

The more universal meaning of right and wrong is expressed in terms of ethical behavior. Ethical also has less burden of meaning from religions and is used less for social engineering. Thus, the definitions I use here are:

> **Morality** is defined here as *a distinction between right and wrong based on local standards of behavior.*
>
> **Ethics** is defined here as *a distinction between right and wrong; based on the organizing principles governing reality.*

Learning to recognize and understand the universal principles that moderate the organization of reality is an important part of the Mindful Way. It is easy enough to ignore those principles in the short term but learning to habitually live in agreement with them is the path of least resistance toward spiritual maturity.

First Ethical Consideration

Useful guidance in ethical conduct came from Jane Roberts' Seth in regard to how a person should interact with others. The advice is simply that *"Thou shalt not violate…."* (47) Seth went on to explain what he intended by violate:

> *An outright lie may or may not be a violation. A sex act may or may not be a violation. A scientific expedition may or may not be a violation. Not going to church on Sunday is not a violation. Having normal aggressive thoughts is not a violation. Doing violence to your body, or another's, is a violation. Doing violence to the spirit of another is a violation, but again, because you are conscious beings the interpretations are yours. Swearing is not a violation. If you believe that it is then in your mind it becomes one.*
>
> *Killing another human being is a violation. Killing while protecting your own body from death at the hands of another through immediate contact is a violation. Whether or not any justification seems apparent, the violation exists.*

Seth explained that there are ways to deal with situations that do not involve killing. He also suggested that:

> *You would not be in such a hypothetical situation to begin with unless violent thoughts of your own, faced or unfaced, had attracted it to you.*

As for the Golden Rule, in Jane Roberts' *The Individual and the Nature of Mass Events,* Session 852, Seth says:

When you are discussing the nature of good and bad, you are on tricky ground indeed, for many—or most—of man's atrocities to man have been committed in misguided pursuit of "the good."

Ethical Treatment of Human Research Subject

The Belmont Report published by the U.S. Department of Health and Human Services gives us another measure of ethics. (51) It appears to be the golden standard for research ethics involving a human subject. The major points from the Belmont Report are that researchers must respect the person, do no harm (beneficence) and provide due benefits (justice). These three points provide guidance in how to define basic ethical principles.

I discuss the question of ethics in research in more detail in Essay13: *Arrogance of Scientific Authority* (Page 205).

Ethical Understanding	Ethical Principles	Ethical Expressions
Do not violate *Seth*	• Respect • Kindness • Do no harm • Justice • Fairness • Suspended judgment • Courage • Discernment	• Just because I can, doesn't mean I should • Mindfulness is a way of life • Citizenship means cooperation • How will my actions affect me and others? • Is it a belief or a supportable understanding? • I will not impose my will on others • Lessons come from new experiences • Contemplation not meditation

Possible Mindfulness Personal Code of Ethics

A Useful Code of Ethics

Here is a suggestion for a personal code of ethics. It should be useful for anyone in any part of society. The idea is to stay with a foundation ethical concept for Tier 1 that sets the tone for the code. Seth's *Do not violate* is an excellent foundation concept. Tier 2 is concerned with reasonably basic principles which complement or further define *Do not violate*. These should be intuitively obvious in the context of your personal progression. Tier 3 is concerned with how Tier 1 and 2 are expressed. Expressions include phrases intended to provide guidance for how to live by the Organizing Principles. Catchy phrases are useful here, as they make it easier to remember the principles.

You will likely want to add Principles and Expressions as you become used to working with the code. Be careful not to overcomplicate it, though. It is important that you can remember the elements so as to apply them as warranted.

Ethical Conduct is a Lifelong Learning Experience

Learning to live by a personal ethical code often means realigning our unconscious mind away from our human avatar's survival instincts and cultural dogma with which we have been conditioned over the years. Such a change in consciousness takes time and attention that comes through experience.

Remember that your conscious expression is first formed in your mostly unconscious mind. That means you have relatively little control of your first response to situations. The control you do have is before the event by consistently intending to act in a mindful way. A feedback expression to yourself after the event expressing how to be more mindful in a specific way helps to reinforce your intended behavior message to your mind. Yes, talk to yourself.

The process of managing expectations of those with whom you share time, in friendship or service, provides opportunities to exercise mindfulness. Unspoken questions and concerns can cast a shadow over a relationship. An Expression for the Principle of Kindness might be *"Citizenship means cooperation"* or *"How will my actions affect me and others?"* In practice, these translate into making sure people know what to expect from you.

Above all, think of a personal code of ethics as a lifelong way of learning. It is unlikely any of us are able to always live up to the ideas represented by a code of ethics. The most important thing is to set our intention to apply the code to our every action.

References and Alternative Sources
Listed at the end of the book beginning on Page 357.

Essay 6
Paranormalist Community
2014

About This Essay

As explained in Essay 9: *Consensus Building in the Paranormalist Community* (Page 129) **Paranormalists** are defined here as *people who experience, study or have a more than casual interest in psychic ability* (psi functioning, remote viewing, healing intention), **healing intention** (biofield healing, distant healing, healing prayer) *and the phenomena related to survival of consciousness* (mediumship, visual and audible ITC, hauntings).

Although individual paranormalists may not accept survival, or even psi phenomena, all are concerned with study or application of the phenomena. The actions of one member of our community reflects on all of us. We all share a need for useful models describing the phenomena and theories as to their nature.

The *Blind Men Describing an Elephan*t parable applies to the paranormalist community in that each sub-community of interest has a part of the answer but is unlikely to have the correct answer without collaboration with the others.

Our efforts as directors of the ATransC to collaborate with other organizations has been only marginally successful. As it turned out, each existing organization had a different focus. There was, and remains, a serious problem of *founder's syndrome* in which opinion leaders tend to emphasize local pride rather than community support. It also seemed that some saw us as a threat. We have never been a threat, as we have never wanted the ATransC to be an umbrella organization.

Organizations based on a more academically trained membership are a different story. While such groups as the Parapsychological Association, (52) the Society for Psychical Research (53) and the Rhine Research Center (54) have different origins, they often share resources. People with an advanced degree are typically culturally conditioned to collaborate and support what they see as their peer community. Someday, I hope to see them turn that human resource to support of the larger community.

It may help you understand this essay if you understand how I model the paranormalist community. It is composed of several subcommunities of interest, primary of which are:

- **Local ghost hunting clubs** – These are often inspired by the popular ghost hunting programs on television. From my experience, they are typically local clubs that organize meetings and local ghost hunts, more for sport than science, but they provide a great introduction to things paranormal.
- **Pop culture-like general interest groups** – Usually supports a wide range of phenomena. In some cases, as with the ghost hunting clubs, member participation rather than science is the primary objective. From my experience, an underlying vision of their founders is to be the dominant organization; baseless claims of being scientific are common.
- **Specialty groups** – The ATransC began as the American Association of Electronic Voice Phenomena (AA-EVP). Even with the current broader focus on all kinds of transcommunication, the ATransC is still best described as an example of a specialty group. Groups specifically focused on near-death and out of body experiences are other examples.
- **Citizen-scientist groups** – There are a few organizations that seek to further understanding of one or more aspect of the paranormal while including both academically trained people and laypeople. Below, I mention the ASCS as an example.

The *citizen scientist* concept has a rather different meaning in the paranormalist community. In most cases, people trained to conduct proper research are either subversive debunkers or too poorly informed to properly apply the scientific method. That has left laypeople to fend for themselves when it comes to developing rational models and practices. I refer to those who try as citizen scientists.

- **Parapsychological groups specifically for Ph.Ds.** – These are professional associations that only incidentally include laypeople. Collaboration amongst similar parapsychological groups is common. From my experience, claims to be scientific are often only partially true, and the science is often inappropriate for the subject and may include false claims of objectivity.

The Academy for Spiritual and Consciousness Studies (ASCS) (55) has been the most supportive of our community's general interest as a citizen scientist group. Struggling to survive, the ASCS is faced with the proverbial three-door challenge. By the time this book is published, the ASCS may have morphed into another group requiring a Ph.D. for full membership, a more progressive version of its current model or it may have faded away.

The point I want to make here is that the paranormalist community is actually rather small with few people inclined to lead. It is very difficult for an individual to find and follow the mindful way without the support of a community which includes wayshowers, seekers and skeptics. So, if you find a group with which you are comfortable, make an effort to support it by being involved. Interact to learn.

You are an important resource for our community. Examine the group you are in or intend to support. Most are beneficial to the community, but organizations that take a silver-bullet approach to collecting feel-good imaginary examples, rather than objectively evidential results, do more harm than good for the community. Test your teacher! Your support without examination will do little for your personal progression or the greater good.

The Paranormalist Community

Think of a community as a collection of people who share an interest in one or more ideas. If you have more than a passing interest in the ideas related to the paranormal, then you are automatically a member of what I refer to as the Paranormalist Community.

For sure, dogmatic skeptics have an interest in these phenomena, but all who outright reject things paranormal are so far off the rational chart that their negative-only approach must be ignored if we are to make progress. However, with that said, more moderate skeptics in our community who claim to study these phenomena, but who are really trying to prove they are delusion, need to be heard. It is by others questioning what we think is true that we become aware of the need to question those truths for ourselves. Else, we become complacent in a bubble of circular evidence.

My focus here is on we who have an affirmative interest in these phenomena. We must all deal with the same facts and the same misconceptions. And even though some of us reject the Survival Hypothesis, we are like the fabled blind men trying to describe an elephant with touch alone. If we do not work together as a community, it is likely we will not learn the actual nature of these phenomena.

Probably the most important unifying idea for the Paranormalist Community is the existence of a subtle field that permeates reality. This is referred to by parapsychologists as the *psi field*.

A second unifying idea is that it is possible to access information by way of the psi field. This can be thought of as *psi-sensing* or *being psychic*. (16)

A third unifying idea is that a person is able to affect objects of reality by way of the influence of intention on the psi field. This faculty is known in parapsychology as *psychokinesis*. (16)

There is considerable discussion about whether mind is a product of brain or if it is independent of brain. The idea that mind is a product of brain but is able to interface with the psi field is generalized as the *Super-Psi Hypothesis*. (56) Increasingly, this group is working under the banner of Exceptional Experiences Psychology. (57)

The idea that mind is independent of brain, and therefore existed before this lifetime and will exist in a sentient form after this lifetime, is usually generalized as the *Survival Hypothesis*. (56)

It is good to acknowledge that there is a group of people working as parapsychologists under the banner of Anomalistic Psychology who seek to prove reports of paranormal phenomena are not real, but only the imagination of the experiencer. Let us refer to their point of view as the *Physical Hypothesis*. (58)

A Divided Community

While everyone in this community is a paranormalist, there is little agreement about how to explain paranormal phenomena, and so, there are divisions in our community. The most obvious division is the mind-body debate. That is, is mind a product of brain or is it separate with brain functioning as a transmitter-receiver for physical senses and motor controls?

From the paranormalist perspective, brain and mind being separate would seem to require that mind survives death of the brain as a still-living personality. Information thought to be coming from discarnate loved ones via mediumship and Instrumental TransCommunication (ITC) would be coming from them as they now exist in a different environment. As noted, this idea is generally modeled as the Survival Hypothesis. (56)

If mind is a product of brain, then there is no survival. The information thought to be coming from discarnate loved ones would necessarily be coming from memories of them held by still living people or from the residual memory of loved ones theorized to remain in the psi field (aka Akashic or life records). As noted, the idea of the information via mediumship and ITC coming only from survived energy or living memory is generally modeled as the Super-Psi Hypothesis. (56)

The community is divided by what I refer to as the Academic-Layperson Partition. When it comes to understanding things paranormal, on one side of this partition are people who seek to understand these phenomena, usually without a Ph.D.-level education, functioning as citizen scientists. On the other side of the partition are academically trained people, always Ph.Ds., who have claimed the scientific high ground under the banner of parapsychology. Parapsychology has three primary points of view:

> The **Physical Hypothesis** point of view holds that paranormal phenomena are delusion, fraud or are mundane, mistaken as paranormal. In this view, the necessary science-based supporting mechanisms for paranormalist phenomena are not established, and therefore any reference to them must be pseudoscience. This is being addressed these days as Anomalistic Psychology. (58)

> In this, consciousness is considered a product of the brain and ceases to exist when the brain dies.

> > For conversational convenience, I refer to this point of view as **Normalist**.

> The **Super-Psi Hypothesis** point of view is the Physical Hypothesis modified with the contention that the physical universe is permeated by a psi field. (56) From this point of view, if not mundane, delusion or fraud, anomalously accessed information is produced via psychic access to residual memory or the mind of still living people. This is beginning to be addressed as Exceptional Experiences Psychology. (57)

> In this, consciousness is either a product of the brain or a psi field phenomenon originating from the brain.

> > For conversational convenience, I refer to this point of view as **Psi+ Normalist**.

> The **Survival Hypothesis** represents the point of view that we are immortal self temporarily entangled with a human for this lifetime, that our conscious self existed before this lifetime and will continue to exist in a sentient, self-aware form after this lifetime. (10)

For conversational convenience, I refer to this point of view as **Dualists**. For my personal study, I refer to the study of survival as Etheric Studies. (59)

A Community Divided Cannot Stand

Roy Stemman wrote an excellent essay about how Spiritualism is contracting while the skeptical community is becoming better and better organized. In *Skepticism: The New Religion*, he noted that organizations such as the James Randi Foundation have developed a unified message and are producing abundant literature aimed at casting all thing paranormal as pseudoscience, and therefore, a danger to society. (60)

The skeptical message is having the desired effect because some governments, including the US Government, have adopted the skeptic point of view, even to the extent of quoting skeptical literature to support policies concerning funds allocation for research grants.

The skeptical community maintains a list of experts in the skeptical point of view who are available for interviews. Their experts need not be experts in a particular science, only in the talking points about how things paranormal are illusion and paranormalists are either ignorant, delusional or frauds.

In comparison, the paranormalist community has no such focus. Because one parapsychologist with a Ph.D. probably looks like the other *experts*, we shudder to think what supposed experts in our field might say to the media. There are no talking points that many of us have discussed and on which we agree. Consequently, even if the person is conceptually correct, it is probable the person's version will be misleading and technically flawed.

Also, because of the super-psi - survival divide, there are too few widely respected people in our community to legitimately represent the survival perspective. It is common during an interview with a specialist in one aspect of paranormalist phenomena to be asked about other phenomena about which he or she has little experience. The response typically leans toward the skeptical perspective.

From the mainstream perspective, there is little distinction between ghost hunting by a club and serious research by qualified, informed practitioners. All of us are culturally conditioned to trust what a Ph.D. has to say over anything a layperson might say. Thus, it is all too common for a Ph.D. to speak with great authority about phenomena for which he or she has little understanding, under the cloak of parapsychology.

We have had more than a few parapsychologists pronounce with great authority that EVP are probably caused by stray radio signals. Yet, we have never had a conversation with them or had our work cited.

Without a well-considered public face, the paranormalist community will always look like ghost hunting television programs and meaningless noise passed off as Electronic Voice Phenomena. People remember the most outlandish and the silliest.

Skepticism is Relative

Each of us is potentially a skeptic. As a rule, the least deviation from mainstream thought is skeptical of the greater deviation, so that a psychologist studying remote viewing will tend to be skeptical of anything to do with survival. In the same way, a person who accepts the evidence of mental mediumship may be skeptical of the evidence for physical mediumship.

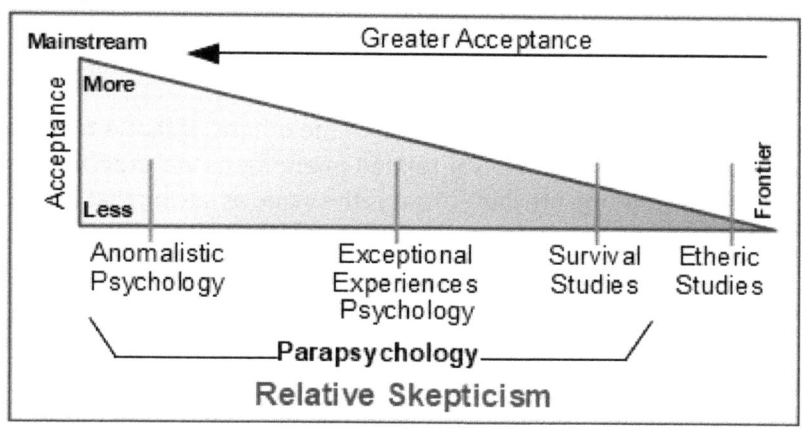

Of course, there is such a thing as healthy skepticism, but the term *skeptic* has been taken over by the zealous, faith-based skeptics who practice a form of scientism.

While we seldom actually refer to our fellow paranormalists as skeptics, the fact remains that some rather well-studied phenomena have been denounced without examination by people who should know better because of their study of less controversial phenomena. Behind the facade of academic authority too often lurks scientism, even in our community.

Gerhard Mayer reported in the *Journal of the Society for Psychical Research* that, based on a recent opinion survey of parapsychologists in Germany: (61)

> *The proportion of responses that disagreed with the statement 'After the physical body dies some part of the person survives' was much lower in the more international PA (14%) than in the GfA (36%) and WGFP (44%) samples. A third of the PA respondents instead displayed an undecided attitude (compared to 8% among the WGFP and 6% among the GfA).*
>
> PA = Parapsychological Association; GfA = Gesellschaft Für Anomalistick; WGFP = Wissenschaftliche Gesellschaft zur Förderung der Parapsycholodies.

A better term for critical thinking is discernment which promotes suspended judgment until more is known. This change of perspective from skepticism, even veiled skepticism, to the suspended judgment of discernment, is perhaps one our first lessons in community building.

Cultivating a Common Culture

Here I argue that the physical is an aspect of the etheric. If that is true, it must be argued that the psi and survival-related phenomena we experience share the etheric as a common attribute. That is the same as saying that all physical experiences share the physical as a common attribute.

If we tried to study gravity without considering the attributes of mass and momentum, for instance, we would always come up with theories that only addressed part of the equation. It is probably true that those theories would be wrong.

That is exactly what is occurring in parapsychology. Some parapsychologists (Normalists) strive to prove all phenomena are physical. Others (Psi+ Normalists) try to show that some may be physical, but many have a subtle-energy attribute they call psi. A very few honestly seek to show that survival-related phenomena are actual evidence of survival.

We are very much like the five blind men trying to describe an elephant. From my experience, it is not possible to study transcommunication without finding an alternative form of space to act as a propagation medium for thought. The psi field satisfies that requirement. Except for the hand-waving Psi+ Normalists must do to explain how information is somehow stored in the psi field, the Super-Psi Hypothesis is technically part of the Survival Hypothesis.

When a Normalist researches exceptional experiences, while deliberately and admittedly ignoring evidence provided via transcommunication, the result must always be wrong in the same way the blind man deciding an elephant as a snake is wrong.

Where to Begin?

The Survival hypothesis is based on the premise that we are immortal self temporarily in this lifetime to gain understanding. To explain that point of view, it is necessary to include consideration of virtually all paranormalist phenomena. Thus, the Dualist definitions for these phenomena are a useful place to begin developing a common culture.

Dualist Concepts Common to the Larger Community

One need not believe in survival to benefit from the more important implication of survival. The lessons learned listed here are pretty high-level, but the conceptual models that suggest their validity apply to the human condition in general, as well as the metaphysics related to the psi field.

Mindfulness: The possibility that we survive beyond this lifetime offers profound implications for how we might live this lifetime to the fullest. The most immediate implication is that what we do now matters now, and if we do survive, certainly hereafter. This translates into a need for mindful living that benefits anyone, paranormalist or not, by teaching good citizenship and ethical values for life.

Living as if life matters: We continue to exist, if not actually, at least through our legacy. Memory is modeled by Psi+ Normalists as a simple thoughtform of residual energy. Memory is modeled by Dualists as Worldview as it is part of and informs a sentient life form. Either way, we continue to have a presence in the psi field that can be sensed by others. With only slight modification, the metaphysics of how that might be can be applied to both Super-Psi and Survival.

The consequences of our actions during this lifetime unavoidably affect our community now and represent our legacy on which we can have substantial influence even after we are gone or transitioned.

Personal responsibility: Personal responsibility means learning to live as a good citizen of the greater reality, but more important, it means learning to make our continued existence as meaningful as possible.

The idea of personal responsibility is important to the entire paranormalist community, because even if our view is that only memory survives, the unifying idea that we are all connected via the psi field means that our actions affect the larger community on both sides of the veil.

Spirituality

A good citizen of the paranormalist community is one who seeks to be a contributing part of the community, rather than an accidental bystander. An excellent definition of *spirituality* is offered by Deepak Chopra in his *Huffington Post* blog: (62)

Spirituality is the experience of that domain of awareness where we experience our universality. This domain of awareness is a core consciousness that is beyond our mind, intellect and ego. In religious traditions, this core consciousness is referred to as the soul which is part of a collective soul or collective consciousness, which in turn is part of a more universal domain of consciousness referred to in religions as God.

Spirituality is about being aware that we are an important part of a universal community, what we do matters here and hereafter and that our every action affects the rest of reality. Yes, most of us are accustomed to thinking of the kind of spirituality traditional religions seek to evoke, but the kind of spirituality I am thinking about is the kind that, when personally realized, leads to greater understanding and the desire to be a good citizen of the collective (the community on both sides of the veil).

If you are familiar with the idea of mindfulness, you will understand that mindfulness is a roadmap to better living. Spirituality can be thought of as an eventual product of mindfulness done right. The characteristics of a spiritual person and a mindful person converge on agreement with the actual nature of reality.

Discernment

A favorite saying of mine is that you should *"Believe what you wish but understand the implications of what you believe."* Being mindful of what is in your worldview determines the world you live in. See James Carpenter's First Sight Theory (16) (18)

There has been a lot of discussion about how our expectations tend to determine what we experience. For instance, the Sheep-Goat Effect, so named in 1942 by Professor Gertrude Schmeidler, implies that people who allow for the possibility of new ideas (believers or sheep) are more likely to experience new ideas. (63) The inverse of this can be seen in the concept of incredulity blindness, a term I use to describe how some people are inexplicably unable to see or hear examples of paranormal phenomena (goats).

In research sponsored by the ATransC and conducted by Mark Leary, Ph.D., listeners were more apt to hear what was expected in ambiguous sound streams when told what to expect. (64) In practice, it is known that the mostly unconscious mind colors what is transferred to conscious perception based on Worldview and expectation. (16)

If perception is the dominant factor in how we experience phenomena. Discernment is the key to seeing phenomena as it really is. That is a skill all of us can learn.

Our mostly unconscious mind processes information with the intention of characterizing it based on familiar references. Once a decision is made, it is very difficult to change our mind. Thus, learning to intentionally postpone making an agree or disagree decision until more information is available helps keep the mind open to new ideas. It also helps avoid making the wrong agree or disagree decision. See Essay 1: *Conditional Free Will* (Page 9) for more information about how we think.

Cooperation

Read Deepak Chopra's definition of spirituality (Page 92) again. Especially notice the way he uses the term, collective. *Universality, collective soul, collective consciousness* and *universal domain* are all terms that describe interconnectedness amongst people.

A common factor in transcommunication is the implication and sometimes outright admission that the communicator is a group consciousness which presents a representative personality to facilitate our body-centric comprehension. Much of what is brought to us via deep trance channel, even our dreams, is the hint that we are part of a collective of personalities. While we experience a lifetime with a body-centric focus, our etheric personality, which is the intelligent core of who we are, is cooperating with fellow personalities to assure our continued education.

This idea of being an etheric personality, temporarily entangled with our avatar body to gain understanding, is admittedly more hypothesis than objectively known. The one persistent message in this concept that you will likely recognize is that we are ultimately a collective under one source.

The way that manifests in practical terms can be seen in how people mature, first as students and then as teachers. Many of us have the compulsion to share lessons learned. For instance, the reason for writing essays like this is specifically to promote cooperation in the community. People like me write them in the spirit of *"those who have come before pausing to help us achieve even greater understanding."*

The idea of a collective of like-minded people cooperating to promote further understanding involves the idea of rapport which is the link of attention/intention that develops between two people as they share ideas. (65) A community is formed by a sense of rapport which integrates many people and only marginally related ideas into a community of common interest. In such a community, educational resources, standards and a common voice to speak to mainstream society can evolve.

But only if we cooperate.

References and Alternative Sources

Listed at the end of the book beginning on Page 357.

Essay 7
Clarity of Communication
2014

About This Essay

It is likely that all of us have had the experience of explaining something to others, only to have them later indicate that they heard something very different. This is especially a problem when discussing conceptual ideas.

Ideas such as *words have meaning,* and *the power of words,* are common themes in my writing. The reason I keep writing about clearly communicating ideas and saying what we mean is that participation in discussions on the Internet has brought me face-to-face with the reality that I am a poor communicator.

It is surprising to me that my focus has recently changed from explaining the need to clearly express ideas to the need to pay attention to what is said. As I try and try to be clear, I see that my listeners and readers are not doing a very good job of hearing what I am saying. One might say they are inconvincible.

There is support for this! In recent years, I have noticed articles in the popular science media about how we do not directly experience our environment. (66) (67) (68) Instead, our mostly unconscious mind senses the information and only lets our conscious mind know what it thinks about the information based on memory. As I explained in Essay 1: *Conditional Free Will* (Page 9), this is not some New Age wistful thinking. It is emerging mainstream understanding that has profound implications for the paranormalist community.

This essay represents one of my first efforts to turn the perspective from teaching the need to be careful what we say to teaching the need to be careful how we listen. Yes, it remains important to clearly state the information, but clarity of communication represents a two-way exchange. It began as something of a rant that was to be titled *Inconvincible,* but after a few rewrites, I calmed down enough to change the perspective from complaining to honest questioning.

Introduction

What our audience hears us say is too often different from what we intend. Probably all of us have had the experience of explaining something, say giving instructions about how to do something, and then seeing the person do it wrong. *"Why?"* you might ask of the person. *"I thought you said to...."* is likely the answer.

The ability to communicate an abstract concept is essential for anyone wishing to share ideas about the paranormal. And, as it happens, virtually every idea about the paranormal is abstract. Lacking a common vocabulary, the words we use are often laden with unintended meaning. What does spirit mean? Is spiritual healing just putting a hand on someone's forehead and praying out loud to God? Are dead people demons if they talk to you? Is that person crazy for hearing dead people or just a witch?

How we answer questions like these depend on our listener's upbringing. A person who has spent most of a lifetime cloistered in a very orthodox religious family will likely hear words like spirit differently than a person raised in a Spiritualist household. A technical metaphysician will answer differently than a causal seeker. On the other hand, a casual seeker will likely be confused by a metaphysician's answer.

While I am no expert in communicating abstract ideas, I do have experience in communicating complex ones. From my experience, the *art of communication* is all about understanding that:

1. People have a style of learning.
2. What people read or are told is different from what they understand, and understanding is always based on their worldview.
3. People pay attention to, or ignore ideas, depending on how the information is delivered, their interest and learning style.

4. A mismatch of agenda impairs communication.
5. Worldview changes in small increments; learning hardly ever occurs in great leaps.

The two messages I would like you to take away from this essay is that sharing information means understanding your audience and your audience's perception requires their attention. It is especially important for mediums, healers and society leaders to be mindful of how their words are received. For instance, mental mediums speak under the cloak of authority as if they are holy people speaking in the name of the dead. Like it or not, that authority is unconsciously attributed to them by many in their audience. The result is that mediums have extraordinary responsibility to manage their comments so as to avoid unintended implications and exaggerated expectations.

> Here, it is important to picture yourself outside of the paranormalist community. Most paranormalists will know what you intend because they have direct experience. We are insiders, but people in mainstream society do not know what you intend and only hear your words in the context of their experience. For that reason, it is safest to assume members of your audience are literalists who will hold you to the literal meaning of every word.

People Have a Style of Learning.

Just as people are born left or right-handed, people are born with basic traits that tend to influence their behavior. Using the avatar model for a person, our entangled self brings a personal style in the form of prior understanding and inherited urges, but the human body also influences our decision-making because of the more dominant human instincts.

Personal Style

Personal styles are cataloged and studied in psychology as a means of understanding human behavior and have been adapted to teach salespeople how to relate to customers. A useful categorization is shown in the Personal Styles diagram below. (45) They are:

Analytical: Thinking, thorough, disciplined; always a student of the subject

Amiable: Supportive, patient, diplomatic, healer and caregiver

Driver: Independent, decisive, determined; always thinking about the next step

Expressive: Good communicator, enthusiastic, imaginative; often the opinion setter

Each basic style is typically further divided so that a person might be seen as a Driver-Analytical or a Drive-Expressive. The point of these styles is that people likely begin dealing with a situation from the perspective of one of these styles. Styles are probably associated with your intelligent core (*I Am This*) as opposed to your conscious self (*I think I am this*).

Astrology

Astrology is based on the assumption that behavior is influenced by the position of the stars and planets at the time of the human's birth. Astrological influences are probably associated with your human's body, as opposed to your immortal self.

Astrology and styles are not being recommended here. I am using them to demonstrate that we display personality characteristics that are evident at a very young age and which tend to shape our lives and how we relate to others.

Of course, the secret wisdom of the diagram is that, by practicing the middle way, we can move toward specific behaviors as appropriate, but return to balance without attachment to the outcome of our actions.

By the way, in the context of personal progression, *detachment* is a term used to indicate an openness to unexpected outcomes. For instance, consider only being offered a sedan when you were hoping to have a sports car. If your actual objective was to have a safe, dependable car, you succeeded in attracting one to you. Perhaps rejecting the sedan would be a poor decision. It is a judgment call to be made at the moment, but one to be made from the mindful perspective.

Being open for possibilities sometimes means following the surprising way. Being attached mean not being open, thus possibly missing important learning opportunities. Our friends on the other side are always seeking to help us. Being attached to an outcome is a form of resistance to that change.

7. Clarity of Communication

Aries **Driver** Self-initiative Intuition	Sagittarius **Driver-Expressive** Where others fear to go Intuitive thought	Aquarius **Expressive-Driver** Visionary Speculative thinking	Libra **Expressive** Cooperative relationships Thinking
Leo **Driver-Amiable** Steward and hope Intuitive feeling	**Driver-Analytical** Mission of discovery Fire	**Expressive-Amiable** Minister to the heart Air	Gemini **Expressive-Analytical** Brings order Logical thinking
Pisces **Amiable-Driver** Self-sacrifice Empathetic	Water **Amiable-Expressive** The great cause	Balance / Earth **Analytical-Driver** Methodical researcher	Taurus **Analytical-Expressive** Teacher Empirical thinking
Cancer **Amiable** Brotherhood Of man Feeling	Scorpio **Amiable-Analytical** Desire and intent Self-control	Virgo **Analytical-Amiable** Discrimination Sensory feeling	Capricorn **Analytical** Functional integrity Empirical

Personal Styles with Astrological Houses
Styles and signs are ways of modeling observed human behavior.
This is not an endorsement of either model.

So What?

A person with a Driver personal style, for example, tends to be impatient with people who are slow getting to the point. While they are important to have around if you want to get things done, they tend to scare Amiables. Expressives help to keep Drivers on track and Analyticals help to keep Drivers on the right track.

For a second example, Amiables connect a community with its purpose as its feeling, sensing aspect; however, that influence must be balanced by the other three aspects for the community to thrive.

The Internet brings people together according to their interests, but there is usually no way to separate people by learning styles. As such, it is common to see all four styles represented in any one thread, often to the dismay of the person who started it. This diversity is important but can be damaging if communicators do not act accordingly.

Selective Understanding

What people read or are told is different from what they understand, and understanding is always based on their worldview. It is becoming increasingly evident to psychologists that we unconsciously process what we sense and only become aware of the results of that processing. The translation from what is heard, felt or seen is based on our worldview. Worldview represents personal reality. It is like a database that is filled by our parents, teachers, local culture and the media. (29)

A hint as to the rules for this unconscious processing of information—what I refer to as the Perceptual Loop—can be seen in a person's personal style, as described above. The Perceptual Loop is an unconscious process that compares a possible understanding of incoming information with what is in Worldview. As modeled, the result of this iterative, streaming process is:

Accept which is presented to conscious self in a form Worldview understands.

Reject which results in the information being ignored.

Conditional acceptance which changes Worldview and presents new understanding to conscious self. This is the ambiguous maybe result we seek in learning new concepts.

See Essay 1: *Conditional Free Will* (Page 9) and Essay 8: *How We Think* (Page 109) for a more complete explanation.

So What?

If the listener is accustomed to hearing about the subject from the perspective of, say a strong religious upbringing, then a statement such as *"I heard from my loved one last night"* may well be heard more in terms of possession. The person may hear the words correctly, but the words will likely be understood from the person's personal point of view (Worldview).

The founder of a system of thought I was once involved in was a technical metaphysician and his way of explaining things agreed with my way of learning. He made his transition and the man who took his place spoke about the same ideas in more *feeling* terms. One phrase I remember from the new leader is *"The Golden Heart."* Well, okay, I have no idea what that means. That was the beginning of the end for my time with that group.

In the *Golden Heart* example, I literally have nothing in my worldview to help me visualize what that means. My personal style leans toward Analytical with emphasis on Expressive and all I am sure of is that *Golden heart* does not register with me.

The point here is that we unconsciously color what we experience to make the experience more familiar. Sometimes this may result in our listeners understanding the message in a very different way than what was intended or maybe even not registering the information at all.

Selective Attention

People pay attention to or ignore ideas, depending on how the information is delivered, their interest and learning style. This is a consequence of Item 2 and is referred to as *Switching* by James Carpenter in his proposed First Sight Theory. (38) In that theory, perception is modeled as an unconscious streaming process in that it produces a flow of impressions to the conscious mind (awareness) as information is received.

It is important to note the incremental nature of how awareness comes to us. A continuing process such as someone talking or information from reading a sentence, emerges into the conscious mind as a gradual *coming to understand*. If the reader is a Driver and the information is written by an Amiable, the Driver's attention will likely switch between trying to absorb the message and wondering why the person didn't get to the point faster.

For instance, it is difficult for a Driver-Analytical to sit through a video to access information. For them, video presentations just take too much time to get to the point. In fact, a five-minute video seldom conveys more information than can be written in a few paragraphs, so unless some animated or reality segment is required, an Analytical and certainly a Driver will probably not sit through a video. There is also the danger that some people will resent the perceived waste of time.

The same can be said for explanatory articles. Magazines intended for popular consumption often begin articles with a too long *"It was a sunny day and I admired the person's home as I drove up..."* kind of dialogue. If an article is a reasonably technical discussion about immunology, for instance, why would the reader be interested in where the immunology expert lives or if the sun was shining?

So What?

Your audience will listen to you with rapt attention but take a close look at their eyes. Are they glazed over? Be aware of your purpose and get to the point. In technical writing, a good introduction, or better, a good abstract may be your only chance to communicate your point. Video may be easier for you, but a few written paragraphs might result in better communication. Presenting a video without some kind of written brief may be a lost opportunity to communicate.

Also, think about what you want from your communication. Do you expect your target audience to refer back to parts of your presentation? Would you like your work to be cited as a reference in an essay? Imagine your audience as you write.

Controlling Switching

Another implication of switching is that we are conditional about what we focus on and therefore what we experience. During conversations, we likely, rapidly move our attention from what we are supposed to be experiencing to whatever we are planning ... maybe lunch. It is important to be mindful of this because it has a lot to do with our ability to learn.

If developing mediumship skills, for instance, the ability to suppress switching will greatly enhance lucidity. While switching is an unconscious process, it is driven by Worldview and that can be changed over time like changing a habit. Intention is a function of conscious self. As we intend to focus our attention, we signal to our unconscious perceptual processes what is important to us, but it takes time to develop that habit.

A mismatch of agenda impairs communication.

This is all about point of view. An example is what I have experienced when trying to explain to a skeptic editor why the Rupert Sheldrake article in Wikipedia is biased. The skeptic's agenda was to make Sheldrake look like a fool, and saying he is a biologist would indicate he is qualified to discuss morphogenesis interfered with that objective. My objective was to have the article describe the man in a balanced manner. No matter what was actually said, the outcome was already determined by the skeptic's agenda and the overwhelming number of skeptics backing him.

Thomas Harris told us in his book *I'm OK-You're OK* that real communication between two people could only occur if both are in an *okay* place in life. (69) This is also true of agendas. What are the two parties trying to get out of the exchange?

Mismatch of agendas tend to produce the phenomena of skepticism. The less what we say agrees with what our listener is expecting from us, the more likely our listener is to reject our words. The result is a skeptical response that emerges into our listener's conscious awareness. Everyone is potentially a skeptic. It depends on how well Worldviews match. See Essay 10: *Skeptic* (Page 163).

People always have *tells* that give them away if only we are paying attention. For instance, it is possible to anticipate people who might righteously attack others, based on their history. The recent attack by several parapsychologists on one of our mediums is an example of a totally irrational response instead of the true and expected levelheaded report indicating that phenomena had been witnessed under stringently controlled conditions. The irrational attack *tells* us the attackers may be threatened by the possibility survival might be proven by the practitioner they were studying. See Essay13: *Arrogance of Scientific Authority* (Page 205).

The investigators had substantial access to the medium during an extended series of séances. The medium later told me he understood that the investigator's agenda was to understand the physical phenomena. Because phenomena were produced despite the stringent controls and distractions, he believed they would produce a positive report. Yet, the investigators wrote reports demonstrating their probably unconscious agenda. In writing the articles, the authors mostly ignored data collected via the prearranged protocol, and instead, focused on innuendo and hearsay from outside of the protocol to support *what if* explanations that were not evident in the data. In effect, if not a conscious one, their agenda was to debunk.

This was predicted, as the lead investigator had telegraphed his real agenda about survival concepts on numerous occasions prior to the study. (70)

So What?

An interesting story portrays a group of friends sitting around a kitchen table, deciding if it is time to drive into town for lunch. During the exchange, there is a lot of *"well, okay"* kind of comments, rather than *"Okay! Let's go!"* comments. At the end of the story, it turned out that none of them really wanted to go, but because they had previously agreed, they felt obligated. In that story, all had the same but unspoken agenda not to go into town. They did not clearly express their position and mistook the position of the others. This story was designed for corporate employee training in office communication.

The moral of the story is to make a conscious effort to let others know what you are thinking. Try to pause the conversation long enough for you to think about what is happening. Being honest (candid) about your objectives will greatly facilitate communication.

Conversely, be mindful of the people with whom you are communicating. I know skeptics have an *"If mainstream science does not say it is so, then it cannot be"* agenda in everything they say and do concerning what they see as pseudoscience. If you have read much of my writing, you will know that my agenda is to teach about survival. When you read my writing, and if you are aware of this, then you will know to look for hints of the underlying purpose of what I say.

Most of us intend to say what we mean but we often have unconscious drivers that shape our selection of words. Take a moment and consider whether you are understanding what is being said and what you are saying. A pause can make a huge difference.

> If you have been traveling the Mindful Way for a while, it is reasonable to think you have developed a sense of how your audience is responding. This may be especially true of conversations before they begin, as people tend to telegraph their agenda in the opening.
>
> If you have a sense that the person has an agenda to which he or she is strongly attached, find a way to gracefully avoid the conversation. Being a teacher takes many forms, of which outright teaching is only one.
>
> I have ignored my own advice on this many times, as I seek to live up to my self-declared responsibility to teach. I have mostly stopped agreeing to examine examples because my response virtually always pisses off the person. It is in supposedly good deeds of trying to explain optional ways of thinking of a person's experience that I too often get a response like: *"Yes, I know you are supposed to be the expert, but I have to disagree with you."* That is typically followed by a long, passive-aggressive response.

If the person exhibits such a degree of naivety, or worse, delusion, change the subject. It is a fool's errand to think you can teach a person who clearly has the agenda to prove beliefs which are contrary to your understanding. Trying and failing tends to sow seeds of self-doubt.

Worldview changes in small increments

Learning hardly ever occurs in great leaps. This is an organizing principle for persuasive expression. Based on First Sight Theory (38) and on the Hypothesis of Formative Causation, (15) Worldview resists change, but does accept small changes if they are an evolution of existing belief or understanding.

In practical application, we tend to readily experience something if it agrees with our worldview. Remember, this is an unconscious process based on our prior conditioning, so if the experience is not recognized by Worldview, it might be outright rejected in the unconscious process. This means it will not be consciously experienced. If it is possible for the perceptual process to imagine how the experience might agree with prior conditioning, that *maybe* outcome can evolve Worldview to accommodate that slight difference.

It is the *maybe* outcome of the perceptual process that results in learning. It is the rejection of information that does not agree with Worldview that leads to skepticism and a breakdown in communication.

My effort to explain the Implicit Cosmology is a good example. (12) Many parts of the cosmology are likely new to you. Because of this, you will likely reject it if you try to absorb it all at once. To facilitate your understanding of it, I incorporate the concepts in my essays. For instance, you may have noticed that I often speak of a person as an immortal self entangled with a human in an avatar relationship. As I read my audience, I expect that the idea of an avatar relationship will make sense and will not be outwardly rejected.

So What?

An interesting aspect of early alien abduction reports was the frequency in which rabbits were involved in the accounts. The ability of the perceptual process to imagine an alternative perception of an experience, to make it agree with Worldview, is thought to result in substitutions in our perception, such as rabbits for little gray aliens. This is a cause of what Spiritualist refers to as coloring. Coloring is usually an unconscious, honest process but sometimes a conscious one.

An interesting example of cultural contamination that applies here is how descriptions of the reported abductors changed from rabbits to the big-eyed gray hominoids commonly reported today. In 1987, Whitely Strieber published *Communion: A True Story by Whitley Strieber,* (71) which portrayed a *typical* gray extraterrestrial. Today, close encounters with aliens can be expected to include the now standard alien description which agrees with the *Communion* book cover. That kind of cultural contamination must be considered when evaluating the usefulness of any report based on perception.

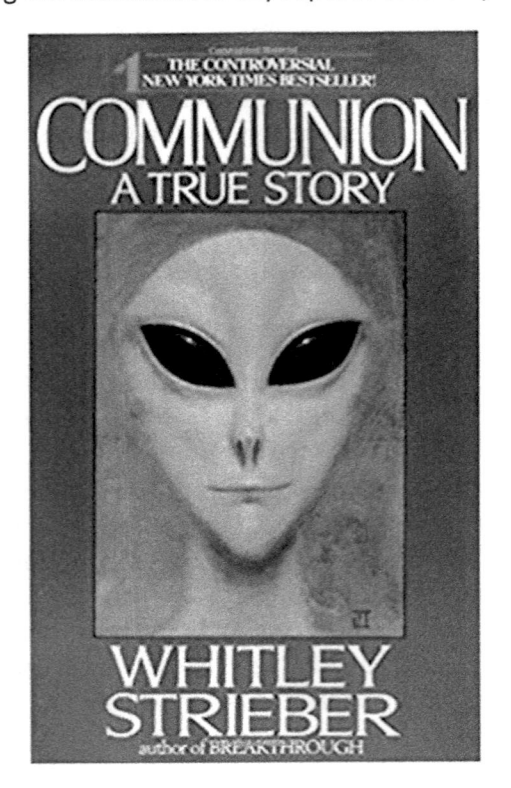

The streaming dream experience looks a lot like our unconscious mind attempting to find a story for sensed information that our worldview will accept. This ability of our mind to tell us a story emerges into our conscious awareness as understanding that is possibly only like but not the same as what was intended.

This is also a factor in what is sometimes referred to as storytelling, in which a person tries to make a supposed paranormal example or mediumistic message make sense by concocting a plausible explanation. A creative story can bring meaning where there is none intended.

If you are training to be a mental medium, be aware that what you sense via psi functioning is possibly translated by your perceptual processes into your personal symbols based on your point of view.

Incremental learning is involved, so for me, it has been a lifelong process of learning to recognize how my unconscious perceptual processes present mediumistically sensed information to my conscious awareness.

Automatic writing, pendulum work, even physical phenomena of the séance room is subject to this translation of original intent into the medium's or an interested observer's understanding. Here, it is important to note that the medium works with the sitter and both contribute to perception. This co-creation of experiences appears to be a factor in all forms of trans-etheric influence.

What I refer to as the *Intention Channel* is the link of influence between conscious awareness as source of intention and our mostly unconscious as servant of intention. In deep-trance mediumship, the Intention Channel between the medium's conscious and unconscious awareness appears to be idled in some way so that the external communicating personality is more able to express its intention on the medium's perceptual processes. This transfer of the source of intentionality appears to be necessary for clear lucidity in mediumship which produces uncolored message and phenomena.

As a final note, it is apparent that many of us are able to spontaneously enter into a relatively deep trance and remain functional. This usually occurs without our realizing the change in state. For instance, when we pause to remember something, we momentarily suspend many of our mental processes in order to increase lucidity. This translates as clearer access to the mostly unconscious perceptual processes. Acknowledging to ourselves that we are likely functioning trance mediums is one way to increase routine lucidity.

Discussion

This essay began as a contemplation about why I am uncomfortable exchanging comments in social media. On the Internet, what we say through our writing is all we are to others. Others seldom tell us they do not understand what we say. They do not see our smile or always understand if we are kidding. This is especially true if there are unexpected cultural implications attached to our words.

The Internet is a most powerful tool for community building. But to use this tool, it is important to be mindful of the underlying dynamics that shape how our words are understood.

This is doubly true of serving others via mediumship. Try to listen to what you say from the perspective of mainstream society. Many people believe in some things paranormal such as ghosts and heaven, but they subscribe to what mainstream scientists teach them. Part of what they are taught is to be suspicious of what the mainstream does not understand.

Especially, pay close attention to the thoughts that came to you while reading this essay. I have encountered substantial resistance to the idea that we only indirectly sense our world, and that what we do consciously sense is colored by our expectations based on memory.

If you are a Spiritualist, or if you have read a lot of Spiritualist literature, you will have been taught a very different point of view about how you interface with your communicators and the world. As a Certified Spiritualist Teacher, I say with the utmost desire you to objectively understand that what you have been taught is likely out of date. It is for you, only, to make sense of the more contemporary science. I say only because your usual teachers may also be struggling to adopt a more rational point of view.

References and Alternative Sources
Listed at the end of the book beginning on Page 357.

Essay 8
How We Think
2014

About This Essay

Perhaps you have heard the phrase *"For those who have eyes to see."* The idea is not that some people have a special set of eyes. It means that some people have developed the ability to notice things about their environment that others tend to ignore. This is an especially important concept when it comes to the Mindful Way.

The earliest reference to the idea I have found comes from Hermes, Line 12 of the *Emerald Tablet*: (28)

> 12. *So the world was created. Hence were all the wonderful adaptations of the One Thing manifested; but the arrangements that follow this great mystic path are hidden.*

The One Thing is the organizing principles expressed by Source. In Line 2 of the lesson, Hermes speaks of the One Power, which is modeled as the Creative Process (19) in the Implicit Cosmology. (12) The Creative Process is a person's attention on an imagined outcome to produce an intended order.

Hermes was explaining that all of reality was formed by way of natural, knowable principles. An important concept here is the need for the seeker to gain sufficient understanding to perceive the secret wisdom of Nature. The phrase *"...the arrangements that follow this great mystic path are hidden"* can be understood as *"This great mystic path is hidden except for those who have eyes to see."*

Hermes was telling his listeners that they needed to understand what he had just told them if they were to understand the One Thing in order to perform the Great Work. Also see Essay 17: *The Heretic Concepts* (Page 285).

For a more contemporary reference, famous Swiss psychiatrist Carl Jung told us that *"Synchronicity is an ever-present reality for those who have eyes to see."* Essay 19: Progression, Teaching and Community (Page 319).

Here are the implications of the principles as I understand them:

1. **The secret of lucidity:** Recognize and learn to live in balance with the organizing principles of reality. Our purpose in this lifetime is to gain understanding about the nature of our world. Doing so can be described as gaining spiritual maturity.

2. **The urge to seek understanding:** Our immortal self has an instinct to actively seek understanding. Even though we make some progress simply by living, ignoring the urge to actively seek understanding can be a source of frustration, a sort of restlessness that brings dissatisfaction with our life if we are not responding to the urge. See Essay 3: *Prime Imperative* (Page 49).

3. **Learn to be a seeker:** There are many different ways to gain understanding. Certainly, the ancient wisdom schools are a tried and proven way, but they are difficult and can take a lifetime to achieve mastery. Simply deciding to do so, and then consciously seeking understanding, is a way, but my experience is that self-guided ways are uncertain. The best way I know to consciously seek understanding is to know how we think, and then learn to integrate that into our daily living. That is what I describe in my writing as the Mindful Way.

   The Mindful Way still takes a lifetime, as cultural contamination is a never-ending influence, but progress gets easier as mindfulness becomes equally never-ending. See Essay 2: *The Mindful Way* (Page 37).

   Finding a teacher, a guru, if you wish, was about the only way to learn about the nature of reality in ancient times. Today, there are many would be gurus offering to teach, but I hesitate to recommend a single one ... including myself. The single teacher as the only way to gain access to knowledge can now be replaced with selective reading of information accessible via the Internet. That is one of the reasons I write so much about community. Today, the Mindful Way can be based on the community as teacher and sufficient discernment of the seeker to distinguish objective understanding from fantasy.

4. **Limited imagination limits experience:** An organizing principle I have found useful is the Principle of **Perceptual Agreement** defined as ***Personality must be in perceptual agreement with the aspect of reality with which it will associate.*** (27) Perception is based on understanding. That is, our access to actual reality is limited by our ability to perceive reality as it is and not as we have been taught. That depends on the correct alignment of our worldview with the actual nature of reality.

When our human is no longer able to sustain us in the physical, we will transition to a new venue for learning; however, we will only be able to enter one which we are able to visualize. Otherwise, our worldview will discard the new venue as being outside of our sense of reality. You and I cannot share the same heaven unless we are in perceptual agreement with the nature of that heaven. See Essay 16: *What is it Like on the Other Side* (Page 267).

5. **Understanding is the gatekeeper of future experiences:** To further explain the idea of Number 4, we live in a personal reality defined by our worldview. That personal reality is a subset of the actual nature of reality. You cannot experience my personal reality unless you and I agree about the nature of reality (perceptual agreement).

    Your degree of spiritual maturity, which is the extent to which your personal reality agrees with actual reality, has a direct influence on your moment-to-moment experiences and continues to influence your experiences as you transition out of this lifetime. Put another way; if you are spiritually immature now, you will still be spiritually immature when you are out of this lifetime. Dying does not make you more mature, only more aware of your immaturity.

From my study, I have learned that the way one develops the *eyes to see* is by developing a more lucid awareness of mostly unconscious mind. Doing so is an expected result of habitual mindfulness.

Another important saying is *You have to do the work!* The best way I know of to do the work is to learn to follow the Mindful Way. That begins with learning how we think.

Introduction

Emerging understanding of how we think is shedding important light on how we might seek to improve ourselves, our psi functioning and our relationship with the community. As it turns out, the body-centric perspective of *I think I am this* is simply not supported by current research. New thought is leading to a more dynamic perspective in which consciousness is seen as a product of attention and expression as a function of intention. These are etheric rather than physical in nature, suggesting that understanding the immortal self-centric perspective might show how to consciously access the true *I am this* nature of who we really are. This essay provides an introduction to this new thought.

First, it is important that you understand that even though I use a few psychological and parapsychological terms, I am neither psychologist nor parapsychologist. As an engineer, I tend to maintain a pragmatic approach to modeling and often focus on different concepts than you might expect from people trained in psychology, so please don't tell your parapsychologist friends that *"Tom said so."*

Terms

Most of the terms I use are intended as they are commonly used, but there are a few for which a more obscure definition is intended. A full glossary of the terms I use for these essays can be found at ethericstudies.org/glossary-of-terms/.

Refer to Essay 1: *Conditional Free Will* (Page 9), for an overview of the model I use for mind. A more thorough explanation is included in *Your Immortal Self*. (1) A brief description of each functional area is included in this essay. Although I repeat these explanations in many of the essays, I tend to describe them a little differently each time and the different approach might help you understand.

About Spirit-Related Terms

Here, I am careful to avoid religious connotation, but some terms common to religion are too useful to ignore.

Spirit: Saying that someone is *in spirit* is the same as saying they are in the etheric. If you are religious, it is the same as saying *in heaven*. Considering the Implicit Cosmology and the anatomy of a life field (Page 12) explained in Essay 1: *Conditional Free Will* (Page 9), I refer to a discarnate entity (sometimes, a ghost) as a personality, etheric personality or immortal self rather than a spirit. The self-aware aspect of immortal self is referred to here as conscious self. Conscious self represents your conscious perspective of reality; the experiencer as a traveling perspective. Think video camera in a drone.

Nature spirit: The formative personalities sometimes referred to as devas are also sometimes referred to as nature spirits. Considering the *Anatomy of a Life Field* discussion in Essay 1: *Conditional Free Will* (Page 9) as a building block of reality, it is probably better to think of nature spirits as just like you and me, but with different intention and from different collectives. As such, I prefer to address their function rather than thinking of them as a special form of personality.

Spiritual: *Spiritual* relates to the high ideals often associated with being very religious. In my writing, *spiritual* is intended in a more objective sense emphasizing understanding rather than believing. Thus, I say *spiritual maturity*, meaning the development of understanding about the nature of reality.

Spirituality: *Spirituality* is the point of view associated with seeking to gain spiritual maturity. It is also a state of spiritual maturity but is relative to how lucid a person has become and the degree to which that lucidity has manifest as compassionate behavior.

Personality

Personality is the immortal part of who we are; our intelligent core. In this model, there is personality, unconscious self and conscious self. When entangled with an avatar, there is also the avatar's body consciousness. I generalize our etheric nature with the term, immortal self.

Lucidity (Page 123) is further explained below in this essay. It is seen as the degree to which we are able to sense actual reality, which is the same as saying how well we are able to consciously manage our Perceptual Loop which is modeled as the perception and expression generator of our mind. That means that part of who we are which is normally thought of as unconscious mind is better described as mostly unconscious. As we gain in spiritual maturity, it becomes more accessible to our conscious self.

I refer to who we are while in a lifetime as a *person* which is ***immortal self entangled with a human in an avatar relationship for this lifetime***. When we are a person, our conscious self normally has a body-centric perspective as if seeing the world through the eyes of the human. Of course, the reason for gaining spiritual maturity is to develop an immortal self-centric perspective.

Just to be sure we are on the same page, *centric* is used here in the sense of our perspective or where we think our real self exists. We naturally assume a body-centric perspective when we enter a lifetime. The objective is to change that perspective to that of immortal self in the greater reality (immortal self-centric) while being able to operate as a person (body-centric).

Worldview

A thoughtform (etheric field) that functions as a database or repository of memory and human and spiritual instincts. It holds our beliefs and understanding which represents (determines) the nature of our personal reality. This model assumes we are born with only a degree of conceptual understanding and an urge to gain further understanding, both inherited from our personality. See Essay 3: *Prime Imperative* (Page 49).

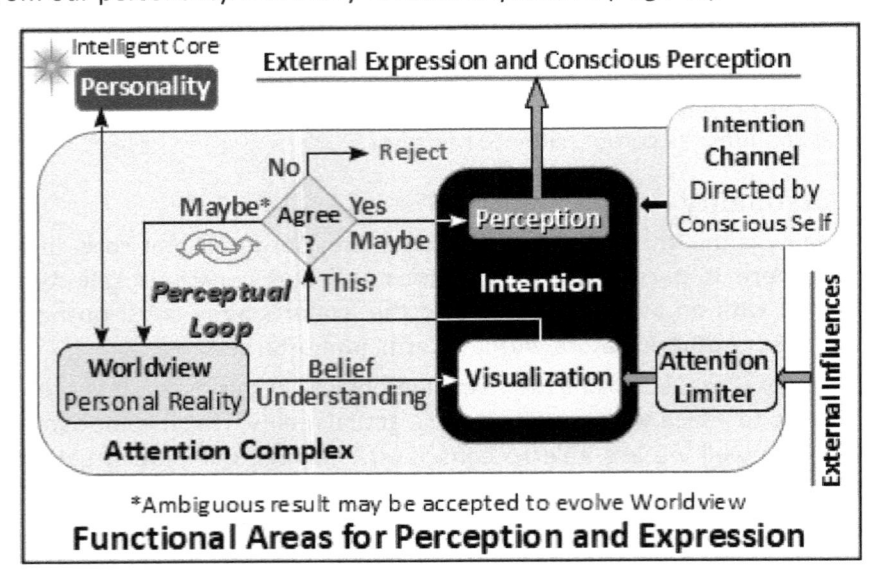

*Ambiguous result may be accepted to evolve Worldview

Functional Areas for Perception and Expression

As shown in the *Functional Areas for Perception* Diagram (above), Worldview represents the standard by which the output of the Perceptual Loop is determined. In effect, it determines what the conscious self is able to consciously experience. Worldview is functionally situated between the Attention Complex and Personality. Even though Personality represents the life field's intelligent core, its ability to experience the physical is limited by Worldview.

In an avatar relationship (a lifetime), Worldview is populated with the body's instincts as well. The avatar influence appears to be dominant in the first few years of a lifetime. Part of the process of maturing toward adulthood is the increasing influence of inherited spiritual instincts and prior understanding of perception.

Perception and Expression

Perception produces external expression. The output of the Perceptual Loop may be conscious awareness of something, a decision to act or an ideoplastic formation (expression) of our perceived reality. Basically, Expression represents unconscious thought coming to conscious awareness based on how it is perceived in the Perceptual Loop. In a way, it is the etheric-to-physical interface. Expression is the function by which we *make the world* as our personal reality.

> The *Ideo-* prefix means idea or image. Ideoplastic is used here to mean the nature of objects formed in the creative process as a mind-to-object expression. In this model, the world we witness through our human's eyes is described as an ideoplastic expression.

In the Implicit Cosmology, all of our expressions are thoughtforms. We assign physicality to some of them and dream-like qualities to others. The expression and expression processes are the same in the Attention Complex (mind). The difference is in conscious self's intention. In either case, the Attention Complex submits the visualized result to conscious self in the manner of *"This is what I said"* or *"This is what I hear."*

Attention Limiter

Both psi and physical information sensed from the environment are filtered, depending on whether or not it is of interest to us. We are presumably immersed in a huge amount of psi and physical stimuli and this process helps to enable the Perceptual Loop to focus on more important input. In terms of digital data, think of this as the router interface: if the Attention Complex is not specifically addressed by the environmental signal, the signal is ignored.

Perceptual Loop

A process in which stimuli are translated into awareness ... or rejected. This is a streaming process in which:

- **Ignore:** Information from the environment is visualized, compared to Worldview, and if there is no recognition, it is rejected, and the conscious self does not become aware of the information.
- **Try again:** The information is visualized, compared to Worldview, and if familiar, it may be modified for repeated visualization. The visualized image will likely pass through the loop several times to be modified to agree with Worldview or rejected.

- **Accept:** If it is a good match with Worldview (personal sense of what is real), it is accepted and sent to the perception function and will be consciously experienced.
- **Maybe:** If the Perceptual Loop produces a *maybe* or ambiguous result, it is accepted and sent to the perception function and allowed to change Worldview.

It is important in cosmology to identify a mechanism by which evolution may occur. Rupert Sheldrake's Hypothesis of Formative Causation includes provisions for *creative solutions* to evolve *Nature's Habit*. (15) In the Implicit Cosmology, an ambiguous result of the Perceptual Loop evolves Worldview. Thus, we can see that a small change or *a little newness* is more apt to cause learning than will a dramatic change.

Attention Complex

As a functional area of the Implicit Cosmology model, the Attention Complex represents the mostly unconscious mind. Personality has individual awareness which is limited to its etheric environment and what is available from its collective. It also accesses Worldview. This information access is also limited by what it can comprehend (perceptual agreement limited by Worldview).

Conscious self is our outward sense of *I think I am this* and *I live in this world*. Conscious self's access to personality, its collective and other personalities is via the Attention Complex. Conversely, our core intelligence's (personality as *I am this*) access to conscious self, its collective and other personalities is via the Attention Complex. Thus, the Attention Complex, as mostly unconscious mind, is the gateway to the rest of reality for our life field.

Attention

Attention is a state of existence which is fundamental to life. It is an automatic process in a similar way that the physical body has autonomic processes controlling such functions as breathing. Both personality and avatar share this functional area.

An important distinction to note is that attention is part of what makes us sentient. Thus, it is necessary for perception or expression to exist.

Besides being necessary for the existence of life in the life field, attention is also part of the influence conscious self has on the Attention Complex (mostly unconscious mind). The only influence conscious self has on the perception and expression processes is the influence of intention. That is enabled, first by attention on the process, and then by the intention to make it so.

Attention on an intended outcome with the intention to make it so is the creative process which must be initiated by conscious self, but the actual visualization of what is intended is developed in the Attention Complex. The result is that we might wish to do something, but the nature of that something is limited by our ability to visualize it, which is limited by Worldview.

How We Think

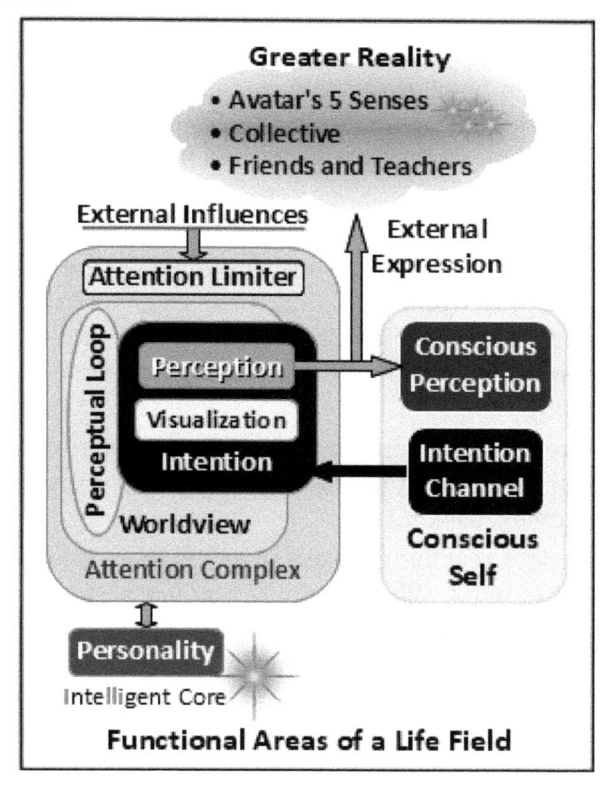

Functional Areas of a Life Field

It is a pretty bold statement to say, *"this is how we think,"* but whether or not this model is technically correct, it is a very useful one for understanding our etheric nature and how we experience our physical and etheric environment. Learning to integrate this model into our daily thinking gives us important tools for improving our intuitive sense and helps us place our feet more firmly on the Mindful Way.

The *Functional Areas for Perception* Diagram (Page 114) provides a block diagram of how we think based on the most current science I can find. All of the functional areas are in the etheric and are better thought of as nonphysical fields which are bound by attention into a system. As such, the Attention Complex represents our mostly unconscious mind.

Here are a few of the consequences posed by the diagram:

- Mind is nonphysical and exists in the etheric aspect of reality. In the etheric, everywhere is here, so that an apparent change in location is actually just a change in perception (change of perspective).

- Conscious self is also nonphysical. When in an avatar relationship with a human body, everything we know about our body's physical world comes to us by way of our body's five senses. I speculate that the information stream is transformed from bioelectric to psi in the brain.

    Still as a streaming signal, the psi form of that information is first processed by our mostly unconscious mind via the perceptual Loop, as informed by worldview. That means signals from our body are processed in more or less instantaneous sample. In that way, we develop awareness of the signal probably a moment after it is sensed by the body.

    The resulting perception of the signal becomes part of worldview as memory and is sent to our conscious self as conscious awareness in a streaming fashion.

    Conscious reaction to those signals is again translated by mostly unconscious mind. While this may produce a streaming output for the body, it appears that the results of our reaction will also be a gestalt thoughtform; a whole thought with meaning, intention and purpose, all as they are moderated by worldview. The thoughtform appears to be expressed into the etheric environment. The process is modeled this way because we know it can be sensed by other personalities.

    If the response is a physical action, it must be transformed from the psi form to a bio-electric signal in the brain. Thus, the brain acts as the etheric-physical interface to operate the body.

- Our Attention Complex receives information from other personalities via thoughtforms developed in their Attention Complex and according to their worldview. In the case of remote viewing, for instance, we see what other conscious selves are registering as physical in their perceptual process.

  Our Perceptual Loop appears able to merge information about the physical environment experienced by other personalities that have been received from their entangled body. Each personality has learned to assign physicality to signals they have been taught to think of as physical. The result for each of us currently in a lifetime is the perception of a physical world and our interaction with it.

  To be clear, we do not see the remote physical scene because it only exists as an ideoplastic construct of other conscious selves. The way we psychically sense a scene is to sense the expression products produced in the Attention Complex of the people who are both aware of the scene and whose worldview is reasonably in agreement with ours.

- Since the Perceptual Loop is a streaming process, environmental information is continuously sampled so that we experience an evolving perception as more information becomes available. In dreams, this may explain stream-of-thought dreams in which one moment suggests the next but does not always produce a logical experience.

First Sight Theory

(Use discernment here. I am explaining this theory as I understand Carpenter's explanation.)

James Carpenter has proposed an important hypothesis he refers to as First Sight Theory. (38) (16) He explains that people first sense the world psychically. As he puts it: *"What if ESP is like subliminal perception? What if psychokinesis is like unconsciously but psychologically meaningful expressive behaviors?"*

He answers these questions by proposing two propositions based on the assumption that: (paraphrasing) everything and everyone, every action in the past, now or in the future, perturbs the subtle energy space that connects all of us. Carpenter argues that research indicates people psychically sense these changes in psi space just as they physically sense changes in physical space.

> By "research indicates" I am saying that Carpenter based his theory on meta-analysis and personal research concerning hundreds of research reports—both mainstream and parapsychological.

According to Carpenter, if that is true, then: (still paraphrasing)

First, people sense their environment psychically as well as with their physical senses.

Second, people process this information unconsciously, and it is the conclusion of this processing that they become aware of and react to ... not what has been psychically or physically sensed or unconsciously considered. A person might psychically sense someone near or far, a person's actions and apparently their thoughts when they are expressed as intention.

> **Note 1:** In this theory, the expression of intention is what produces a change in psi space which can be sensed by others. Simply thinking of something does not appear to produce information which is detectable by others.

> **Note 2:** The result of the unconscious perceptual process is described as formatting of *"experience and action."* This is the perception functional area in the Functional Areas of Perception diagram (above). In this context, the person does not directly experience information from the environment. Instead, the person becomes aware of the information after it is formatted in a way that agrees with Worldview.

Carpenter proposes thirteen corollaries which amount to a decision tree defining the Perceptual Loop in the above *Functional Areas for Perception* Diagram (Page 114). I will spare you the list here, but it may be useful for you to read the Perception Essay at ethericstudies.org/perception/. (18) Also refer back to Essay 1: *Conditional Free Will* (Page 9).

The diagrams here are intended to reflect the essence of Carpenter's theory, but as modified based on other parapsychological literature and lessons learned from transcommunication (physical, mental, trance-channel forms of mediumship and Instrumental TransCommunication which includes EVP). The concepts are presented as the Implicit Cosmology (12). The application of these concepts is described in Essay 2: *The Mindful Way* (Page 37).

Implications of Unconscious Preprocessing of Thought

If the above information is reasonably correct, then our conscious self which is the *I think I am this* part of who we are, and what we sense around us, are the product of unconscious processes of which we have only indirect control. This means that we sense a lot more about our environment than that sensed with our five physical senses. In fact, we are likely psi-sensing stimuli much before our physical sensing.

Paying Attention

If we do not care about something, if it is uninteresting, then it will likely be unconsciously ignored. Of course, our body is constantly watching for survival-related input, so anything potentially scary, good to eat, sexual or related to territory will be first unconsciously reacted to, and if persistent, will enter conscious awareness. For instance, adrenalin may flow in our system some time before we become consciously aware of a looming threat.

An assumption of this model is that we are born with an urge to gain understanding. (23) In effect, that is a spiritual instinct; however, contrary to body instincts which compel the human to act, it appears we must consciously decide to respond to the urge to gain understanding. It may be possible to live a lifetime without responding to our spiritual urges. However, it has been my experience that ignoring them can result in a sort of restlessness as if our friends in spirit sometimes coax us to pay attention. See Essay 3: *Prime Imperative* (Page 49).

This is not a matter of self-determination (free will). It is a matter of finally gaining sufficient presence of mind to realize there is a need. Yes, we are urged to act in small ways, but seldom in a way that leads to conscious pursuit of spiritual maturity. One way we can take conscious control of our progression is by learning to pay attention to our world with the intention of gaining understanding. It is the expression of intention that counts.

In the ancient wisdom of the Cabala, the *Great Work* consists of transmuting the relatively immature mind (lead), into a more mature mind (gold). (30) One of the techniques for teaching this is to teach the seeker to learn as much as possible about everything. Learn to pay attention!

Here, I should add the wisdom of detachment. It is important to be interested in what is happening around us while not being attached to the outcome. Be open to unexpected possibilities. This translates as suspended judgment when witnessing things paranormal. Avoid expecting immediate understanding. The mind will try to shape what is being observed into something familiar, so not needing immediate understanding teaches the unconscious mind to allow what is observed to unfold naturally. The same goes for creating an intended reality. Seek an outcome but remain open to unexpected results.

> I need to emphasize here that our mostly unconscious mind is a world-class storyteller. Always be aware that what we are consciously thinking is a product of what we already know. This means that what we think we correctly understand about an unfamiliar situation may actually be a fabricated understanding based on similar but not the same references.
>
> On a personal note, too often when people bring a report of something paranormal to me, they become angry with me if I do not agree that it is paranormal. It appears they are already set in their mind that it is paranormal and has a specific meaning. That is a form of attachment and prejudgment that defeats learning. It also makes teacher disappear.

Worldview Can Be Taught

Like a mental database, our worldview is *filled* by family, teachers, schools, church, and probably most importantly, the media. A *red flag* concept of the Mindful Way is that we share this worldview with our human avatar. At birth, it is mostly the instincts of our human that governs our behavior. Maturing means learning to moderate the influence of these instincts with understanding. Still, there are adults who are mostly controlled by their avatar. It takes a community because we are not born already understanding how to become a good citizen. It is taught.

It is common to know people who are ostensibly like-minded, but who report rather different experiences from a shared event. This is mostly because of different Worldviews, so that for example, a person with a strong orthodox religious upbringing tends to experience or interpret phenomena with religious overtones while a Spiritualist is more apt to see the same phenomena in terms of interaction with their friends on the other side. Some people see spirits everywhere while others see Natural Law at work.

Interestingly, none of these points of view are necessarily incorrect, as each is based on a combination of good sense and belief. If we normalize them by translating them into fundamental concepts, it is often evident that the beliefs are based on the same concepts, only with rather different attribution. For instance, if we let Jesus be a wayshower instead of a savior and let God be the reality field rather than a father sitting in judgment, the common characteristics become more evident. Angels, guides, loved ones, inner teachers and devic entities all present themselves to our unconscious in much the same way. How they emerge into our conscious mind is determined by how we have been taught to look for them.

Temperament

Since ancient times, observers of human behavior have noted that people tend to fit into personality types or styles of behavior that govern how they interact with the world. This concept is explained in the "People Have a Style of Learning" (Page 97) Section of Essay 7: *Clarity of Communications* (Page 95).

Lucidity

Working with Worldview, the Perceptual Loop acts as a filter to restrict what part of the *raw* psi input becomes available to the conscious self. Learning to manage this process is important in developing a clear channel between conscious self and etheric personality. A person who demonstrates psychic ability is seen as one who is able to sense more of the Perceptual Loop process than the average person. This ability is referred to as lucidity.

The threshold between unconscious and conscious awareness is referred to as liminal. According to Carpenter's last two First Sight Theory corollaries: (38) (16)

Extracting from Carpenter's explanation of the **Inadvertency and Frustration Corollary:**

> Information gathered via psi is not available to conscious experience but does contribute to the formation of conscious experience by the arousal of anticipatory networks of ideas and feelings (assuming that they are heavily weighted, afforded slow switching and approached with the intention of assimilation). Because of this arousal, their action can be glimpsed consciously only by observing thoughts, feelings and behaviors that are inadvertent; that is, not intentional and not obviously caused by any current experiences. Someone who has become skillful in interpreting them is thought of as relatively psychic.

Extracting from Carpenter's explanation of the **Liminality Corollary:**

> The arousal of anticipatory networks of ideas and feelings resulting from unconscious psi information may be considered liminal ones, in terms of the boundary between conscious and unconscious thought. Habitual

interest in liminal experiences facilitates expression of psi processes (openness), leading to unconscious reference to psi material (and other streams of unconscious material). A more positive, open, secure state of mind will tend to facilitate reference to a broader spectrum of contextual, potentially liminal experience. In other words, habitually paying attention to subtle information emerging from your unconscious can lead to more direct awareness of what has been psychically sensed from psi space.

This means that a person wishing to develop psychic ability should learn to focus on inner thoughts, think what is sensed is potentially useful and expect more information. It is important to note here that no distinction is made between mediumship and psychic functioning. From the perspective of perception, the mechanism of transcommunication is the same and only the attribution is different. That is the theory.

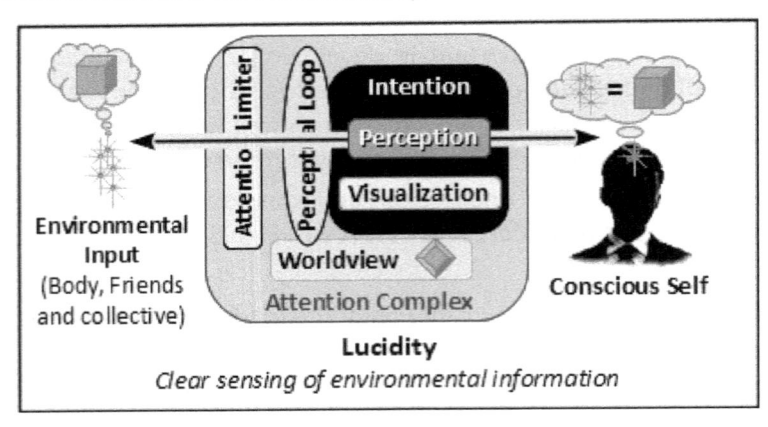

Lucidity
Clear sensing of environmental information

Instead of seeking a meditative state of no mental chatter, every thought should be invited and at least briefly examined with interest (contemplation). I like the meditation in which I mentally stand at the opening of a vast cave. The world full of light and solidity is before me, while fragments of thoughts flitter toward me out of the cave as if they are butterflies. I turn to face them, and as each thought comes, I briefly attempt to make sense of it, before sending it back into the cave. I mentally follow the thoughts that are especially interesting, deep into the cave, seeking to understand their source.

It is the source I am most interested in, and as I seek to visualize that from the temperament side and not the astrological side, I intend that my unconscious mind will learn what is important and what is not, based on reason and not on habit.

Information from our friends on the other side comes to us via a mind-to-mind exchange by way of our Attention Complex. As illustrated in the Lucidity Diagram (above), information from the etheric is still filtered by Worldview, but that filtering is moderated by understanding so that greater understanding results in greater lucidity. Ideally, the experiencer (conscious self) learns to sense the difference between that which is cultural contamination and that which is from the environment. For me, it is often the surprise factor that hints of something outside of my musing.

Hyperlucidity

Hyperlucidity is marked by the tendency to find phenomena *everywhere* despite considerable testimony to the contrary by peers. The concept is adapted from occasional reports of odd behavior exhibited by some people when they visit an emotionally charged place (Paris Syndrome, Jerusalem Syndrome) or experience great beauty (Stendhal Syndrome). (72) (73)

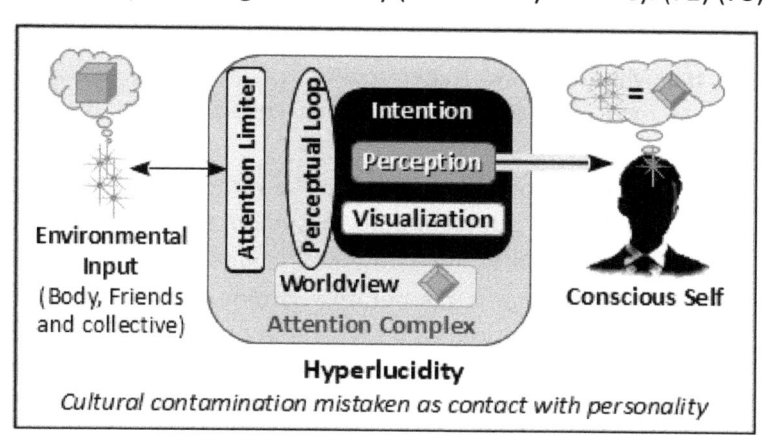

Hyperlucidity
Cultural contamination mistaken as contact with personality

In the Hyperlucidity Diagram, the experiencer (conscious self) imagines what is being perceived to be from transcommunicators while in fact, it is just from the physical environment or internal mental musing guided by cultural teaching.

The key to understanding hyperlucidity is to look for hints of magical thinking. If a person thinks ability comes upon them as if as a gift, rather than by way of study and practice, then an observer should first look for the objective evidence. If that is available, the next question concerns the validity of that evidence. Can it be independently experienced?

This concept provides a possible explanation for the abundance of not so evidential evidence displayed on the Internet. From personal experience, I know that mental mediums can be very evidential, but it takes practice to learn to distinguish between mental preconception and thought emerging from personality. One of my favorite talks given in our Spiritualist Society was by Steve Crow, titled *You have to do the work*. A person tending toward hyperlucidity is typically one who first comes to these phenomena as an expert.

This is a difficult concept because there may be a tendency to use it as an argument against the existence of transcommunication. Instead, hyperlucidity should be looked at as an enthusiastic approach to the study of things paranormal that is not balanced by discernment and education about the subject. As a behavior, it will eventually moderate as the person become better informed.

> I know I tend to harp on this but forgive. Hyperlucidity appears to be the greatest roadblock to gaining understanding you will encounter on the Mindful Way. Hyperlucidity is a pervasive syndrome that inflicts most people new to serious consideration of paranormal phenomena.
>
> Hyperlucidity often shows up in the form of pronouncements of truth based on unsubstantiated, assumed *truths*. The effect is to reject the offer of alternative thought from others in the community.
>
> When I say that a cooperative community requires a candid exchange of information, I am saying that personal pet theories must be held in abeyance in order to consider new ideas. Candid responses are those that reveal honest consideration. Discernment means recognizing the influence of our inner judge when it poses as all-knowing.

Discussion

This essay rather briefly covers a lot of material. While much of it may be new to you, even contrary to what you have learned, it is supported by quite a lot of solid research by many very knowledgeable people.

Psi field, psi functioning and first sight are concepts that may well initiate a global mind change. It is for you to decide if you are willing to join the early adopters on the path of mindfulness in the pursuit of greater lucidity. The way is not easy, but reward comes early.

The Way of Progression
Through community comes knowledge
Through teaching comes understanding
It takes a collective

Take advantage of opportunities to discuss these concepts. Doing so helps their integration into your worldview. What you have to say is important because sharing your unique perspective, observations and experiences will help the rest of us.

References and Alternative Sources
Listed at the end of the book beginning on Page 357.

Intentional Blank Page

Video-loop ITC Image collected by Tom and Lisa Butler.

In the color version, it is clear that this is a man wearing a brown vest. It is possibly leather with fur lining around the arms and neck. He may have a mustache. Color seems to be true in many of our examples, and here, we see a blue-skinned man.

Essay 9
Consensus Building in the Paranormalist Community
2017

About This Essay

In Essay 1: *Conditional Free Will* (Page 9), I tried to explain the limits of our ability to say what we intend. When I speak, I intend to speak the truth, but in actuality, I am only able to speak the truth as I understand it. We are all the same in that regard; however, I am willing to argue that I am more able to speak the truth than the average person. The reason is that I have spent much of this lifetime trying to learn how to see the truth ... Nature as her actual self. This is something you can do, as well.

When I write an essay, my motivation is usually to explore a concept that has been more frequently coming to my attention than usual. This essay is a good example. I had been shown one example too many for why I am not doing a good job of explaining myself.

One good example is my failure to organize an ethics panel. Five or six opinion setters in our community had agreed to participate, but when it came down to actually doing the work to get things started, none were to be found. They wanted to be part of it, so I must assume I failed in some way to motivate them into action. I blame myself because such disappointments are common with my bright ideas.

Another example is the many hours I might spend explaining the Implicit Cosmology to people, only to find months later that their actions argue that they did not believe or understand a word of it.

The crowning disappointment has been the public response to *Your Immortal Self*. (1) The book was not intended to make money, but sales are the only metric I have to tell how well it is received. I do watch to see if people comment about it or list it in recommended reading. It did show up on a couple of lists but fell off one in weeks.

I sent it out to a few people I considered important opinion setters in our community; however, despite kind words, I have the sense they did not actually read the work. Two did give me excellent reviews, which probably accounts for those who have purchased it.

The most telling, and important comment I received was that the book was intimidating. Others complained that it is too long or too complex. These comments are important to me and help me understand the need to learn how to more effectively communicate.

While new, when compared to, say momentum or osmoses, the concepts are not all that confusing. It is just that I do not stop explaining until I have addressed all of the variables. There are a lot of variable you can expect to encounter when trying to apply the concepts, so yes, the story is probably too long.

This book was to be around 250 pages, but here it is at nearly 390. So, I did it again. And despite my efforts to be more reader-friendly, it will likely be seen as intimidating. The moral of my little story is that our community needs people who can convey these concepts in more reader-friendly ways. It is by that skill that we will see a useful consensus about the nature of our reality develop.

This essay began with the title of *Inconvincible*, in the sense that some people cannot be convinced about new ideas. But, every time I started the essay, it quickly degenerated into a rant. The same thing happened with Essay 14: *Open Letter to Paranormalists* (Page 223). All of my attempts turned into complaining. I realized this and began again, only for the writing to drift into another rant. Being me is frustrating sometimes. I suppose we can all say that. The solution is not easily found when I am busy being frustrated.

Sometimes I wake up in the morning with insights about whatever I am working on at the time. The change in approach for the essay usually begins then. It still took me many restarts, but with each, the perspective gradually changed from me complaining, to more constructive observations with possibly actionable suggestions.

This is where I explain that I have help writing these essays. Well, yes, I clearly do, but the trick is learning to listen to that help. My objective is to trust the urge I sense to write this or that in the same way, I trust advice from an old friend … with the understanding that it is me who must accept responsibility for what I write.

This is to say that this essay began as a rant but ended up, I think, as a useful guide for working as a cooperative community. If you like it, thank my friend Mr. Writer. If you do not, then let me know in what way and I will try to make it more useful.

Introduction

Longtime observers of the paranormalist community may have noticed considerable change in the community over the years, but with little apparent progress in understanding the nature of these phenomena. All of us are in the same boat, *Paranormal*. In a very real sense, we are dependent on one another, as scientists need practitioners to produce examples for study and practitioners need scientists for guidance in how to work with them.

If progress requires cooperation amongst citizens of the community, how can cooperation be improved? In my view, the answer has everything to do with how we mentally process information and the nature of our spiritual instincts.

Shared Perception of Reality

If our purpose is to gain understanding about the nature of reality, and if there is one actual reality, it seems clear that each of us is converging on the same understanding. If this is true, our differences are ultimately temporary.

I say we are converging on the same understanding because it appears that understanding, itself, is a relative concept. An experience might be new to us and our understanding of it only slight, but that understanding becomes more precise as we have other, similar experiences.

And so, it might be true that none of us are completely right in our opinion because our understanding is not complete. Even so, it is arguable that some of us see reality with greater lucidity than do others. If so, the sages were right all along in saying that *the truth is evident for those who have eyes to see.*

In the end, it comes down to the simple truth that the present is created from the past. That is, unless we have developed the presence of mind to examine the validity of what we perceive, with the intention to better align our understanding with the actual nature of reality, our perception will be entirely based on memory of past experiences.

You may know this *presence of mind* as mindfulness and the path we must travel to gain it as the Mindful Way. See Essay 2: *The Mindful Way* (Page 37).

> This is a good place to discuss karma. Karma is commonly described in New Age circles as a system of merits we earn through good deeds

and demerits we must work off by making amends. Karma is said to follow us from one lifetime to another.

As it is described in the *Katha Upanishad*, (74) karma is of our physical nature, as opposed to our spiritual nature. The term is used in the sense spiritual lessons we are intended to learn in a lifetime. I see a close comparison between the karma of Hinduism and Buddhism to the Prime Imperative I discuss in Essay 3: *Prime Imperative* (Page 49).

In both systems of thought, we are born into this lifetime with the urge to gain understanding about a specific aspect of reality. Seeking the lessons is our purpose for existing. As I understand the Upanishads, a *karma* is one of those bits of understanding we are intended to gain in this lifetime.

So, you can see, karma can be paraphrased as imperfect understanding about the nature of reality that is intended to be resolved in this lifetime. It translates in practical terms as the better you understand reality, the less difficult the remaining lessons (karma).

Community

As I will explain in *Spiritual Anatomy* (Page 143), below in this essay, one of the more important tools for the Mindful Way is the cooperative community. Participation in a community enables self-evaluation. The process of composing what to say helps us organize our thoughts. An honest response from others in the community helps us develop an idea of how sensible our understanding has become.

An underlying assumption of a cooperative community is that its members return candid responses. They need not agree, or even be rational, but they do need to be candid by speaking up about why they agree or disagree and their questions. It is when people do not return honest feedback that the cooperative community fails.

Paranormalists and Paranormalist Seekers

Paranormalists are defined here as **people who experience, study or have a more than casual interest in psychic ability** (psi functioning, remote viewing, healing intention), ***healing intention*** (biofield healing, distant healing, healing prayer) **and the phenomena related to survival of consciousness** (mediumship, visual and audible Instrumental TransCommunication (ITC), hauntings). However, being a paranormalist does not mean a person is consciously seeking to gain understanding. For instance, to a casual observer, ghost hunting clubs appear to be more like sporting societies than study groups.

As you read this essay, keep in mind that I am addressing it to paranormalists in general, but when it comes to questions of increasing understanding, I am specifically directing my comments to seekers. Here, a *seeker* is **one who consciously responds to the urge to gain understanding about the nature of reality.** (23)

> As an aside to the skeptics amongst us, being a paranormalist does not mean all paranormalists think these phenomena are actually paranormal. In fact, skeptic-ism is relative so that the further a concept is from mainstream thought, the more likely it will be rejected by a person who studies these phenomena. For instance, anomalistic psychologists (58) reject psi and survival phenomena, while exceptional experiences psychologists (57) may accept psi but reject survival phenomena.

The cohesiveness of a community is based on its member's common understanding of the concepts which define it. Greater agreement in understanding tends to attract improved understanding, while decreasing agreement tends to disperse the binding influence. In effect, unlimited freedom of opinion with little or no collaboration produces a chaotic environment which tend to defeat efforts to develop a cohesive foundation of understanding.

In this sense, the paranormalist community is fast becoming a failed community. Even as more is understood as sound theory, we have less and less general agreement about the nature of the related phenomena. Belief rather than objective understanding, profiteering and pet theories dominate. It seems that everyone has become their own expert, learned leadership is often anti-psi, but predominantly anti-survival.

Roadblocks to Idea Sharing

As I have said, establishing the consensus necessary for community norms requires members of the community to be candid about expressing their point of view and open to the possibility of seeing other people's point of view. Consensus building is typically an ongoing process as opinions converge toward a common view. If a common view is not found, the community can be expected to splinter as people who do agree, self-organize into sub-communities of interests. This appears to be our current state.

A Fractured Community

The old blind men trying to describe an elephant parable applies here. From it, we learn that a coherent explanation of the phenomena may only be possible if all the sub-communities of interest work together. As it is today, some sub-communities disagree with the others to the point of actively attempting to debunk their view of the phenomena.

The dominant sub-communities are:

1. **Parapsychology:** Academically trained, usually with doctorates, concerned with theorizing and conducting research, primarily considering the human nature-based response to phenomena, especially psi field-related phenomena. Often, research is designed to explain why people believe in these phenomena rather than to develop metaphysical models.

2. **Citizen Scientists:** People who are involved in the study of these phenomena on the layperson side of the Academic-Layperson Partition. These include hauntings investigation as opposed to ghost hunting. Occasionally apply well-considered, protocol-based investigation of mediumship and ITC. More an emphasis on survival phenomena.

3. **Mediumistic Practitioners:** People who apply techniques to induce transcommunication, such as mental and physical mediumship, automatic writing and ITC. Healing intention is included here. Usually an assumption of survival, but also presumably psi-only practitioners such as remote viewing and healing intention

4. **Hobbyists:** This community of interest is primarily composed of ghost hunters. There are hundreds of ghost hunter clubs which are usually operated as social clubs with haunted place *investigations* as outings. They usually employ quite a lot of technology to collect *evidence*, including recording for EVP.

5. **Seekers:** People who are interested in understanding these phenomena as it relates to their true nature and relationship with reality. Such concepts as mindfulness, personal improvement and human potential apply to this community of interest.

These sub-groups tend to be interdependent so that parapsychology depends on the other groups for access to phenomena, and the other groups depend on parapsychology for learned guidance about the nature of the phenomena and possibly how to work with them. This happens to some extent; however, my observation has been that all five groups mostly ignore the others. They have developed their own stories about the nature of these phenomena; often many different stories within a group.

Thus, each group tends to insulate itself from the others. The effect of this is self-review which leads to a possible bubble of baseless beliefs used as the standard within the group to *peer review* new thought. This internal consensus tends to be a roadblock for the development of a global consensus. The effect is more like the various belief systems developing into a sort of contemporary form of religions competing for the ownership of actual truth. As we have found amongst orthodox religions, belief-based competition for truth assures that no global consensus will form without individual acceptance of the need for change.

A common story is necessary for progress to be made in our understanding. Lacking it, the paranormalist community is rapidly contracting into isolated subgroups. For the most part, the way is left for superstition and scientism to overshadow objective study of these phenomena.

My objective here is to understand the dynamics of this dispersal and seek ways to find a common story. The obstacles I see, those behaviors I think are at the center of this failure to communicate, include:

1. **People tend to be protective of their examples of phenomena**

    Sharing examples of phenomena is an important part of social media for paranormalist. On the surface, sharing is an effective way to teach one another about the nature of the phenomena and encourage further attempts to collect more examples.

    However, many shared examples are simply not paranormal. Some are artifacts of whatever technology was used. Many of these technologies are poorly understood and most of us do not know the many ways the technology might fool us. For instance, photographic latency can produce ghost lights and bright objects and particulates can produce orbs which are often described as evidence of ghosts.

    A witness may not think the example is evidential, but it has become something of a social norm to praise the example, perhaps to support the person's emotional need for validation or the witness not wanting to be shunned as a doubter. The witness, failing to be candid about the example, effectively reinforces the exhibitor's erroneous belief. No learning occurs and community understanding about the actual nature of the phenomena is further devolved into belief and away from objective thinking.

While praising an example that probably is not paranormal may seem like a kind gesture, it is more likely the witness does not wish to argue with the exhibitor. We have had so many angry responses to our reports that examples are probably mundane, that now, we hardly ever agree to examine examples from the public.

Our perspective on this is explained at the top of the ATransC.org website front page:

When encountering an extraordinary event, seek first to find commonalities with other extraordinary events and known science before assuming something new. Always error on the side of the mundane.

2. **Delusion**

Delusion is defined as **belief in something despite evidence to the contrary.** This may be a touchy subject, and certainly, I am not qualified to speak of it in academic terms. While not intended here as a derogatory comment about a person's sanity, it is intended as a red flag for experiencers and witnesses. This human condition is described in the Implicit Cosmology (12) as hyperlucidity and is considered a form of erroneous perception that is self-correcting with education. See "Hyperlucidity" (Page 125) in Essay 8: *How We Think* (Page 109).

There are a number of common forms of delusion we encounter that make discussing examples difficult. Typically, it is the result of naturally occurring perceptual artifacts that become obvious when how we think is understood. In one form, it is natural for the mind to attempt to find meaning in ambiguous sounds. This usually occurs when a person is distracted or hypnagogic (not fully awake). For instance, always wanting to be the judge, our mind will try to hear the mostly unnoticed hum of a motor as a distant conversation, radio station or music ... a familiar sound. Sounds other than the motor's hum cannot be recorded, making recording a good test.

This auditory artifact can be reported in audio recording when the experiencer thinks there are countless EVP, often evil, in a recording. Yet, others do not hear the same. We conducted a study in which online listeners were asked what they heard in soundtracks we specifically explain had no speech. A surprising percentage of people insisted there were spoken words present. (75)

Regretfully, it is necessary for paranormalists to make themselves aware of the potential for delusional experiences when experiencing phenomena or witnessing examples. We are all subject to delusional episodes, but it is important to recognize that some people should avoid working with anything paranormal because of a tendency toward erroneous perception.

3. Pet theories

When we assumed leadership of the ATransC in 2000, a frequently expressed theory was that the only real EVP were those found in the subsonic region of the audio spectrum. Later, some people insisted the only real EVP were in the ultrasonic region. Reversing the audio track to find EVP was popular back then, and some insisted that the only real EVP were those found by reversing the soundtrack. Over the years, it has become evident that the unseen communicators will speak where we are listening.

To our knowledge, none of those theories were tested under controlled conditions. None have stood the test of further experience.

Prior religious training can cause people to believe their unseen communicators are stuck or earthbound. The same can be said about the way religious beliefs can lead to fear of encountering demons, possible possession and the danger of evil spirits. We know that perception is influenced by prior belief, which I refer to as cultural contamination. Because of this, the *proofs* offered by these theorists are suspect. Such faith-based beliefs can make these phenomena unnecessarily scary for the average experiencer and help to defeat our community's forward progress.

The idea of a pet theory also comes up in the form of *"Spirit told me."* We see evidence of spiritualism in this argument, and even though we are both ordained Spiritualists, certified National Spiritualist Teachers and certified mediums with the National Spiritualist Association of Churches (nsac.org), there seems to be no way that we can convince such seers that they likely colored the message from spirit to agree with their worldview.

It needs to be noted that a pet theory is something of a gift in the sense that it cannot exist unless the theorist has been able to see beyond the dogma of our community. However, a theory cannot move us forward unless people are willing to challenge it by asking why. It also cannot move us forward if the theorist is unable to explain it, or unwilling to engage in extended, candid discussions about its merit.

Just as a person manufacturing an ITC device has the ethical responsibility to conduct studies to establish that the device works and to document how artifacts occur, so does a theorist have the responsibility to establish a foundation of evidence and supporting precedence.

The usual reaction when we question a pet theory is anger. Again, this is a form of being protective of one's phenomena. We have also noted a degree of arrogance in theorists as if our degree or experience does not warrant consideration of our input. A theorist telling a witness his theory is right (period!), thus ending the conversation, is perhaps the most common way learned discourse is defeated in our community. The antidote is for the conversation to continue with a well-considered *"Yah, but"* or *"Consider it this way."*

I am potentially guilty of this *"My theory is better than yours"* syndrome. Knowing the potential of this problem, I spend a great deal of time examining my models and trying to normalize other models so as to compare apples to apples.

4. **The *Silver Bullet* Syndrome**

Transform EVP are the original form of transcommunication and remains the form best understood by way of practice and research. They appear to form under the influence of intended order on chaotic audio noise. The problem is that transform EVP can be a difficult means of contacting the other side. Not everyone is successful and only a relative few collect clearly understandable examples.

On the other hand, opportunistic forms of EVP are supposedly found in the output audio stream of radio-sweep, EVPmaker and most technologies that use detected changes in environmental energy to select from a pre-recorded speech database. The resulting sound stream consists of bits of speech, some of which are supposed to be arranged to produce the phenomenal message. The bits of sound are in every output. All that remains is for the practitioner to figure out how to make the resulting babble seem like a message.

In practice, we see that the majority of people trying EVP are willing to force those bits of sound into expected meaning, whether or not it is actually present. The result is that opportunistic techniques have become so popular that they have pretty much pushed the old fashion and harder to use transform technique into obscurity. In fact, much of what is being reported as EVP today is probably not EVP.

After considerable study and honest attempts to see how opportunistic technique might produce EVP, (64) (76) (77) (78) we have finally *officially* set it aside as a technique that causes too many false positives to be of use. Our policy now is to explain to people that, for teaching or research, we no longer consider supposed EVP produced by opportunistic techniques, including ghost or spirit boxes and some of the new apps.

We routinely suggest that we would be happy to revisit this policy if someone presented us with meaningful research supporting opportunistic techniques. The people who should be conducting such research are those selling the equipment. As it appears, the profit motive mentioned next in Item 5 appears to be the reason the vendors work so hard to convince people the equipment works.

Again, the tendency of people to be protective of their examples of phenomena appear to be the problem. People who openly discount examples, no matter which phenomenon, are likely to receive an angry response, or be shunned as a non-believer. From my perspective, the transform-opportunistic EVP divide has had a devastating effect on the forward progress of our community. At root is a widely shared belief in something that has little or no objective support.

5. **A profit motive**

Some people have developed a theory to explain some or all of these phenomena. While many are only motivated to prove they are right, it seems the most aggressive are those who have written a book, and now benefit from the book being right. We see this in a number of opinion setters in our community, as they have found an audience to agree with their views and now seek to profit with books and website advertising revenue.

The people promoting opportunistic EVP tend to be making money by getting people to purchase the equipment. This is also a potential problem when people promote an event to pay the rent. We have attended a few excellent mediumship demonstrations held at private venues, even hosted a few, but some we attended bordered on deception. We have been kicked out of a séance facility in France because we were *too negative*, according to the sponsors. In fact, we were challenging the promoters for disingenuous representation of the phenomena. It would have been unethical had we not.

The profit motive usually consists of a person trying to sell a book. But many researchers depend on donors to fund their university department or at least a project. An interesting new development is academics seeking funding under false pretenses; perhaps ostensibly to study survival phenomena while intending to debunk it, or seeking funding to develop old, failed ideas as if they are newly invented.

People should be paid for their work. Lisa and I originally sold signed copies of the book we wrote for the Association, but finally stopped because of the personal cost and time. Without a little profit motive, service to the community has a limited lifespan for many of us. From my experience, a mental medium stands spiritually naked before a client. I for one have found the experience uncomfortable beyond my willingness to serve.

The years people spend developing ability ... in this lifetime or others, warrants reasonable reimbursement, else they should not be expected to be available to serve without the same sort of scheduling and remuneration one would expect for a professional such as an attorney or medical doctor.

A similar situation exists with EVP practitioners. Not everyone is able to successfully record for EVP and only a few we know of are able to contact specific people with any confidence. Like mental mediumship, there is always a risk that the practitioner must report a failed attempt. It also takes considerable time and effort to properly examine the recordings and compassionately deal with the sitter. Thinking someone should do these things for free is the same as thinking a therapist should work for free.

A belief-based assumption used to justify many accusations of fraud is that mediumship ability is a God-given gift that should be used in service to others for free. To be clear, psi sensing of any form is a characteristic of life. It is not a gift to be taken away because someone thinks it is being abused. It is an ability which can be developed and is applied according to the person's worldview.

6. **Academic-Layperson Partition.**

Parapsychologists typically hold a Ph.D. and operate under the cloak of that authority. Members of the other communities of interest typically do not have an advanced degree, or if they do, do not operate under its authority. The expected social order is that people with an advanced degree provide learned guidance to people who are not as well informed about their specialty.

We have observed, and personally experienced, a tendency of parapsychologist Ph.Ds. to avoid discussing theory with laypeople. Instead, much of the parapsychological literature consists of one-sided studies that, with the author's apparent assumption of witness delusion or ignorance, too often discount reported phenomena.

Given the lack of informed discourse amongst the laypeople of our community, it is easy to understand how the Academic-Layperson Partition came about. However, almost every effort we have made to bridge that divide has failed.

When a sub-community is closed to discourse with the other sub-communities ... all trying to describe the same elephant ... the greater community either fails or finds a way to compensate for the missing input. That appears to be happening now as the parapsychological community becomes less relevant in the objective study of transcommunication and survival.

Importance of Sharing Ideas

Two realizations began to emerge as I studied these phenomena and sought to explain the related concepts to others. The most important is that the phenomena are a symptom and not the end objective. This is important so let me say it a different way. EVP, for instance, is a tool with which we can study our spiritual nature and the nature of the greater reality, which is our true home. Yes, having an EVP example containing the voice of a loved one is reassuring, but that should not obscure the fact that our dead loved one is not dead at all, only transitioned to a different aspect of reality. If we miss that point, then all else is just cultural pabulum.

If the implications of these phenomena are as we think, they are profound. They suggest that living a full life means seeking to understand the greater reality and living in accordance with its organizing principles. I describe this realization in terms of mindfulness and explaining the implications of this realization has become the underlying theme of all I do these days. You may have other ways of describing this way of knowing. The important point is that the underlying concepts amongst the different ways of visualizing this idea must be the same if they are real. It is the common underlying concepts for which we must find a consensus.

The second realization is that explaining the first realization is more difficult than I expected. My wife Lisa and I have developed many ways to bring these phenomena to the public. Of the 129 issues of the newsletter published by Association TransCommunication, Lisa and I published the last 54 between 2000 and 2014. We estimate each issue has been read by some 1000 people. The website routinely receives 700 to 900 unique visitors a day. The book we wrote, *There is No Death and There are No Dead*, (79) has made the association more than $34,000, which means nearly 7,000 people have read the book. Over the years, I have personally answered thousands of questions from website visitors. Yet, there is little evidence in the paranormalist community that we have existed.

It is probably human nature to gather all the information about a subject of interest and synthesize a personal point of view that does not normally highlight any one source. Assuming that is true, it is reasonable to think our contribution to the community has been considered and merged with all the rest. The problem I see, however, is that much of the belief popularly held by members of our community appears to contradict what we have written. If we have had an effect, it is a very subtle one.

Again, this apparent difficulty in getting ideas out to the public is not unique to us, but it is a problem that must be resolved if we are to see understanding about these phenomena progress. So, the question is, how do the opinion setters amongst us help us develop stories about these phenomena that are both objectively meaningful and realistically useful? I do not have an answer, but the following point to consider might help in our efforts to collaborate with others.

Spiritual Anatomy

To understand ourselves and others, we must first understand our spiritual anatomy. (80) By spiritual anatomy, I mean who we are as immortal beings and the functional areas involved in developing unconscious and conscious perception. This is important because, considering the more objective theories of today, we can see that it is probable we experience the world as we were taught, rather than experiencing its actual nature. See "Anatomy of a Life Field" (Page 12) in Essay 1: *Conditional Free Will* (Page 9).

This section provides a brief overview of the factors which appear to determine how we form opinions. All of it is discussed in more detail in *Your Immortal Self: Exploring the Mindful Way*. (1) (Also see ethericstudies.org/concepts/.) If you have read other essays by me, you will know that some of this material is often repeated; however, I tend to tailor explanations for the subject, so please take the time to read the version presented here. "Anatomy of a Life Field" in Essay 1 also has a comprehensive introduction to these concepts.

Objective Thinking

For the sake of discussion, it is necessary to make a distinction between that which seems to be less in alignment with actual reality and that which appears to be more in alignment. The American Association of Electronic Voice Phenomena (AA-EVP) was founded in 1982 by Sarah Estep to **Provide objective evidence that we survive death in an individual conscious state.** When we assumed leadership of the AA-EVP in 2000, we were determined to maintain a focus on the *objective* aspect of Sarah's goal. The Association is now the ATransC, and a review of the ATransC.org website will show that we have tried to remain objective to this day.

At the bottom of each page of ATransC.org is a link to the *Our Pledge to You* page, on which we say:

> **We pledge to do all we can to provide the most accurate and up-to-date information about all things etheric. While we do not know what will be seen as true in the future, we will attempt to identify what on this website is supported by empirical evidence, what is speculation and what is common knowledge.**

The pledge is important because we have found that an objective point of view is essential to ethical disclosure of information about the nature of reality. Otherwise, our audience has reason to think our work is based on belief.

As it is used here, **Objective** is intended as **understanding based on empirical evidence supported by sound theory and undistorted by emotion or personal bias.** Considering our goal, our Pledge and the need to support those in need of assurance, it is important that we strive to provide only the most ethical and reasonable explanations about these phenomena.

Of course, I have been way off in the past about some concepts. For instance, we were very supportive of some audio and visual ITC techniques in the past, but which I would hesitate to support today. Naturally, our understanding is also evolving. That is why I have learned to provide references and links to additional information in my writing. This is not to say the additional information is correct. It is to say that the more you know about the subject, the better prepared you are to decide for yourself. It is for you to do the work of exploring the references and links.

Rational Personal Responsibility

When I say that the present is decided by the past, I am referring to the concept that our worldview is the filter for our perception. Our worldview is a mostly unconscious database populated with memory, cultural training and spiritual instincts inherited from our personality (Core Intelligence) and the collective of personalities with whom we share a common source. Since we share Worldview with our human, our human's instincts are also involved. If we do not stop and think about our actions, Worldview pretty much determines how we perceive our world.

Rupert Sheldrake referred to this as *Nature's Habit*. (15) This is another way of saying that we do what we have always done unless we make a conscious effort to change. Making that effort is the essence of mindfulness, and what I think of as rational personal responsibility. Not making the effort is a definition of the average person. For instance, I argue in my *Irrational Nature of Gun Ownership* Musing on EthericStudies.org that owning a gun is more a surrender to our human survival instincts than it is the expression of a rational mind. (81)

The idea of rational personal responsibility is nicely summarized as the first item in the *Ethical Expression* Column of the *Possible Mindfulness Personal Code of Ethics* Table (below): *Just because you can, does not mean you should.* (82) Communities are developed by people cooperating to support a common set of ideals. Cooperative communities are torn apart by people disregarding such ideals in favor of personal wants.

Ethical Understanding	Ethical Principles	Ethical Expressions
Do not violate *Seth*	• Respect • Kindness • Do no harm • Justice • Fairness • Suspended judgment • Courage • Discernment	• Just because I can, doesn't mean I should • Mindfulness is a way of life • Citizenship means cooperation • How will my actions affect me and others? • Is it a belief or a supportable understanding? • I will not impose my will on others • Lessons come from new experiences • Contemplation not meditation

Possible Mindfulness Personal Code of Ethics

Personal Responsibility as a Process

The Essay 1: *Conditional Freewill* (Page 9) includes a discussion about the implication of mostly unconscious processing of perception on our ability to personally account for our actions. Our ability to assure that what we say and do what we intend is limited by our lucidity. Here, **lucidity** is used to mean **the extent to which we are able to sense the actual nature of environmental information before it is colored by Worldview.** Greater lucidity means greater ability to be responsible for our actions; personal responsibility or self-determination. This is in contrast to **hyperlucidity**, which is a useful term for **the problem of thinking we are sensing clearly, while in fact, we are only seeing a story created by our worldview.**

The most important points for you to consider when evaluating your next action is that the environmental information you just received into your mind, possibly a biological-to-psi signal from your body's five senses or a psi signal from a friend in spirit, will have been changed from its original form before entering into your conscious awareness.

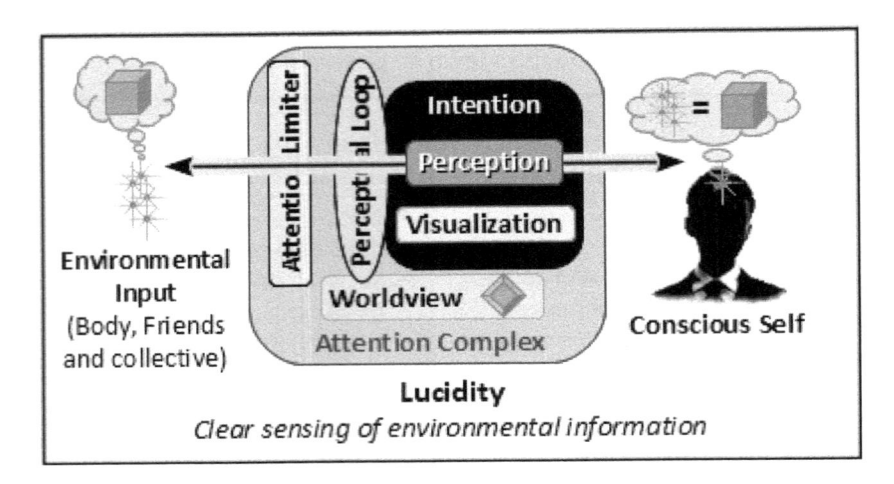

Lucidity
Clear sensing of environmental information

The determining factors for what you finally sense is what is in your worldview and the intention you previously expressed to your mostly unconscious mind. Your worldview is shared by your human body and your spiritual self. That means what you were taught in school, what you learned from that horror movie last night and what your body thinks it is supposed to do to maintain its place in the pecking order or to assure continuity of its kind, are all factors in forming the information that comes to your conscious mind each moment of your life.

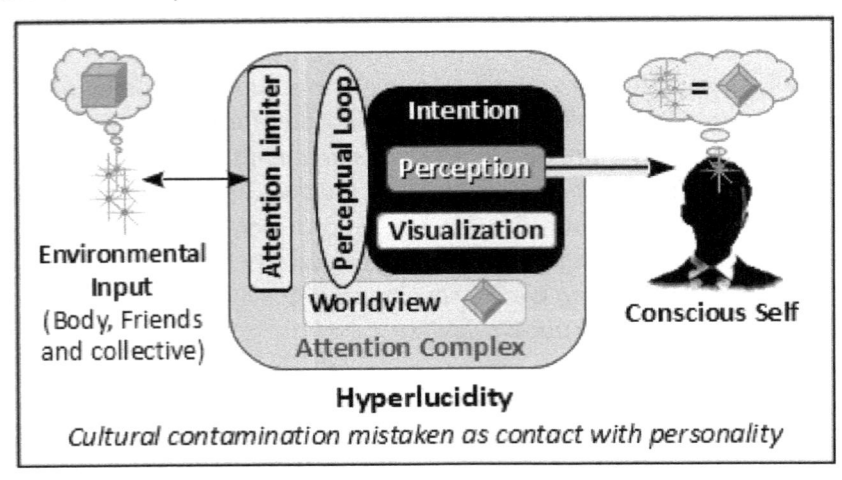

Hyperlucidity
Cultural contamination mistaken as contact with personality

The one conscious influence you have on your mind's coloring of information is the expression of intention to experience things as they actually are. The day you realize the need for this conscious influence on your mind is the day you step onto the Mindful Way. It is how you take control of your self-determination. Expression of intention is a lifestyle behavior, so if you are just beginning, give your mental processes a while to respond.

> As an aside, pay close attention to the statement, *realizing the need*. You can think to yourself, "*Of course, I realize the need*," but that is not the same as fully comprehending the need as an actionable idea.
>
> One of my first talks to a Spiritualist group was about the winning attitude. In it, I explored the difference between gamblers who want to win but do not and those who seem to win all the time. Both people are the same except the winners have learned to change the desire to win into an attitude that winning is expected.
>
> *Realizing the need* is like that. The perception and expression processes of the Attention Complex represent the creative process. The *creative process* can be defined as *attention on an imagined outcome to produce an intended order.* It is decisively influenced by your prior training, so that the created outcome will always be modified to agree with Worldview.
>
> If you were raised to doubt the existence of a greater reality, and to believe that your body is who you are, simply wanting to gain spiritual maturity will not be enough. It is necessary to teach your worldview to accept the necessary concepts. Again, that is the Mindful Way.

Understanding is Relative

One of the organizing concepts for Personality is the **Principle of Understanding: Perception of reality as it is and not as it is believed to be, with emphasis on underlying principles.**

Experience becomes understanding as a person aligns personal reality with local reality. A fundamental concept in the Implicit Cosmology is that personality exists to experience reality, gain understanding from that experience and return that understanding to the Collective. Understanding is seen as being a quality of personality and is, therefore, the quality determining perceptual agreement. Extent of understanding is a measure of progression.

Note that local reality is a special instance of actual reality. Actual reality is as it is moderated by organizing principles which are perceived by Source (expressed by Source). Local reality is the expression of actual reality as it is understood by those personalities said to be collectively imagining physical reality to create a venue for learning. Based on the Principle of Perceptual Agreement, perception of personal reality converges on local reality as it is visualized by those who maintain this venue. Thus, it can be said that *"Experience becomes understanding as a person aligns personal reality with local reality."*

(Principle of Perceptual Agreement: Personality must be in perceptual agreement with the aspect of reality with which it will associate.) (27)

The Nature of Understanding

A distinction is made in the Implicit Cosmology between information, knowledge and understanding:

Information

Information can be compared to raw data. In the context of mind, it is sensory input from the environment brought by the physical senses and countless psi signals. As it is received, information is undifferentiated and is only sensible when organized into some context. Think of the bits and pieces of information coming to us every moment as data points which must eventually be integrated into a sensible image. Much of what people think they know is just information.

Knowledge

When information is considered, and the related concepts are understood, the resulting comprehension is considered knowledge. To be knowledgeable about something is to comprehend related concepts so that they can be reasonably well visualized. This is the key to the nature of knowledge. It must be comprehensible enough to be visualized as a concept.

Here, **visualize** is used in the general sense of **being imaged in some way in the mind**. That might be sufficient recognition of an odor to name its source or formulation of many seemingly unrelated concepts into a form that can be described to others.

Once again, consider the old story of blind men examining an elephant: if each man represents a concept (leg, tail and such), then knowledge would be the ability of the men as a group to be able to describe an elephant. Knowledge is a combination of information, understanding concepts and comprehension of a global sense of meaning.

Understanding

In the Implicit Cosmology, perception is modeled as the outcome of a process that begins with encountering information that is external to the mind. We, as etheric mind and physical body, are immersed in a multitude of environmental signals. All come to us as a psi signal, some from the body's physical senses, the rest from other personalities in the etheric. These signals are processed in our Attention Complex where they are filtered according to our interest. Only some reach our conscious awareness, but always as they have been translated by Worldview. (29) (Refer to the *Lucidity* Diagram (Page 146)

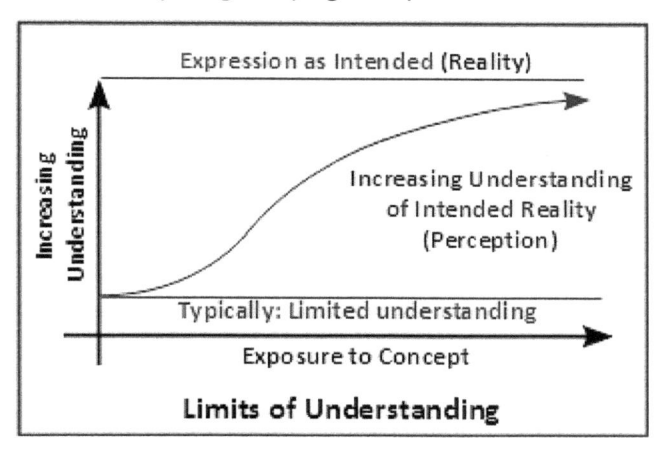

Limits of Understanding

(Chart: Y-axis: Increasing Understanding. X-axis: Exposure to Concept. Curve rises from "Typically: Limited understanding" toward "Expression as Intended (Reality)", labeled "Increasing Understanding of Intended Reality (Perception)".)

The degree to which understanding agrees with reality is also a function of our worldview. For instance, if we have been taught to be prejudiced about something, and we retain that prejudice, we will have that prejudice reinforced by the tone of the information coming to our conscious awareness. Unless we have become aware of the need to examine that perception, we will likely take what emerges from the unconscious perception process without question. Here, mindfulness as a conscious effort to see things as they are intended becomes important in helping us see the world as it is, rather than as we are taught to think it is.

> I want to stress this point. Something is not true because we think it is true. Take, for instance, a person who has been raised in a family that is prejudiced toward minority farm workers. When he sees several in a field tending a crop, he likely sees inferior people doing only work for which they are capable. He thinks this is true because

that is his honest perception ... because that is what he has been taught to think is true and not what he has come to understand by examining the evidence.

However, it is also possible the workers are doing the limited work they have been able to find. It is entirely possible some of them are working their way through college. Others may be the parents of his children's future boss. IQ does not matter much when the infrastructure of society is not working for you.

Many who have considered alternative theories about reality first think of their own story about reality. What they think is true dominates what they have yet to learn, and so, it is easy to discard the new idea with the assumption that current understanding is superior.

Whatever you think is correct, you are probably out of date, and therefore, incorrect. Take the time to rethink your assumptions. Doing so is a most important step on the mindful way.

Understanding is not an absolute. As is shown in the *Limits of Understanding* Diagram (Page 149), correct perception (understanding) is typically limited on first exposure to a concept. As a mindful person seeks to better understand the concept, understanding approaches (converges on but does not reach) perception that agrees with reality. This is an obvious effect when, say, a new principle in science is introduced, but it is less obvious when the process of overcoming prejudices learned in youth and fears foisted upon the avatar relationship by human instincts are considered.

As we move through a lifetime from youth to old age, things held to be true populate our worldview and tend to reinforce future perception. A conscious effort is required to break this cycle. Perhaps even more distressing is that we carry these false truths into transition from this lifetime. The result can be a complex, reportedly emotionally painful period of *getting well* while Worldview is better aligned with reality.

To emphasize the point of these paragraphs:
- We experience what we expect to experience.
- How well what we experience agrees with the actual nature of reality depends on the degree to which our understanding agrees with reality.

- Our understanding is held in Worldview.
- We can only manage our worldview with the habitual, conscious expression of the intention to see reality as it is, rather than as we have been taught.
- If the people with whom we are trying to build a consensus do not understand this point ... more importantly, if they are not practicing mindfulness, it is unlikely we will find common ground for agreement, only mutual benefit for compromise.

Cooperative Community

The **Cooperative Community** Organizing Principle is defined in the Implicit Cosmology as ***an effort to express understanding is necessary for progression. A person is attracted to communities of like-minded people cooperating to facilitate progression.*** (27)

Communities of like-minded people help to satisfy our human's social need. They can be considered cooperative communities if they enable members to express ideas and receive meaningful feedback. The act of ordering thoughts before speaking helps us better understand our point. Learning how others respond helps establish a sense of reasonableness. Without purposeful consideration of what we think about something, it is difficult for us to transform information into understanding. It appears to be understanding that we take with us beyond this lifetime.

A saying I like based on the Cooperative Community Organizing Principle is:

The Way of Progression
Through community comes knowledge
Through teaching comes understanding
It takes a collective

In a cooperative community, it is the responsibility of the individual to take the initiative to either comment or respond based on point of view. Being correct is not the idea. It is the preparation to express an idea that is so powerful. In this way, the speaker is both student and teacher. As an old adage goes, ***our lot is to learn, and having learned, our lot is to teach.*** Each member of a cooperative community fills both the role of seeker and that of teacher, simply by interacting with other members of the community.

An important point of this is that the listener does not need to be convinced by the speaker. Much of my current understanding has come from attempts to explain the reasonableness of paranormalist thought to other Wikipedia editors; a challenge at which I completely failed. The real value comes from composing our thoughts into sensible expressions of what we think is true, and the candid response from our listener. This process fails when the listener does not provide a candid response. Wikipedia editors are very candid.

Making a Cooperative Community Work

A person automatically becomes a member of a community by having the same or similar interests as held by other members of the community. A cooperative community, which is typically a subset of such a community, develops when members of the greater community begin to share ideas and give considered feedback to other member's ideas.

On a global scale, paranormalists are in the same community, but not all paranormalists are open to the exchange of ideas. The test is if people make meaningful comments about things related to the subject at hand and if others respond in a considered fashion. Communication does not begin if people lurk in silence or only offer flippant remarks.

Words have consequences. An offhand, poorly considered remark brings little value in promoting understanding amongst other members. Perhaps more important is that the harm ill-considered comments do to both the exchange of ideas and the speaker's worldview is substantial. For flippant speakers, it sends a reinforcing signal of intention to their unconscious mental processes, telling it to accept cultural contamination without review.

Assuming the person you are talking with does intend to be like-minded, there are points to consider when sharing ideas that might improve chances of reaching mutual understanding. A few points to consider are suggested below. Just use them as guides, as every situation is unique.

Social Media

As it turns out, social media does not support very effective cooperative communities. Yes, anyone can voice their opinion, but it is more of a *talking at* one *another* kind of exchange. From my experience, most people who comment, do so in not very well considered statements that invite little conversation. A good deal of the time, I cannot figure out their point, the comments are so brief or obscure.

Interestingly, it can be argued that social media works as a mechanism for suppressing considered discussion. Is this creating a generation of people who are no longer able to form a meaningful consensus?

Discussion boards have their shortcoming, but an important feature is easy participation in a thoughtful discussion. A second benefit of discussion boards over Facebook-like websites is that the discussion remains for a time so that others can learn. This is why the ATransC Idea Exchange has been maintained. (atransc.org/forum/) Please feel free to take advantage of its capability.

> Access to the Idea Exchange is free, but we now require members to use their real name. Using a pseudonym is a lot like wearing a mask. If you have read the *Lord of the Flies*, (83) you will understand my point when I say that part of personal progression is having the courage of our convictions. It is also important to speak as our real name if we ever hope to become an opinion setter. Building our reputation as a thoughtful person requires name recognition.

Tell Me Three Times

A quote I remember from many years ago, I think from a Native American Shaman, addresses our difficulty in comprehending information. I remember it as: **Tell me three times; once for my head, once for my heart and once for my spirit.** The telling is not simple repetition. Here, telling the *head* is an analytical explanation. Telling the *heart* is explaining why it matters to the person. Telling the *spirit* is an appeal to the underlying moral and ethical meaning of existence.

Trice-telling can be paraphrased here as recognition that a conversation with a person is really a conversation with his or her mostly unconscious mind. As a functional area for thought, the mind has considerable resistance to change. It must be changed in small increments, and depending on how the person learns, the conversation must be addressed to the person's learning style if it is to get past the Attention Limiter. (Review the "Attention Limiter" (Page 17) portion of Essay 1: *Conditional Free Will* (Page 9).)

A change of mind becomes most difficult when the listener's mostly unconscious mind has made a decision. The result of the decision likely becomes part of Worldview to be used to measuring the acceptability of the next bit of information. A possible way to avoid this self-imposed listener roadblock is to avoid speaking in absolutes. Ambiguous information is more apt to be accepted so long as it is not evasive. This may be why some people learn best via storytelling in which the point is part of the storyline.

Learning Styles

The way people approach learning has a lot to do with the way they do best in conversation. I am a good example. Watching a video to find out what a presenter said is the worst possible way for me to learn. The information-per-second delivery rate of video is very low compared to me scanning the abstract and reading the concluding remarks in a PDF file.

Considering the "Personal Styles" model (Page 97) in Essay 7: *Clarity of Communication* (Page 95) (Analytical, Amiable, Drive, Expressive), it is reasonable to argue that a quarter of our population has the same impatience with slow delivery of information. Conversely, an amiable personality might find a video more humanizing, and therefore, better for comprehension. A solution I have often looked for when being referred to a video is if the author has included a brief written synopsis on the same page as the video.

My writing style is technical but detailed. From conversations with others, I gather it is a difficult style for reader comprehension. When seeking to develop a consensus of understanding about something, it is important that others in the discussion explain how they understand the point, but in terms that makes better sense for them. A successful cooperative community is one that has active representatives of all learning styles.

Temperament

You may remember that a basic question in psychology is whether temperament is inherited from a family member (nature) or learned while a child (nurture). Considering Sheldrake's Hypothesis of Formative Causation, (15) temperament can be at least partially modeled as Nature's Habit established by preceding generations (nature). A refinement of Sheldrake's hypothesis offered in the Trans-Survival Hypothesis (10) is that the entangled personality contributes the urge to gain understanding to a person's temperament, while the human instincts contribute the urge to assure survival of the species.

Formative Causation and the Trans-Survival Hypothesis leans understanding of the nature-nurture question toward nature, but cultural contamination has a more contemporary influence on temperament that cannot be ignored. What we are taught introduces a strong, often repeated influence on our temperament, but that is biased by the nature side of the equation. We translate cultural influences (nurture) according to our nature.

With that in mind, it makes sense to define **temperament** as *a predisposition to understand experiences*. And, *a person's temperament is based on previous understanding inherited from Source, personality and the personality's and human's collectives, as it is modified by intended understanding.* If a life field is not entangled with a personality, its temperament is likely inherited from its collective.

A person's point of view is how a person expresses Worldview as it is biased by temperament. For instance, a person with a strong Driver temperament might hear *"Be responsible for your family"* from cultural upbringing as a command to keep family members away from worldly influences at all cost. A community may teach its citizens to be deeply afraid of demons, but if the temperament of a member of that society is to question authority, it may be natural for that person to turn that fearfulness into curiosity as to why demons are fearful. Upon discovering there is nothing to fear, the person's temperament might help to modify Worldview's influence, permitting the person to be more accepting of the unknown.

Villager-Explorer Effect

This is more commonly known as the Sheep-Goat Effect, but I do not much like being called a sheep. In 1942, Professor Gertrude Schmeidler identified a correlation between people scoring high on a belief in the paranormal survey with their psi functioning scores. Conversely, a poor belief score correlated with a lower than chance psi function score. She referred to this as the *Sheep-Goat Effect* with *believers* as the sheep. (63) As noted, for my self-esteem, I am referring to the *believers* as *explorers* because of their willingness to explore new ideas. Conversely, I am referring to disbelievers as *villagers* because of the more conservative attitude of people who follow cultural norms and depend on maintaining the status quo.

This concept implies that people who allow for the possibility of new ideas are more likely to consciously experience new ideas. The inverse of this can be seen in the concept of incredulity blindness, a useful term to describe how some people are inexplicably unable to see or hear examples of paranormal phenomena. Discourse 10: "Perceptual Agreement" in *Your Immortal Self* (1) addresses this concept in detail. (Or, see: ethericstudies.org/perceptual-agreement/) (17)

Suspended Judgment

If you are following this progression for how we relate to our world, you will notice that there are many factors conspiring to influence the way we experience life, most of which are beyond our direct control. As a general statement, instincts must be recognized, acknowledged and managed; once learned, beliefs are difficult to change, but incremental change is possible.

The only real influence we have on our temperament is the conscious expression of intention to be different. An old Zen Buddhist saying is, *"Before enlightenment chop wood–carry water, after enlightenment chop wood–carry water"* This might be paraphrased as *"Question for understanding; after understanding, question again."*

The most powerful technique I know for remembering to question is to habitually practice suspended judgment. Even if a person is skeptical, suspended judgment as an openness to the possibility of the unknown offers far more opportunities to experience phenomena. The Attention Complex which supports perception is designed to make an *agree* or *disagree* decision about what is real and what is not. Once a decision is made, it becomes part of Worldview, and is, therefore, difficult to change.

The perceptual processes in our mostly unconscious mind are also designed to characterize what is real in terms familiar to the person. In this same model, intention expressed by the conscious self is the means by which the perceptual process can be evolved. As such, following the Villager-Explorer Effect, the intention to withhold judgment about experiences enables more experiences. Discourse 9: *Perception and Expression* in *Your Immortal Self* (1) addresses this concept in detail. (Or, see: ethericstudies.org/perception/) (18)

Knowing Enough to Judge

My engineering training has taught me to beware of absolutes. In engineering, there are always tolerances of accuracy which must be acknowledged and possible unknowns for which preparation must be made. Even in metaphysics, I might weave a good explanation, but it is always with the awareness that this is what I understand today. Saying I know for sure potentially closes the door to further understanding.

An example that comes up way too often is the way people pronounce that someone is a fraud. Saying someone is a fraud is potentially a personal violation (Seth: *"Do not violate"*) and potentially threaten that person's wellbeing. Yet, in fact, none of us know enough to say for sure how transcommunication might manifest. At best, all we can do is seek better controls.

Consider the situations in which a physical medium has been caught moving about the dark room during a séance. Of course, this appears to be clear evidence of fraud. However, I have on many occasions, witnessed convincing demonstrations of matter-through-matter during séances. For instance, a plastic strap binding the medium's arm to the chair or an apport.

I understand how deep a physical medium's trance can be. There is nothing in what we know today that says the medium's controlling personality cannot release the medium from the binding straps, move the medium about like a trance puppet and then return the medium to the chair and reattach the straps; all without the medium's awareness. We have seen the effects of this too many times to think fraud is possible. There are just too many sitters trying to sense every movement ... even in total darkness.

Unless such possibilities have been accounted for, no one has the intellectual or moral authority to make accusations of fraud. The best remedy I know is to have sitters maintain physical contact with the medium while the lights are out. It might also be helpful to have a candid discussion with sitters and medium after the medium has recovered from the séance. Then, it would be appropriate for sitters to ask the medium why he or she seemed to be walking about. It is a fair question.

Without the availability of our practitioners to demonstrate these phenomena, the rest of us might never have the opportunity to witness proof of our immortality, make contact with a loved one or even learn the necessary skills for ourselves. Without them, research would be impossible. Attacking them as frauds when too little is known for us to say for sure, only serves our debunkers.

Justified

Over the years I have studied metaphysics, I have found no evidence of evil or demons. For sure, I have encountered people behaving badly, but in every case which I have had the opportunity to examine, the person has in some way, felt justified in his or her actions.

This is an important perspective to have when considering why people behave as they do. There is little sense in simply concluding they are wrong. The need is first for us to examine the sensibility of our point of view, and then seek to understand the other person's motivation.

To be sure, this is not a *turn the other cheek* philosophy. I am not speaking of how to act with someone who poses potential harm. Even though a person may feel justified in threatening actions, it is still necessary for us to react defensively. A person wishing to carry a concealed weapon, for instance, is a person with a fearful personality. We have every reason to avoid them, whether they are justified or not.

My focus is on the frequently expressed fear of evil we encounter amongst paranormalists. Above all else, remember that it is not what happens to us, but how we react to what happens to us. If we can understand why a person behaves or thinks as they do, we are well underway to understand how to deal with the situation.

Essay 1: *Conditional Freewill* (Page 9) is an attempt to identify the characteristics of free will, what of it is real and what is predetermined. The essay also includes a reasonably well-detailed examination of how we think, which should help you integrate what is in this essay.

> It is important to have a clear sense of the evil concept. I say *evil*, and I can almost see my listener's mind go straight to divine evil, as in demons and the Devil. The secondary reaction is that there is no defense against such evil: *we must surrender to our fate, so it is better never to attract evil in the first place.*
>
> Learn to consider the usual meaning of the words you use. Consider the thoughts your words will trigger in your listener's mind. I am not making a distinction that does not exist when I say *there is no evil, only people behaving badly*. I am trying to communicate understanding by avoiding triggering a faith-based response.
>
> As a practical matter, *people behaving badly* are people making conscious decisions for which they feel justified. *Evil people* implies puppets of God who are unable to be any other way. A fundamental concept in metaphysics is that we have self-determination. If so, being a God puppet is a violation of self-determination.

Belief without understanding is Faith

Here is something of a case study. A person contacted me wanting someone to examine a message she found in her cell phone voice mail. The girl told me she thought it might be from her transitioned boyfriend. She explained that she had called the telephone number shown for the recording, but the person there knew nothing of the call. (By implication, the call was, therefore, paranormal.) Because my responses so often anger requesters, it is my practice these days not to examine evidence; however, she was insistent, and I agreed.

Here is where I usually get in trouble. If the person wants to believe something is said in the recording, then I know enough not to get involved. But, I assume people ask us to examine their example because they truly want to know what is said. The motto for the ATransC website includes the phrase *"objective evidence that we survive."* That means we attempt to base our examination of examples on objective evidence, rather than belief. This includes resisting the Phantom Voices Syndrome. (75)

The short audio consisted of a young male's voice which was interrupted so that one heard fragments of voice (about 10%) separated by longer periods of quiet (about 90%). I could make out a low-volume swishing sound in most of the quiet segments. Even with ten or so tries, I was unable to make out what was being said. I reported as much to the girl. True to form, she was very indignant and scolded me, saying *"I thought you were a believer."*

If you listen closely to a radio-sweep sound file, you will sometimes notice a repeated, low volume swishing sound in the quieter segments. That is the sound of the sweep locking on to one station after another in quick succession. In areas in which there are few radio stations, the sweep will have many quiet segments with the occasional fragment of voice or music. The sweep is typically a few seconds long, so a person speaking on one station with no voice or music on the others, will produce exactly what I heard in the girl's example.

In a subsequent email, I suggested that the girl might have been spoofed. I was trying to warn her to use extra discernment. Instead, she said that others heard her name and that the voice had made a cryptic statement which she could not understand. (That, by the way, should have been her first hint that the file should be discarded.)

Feeling reasonably sure the recording was not paranormal, I did a little research about how a telephone number can be spoofed. As it turns out, there are apps one can use that do just that. It is apparently easy and free to put a message in someone's voice mailbox which appears to have come from a different telephone number. Receiving an *"I did not make that call"* when she called the number to ask, proves nothing.

Objective or Belief

The possibly spoofed call illustrates an important problem for cooperative communities. There need not be agreement in any one exchange, but there does need to be rational feedback. Remember I said that we are converging on the same understanding? The intention to gain objective understanding is the key characteristic of a paranormalist seeker. As one myself, for me, a like-minded person is one who is also a paranormalist seeker.

If a person prefers more of a belief or faith-based point of view, a like-minded person might be one who prefers the titillation of being spooked by the unknown. For instance, the ATransC is not a place to find belief-oriented people. I say this because we have suspected one of the reasons for diminishing membership in the ATransC has been our determination to remain objective. It is apparently a lot more fun if people can just enjoy spooky examples, rather than trying to learn if they are only an artifact of the technology.

When you consider with whom you might share your interest in things paranormal, and assuming you wish to seek objective understanding rather than belief, make sure who you chose is interested in the same approach.

> The ATransC is no longer a membership organization. Participation is via correspondence, the website, participation in research/studies, signing up for the *Occasional Update Email* and discussions in the Idea Exchange.

Summary

This essay evolved from my growing frustration about how easily people turn away from an objective perspective and toward a belief-based view of their world. At the same time, some people seem to avoid situations that might help them better understand their world. In a real sense, no matter what, some people are simply not able to or do not wish to be convinced about ideas that contradict their beliefs.

The fact is that we know enough these days to follow an objective way of progression. The book for which this book is a companion, *Your Immortal Self,* (1) includes a model designed to help people understand the nature of transcommunication. I am a reporter and do not claim to be the inventor of the model, but since I have found that it applies to so many different forms of phenomena, it seems reasonable for me to think that it has lasting merit. Even though it has not been reasonably vetted by others, I feel comfortable sharing guidance about these phenomena which is based on the model and personal experience.

It is arguable that the paranormalist community is in contraction. (60) Rather than well-informed seekers, we see mostly debunkers or people who prefer explanations based on quantum mysticism or religious doctrine. In a very real sense, sincere paranormalist seekers are a very small minority.

As a minority, our ability to study these phenomena is not assured. (84) Keep in mind that skeptics have defined pseudoscience as a danger to society and have gained wide acceptance in the idea that these phenomena are pseudoscience. Some parts of the US Federal Government agree. (85)

We are all born with the urge to gain spiritual understanding. I suggest that the more we respond to that urge, the happier we can be. In the long view of our immortality, resistance is self-destructive. In the end, it is an individual responsibility, but if you want to just believe, do the rest of us a favor by telling us up front so that we can stay out of your way.

Consensus building requires the desire to find common understanding.

References and Alternative Sources
Listed at the end of the book beginning on Page 357.

Intentional Blank Page

Video-loop ITC Image collected by Tom and Lisa Butler.

You should be able to see a cow looking toward your right shoulder. In the color version, it appears to be a young brown cow with a white face.

Essay 10
Skeptic
2016

About This Essay

This essay is probably something of a surprise, in that the essays thus far have been all about concepts, mind and implications of the way we think. This essay is decidedly about our community rather than concepts. I write about community issues because the dominant factor in how well we can research, study and practice working with these phenomena is the cooperative nature of the paranormalist community.

Imagine yourself at the very farthest reaches of thought where survival is studied. From that rarified space, imagine yourself facing toward the center of society where mainstream science is king. Everyone between you and the center is likely skeptical of your study.

Most parapsychologists studying psi phenomena do not accept survival as even a possibility. Yet, those who stand closer to the center studying mental aberrations are skeptical of psi phenomena and probably disdain those who think survival is possible. Skepticism favors the status quo established by the dominant mainstream segments of society. People with a Ph.D. tend to think of those without a Ph.D. as the intellectual unwashed.

Just about every parapsychologist has written about the problem of skeptics who try to debunk anything paranormal. This essay was written quite a while ago as my effort to make sense of the way Wikipedia has been dominated by skeptics. As you will see in Essay 13: *Arrogance of Scientific Authority* (Page 205) and Essay 14: *Open Letter to Paranormalists* (Page 223), the ethical shortcoming and anti-survival prejudices of many parapsychologists have inspired more confrontational essays by me.

There are two kinds of skeptic. The obvious skeptics are found amongst Wikipedia editors and mainstream scientists. They believe in the status quo of knowledge as it is taught in most universities and by most scientists.

Organized skepticism has been effective in making it difficult for seekers to gain spiritual progression. They scare away young, would be parapsychologists and the funding necessary for proper research.

However, the obvious skeptics are not really the problem. The problem skeptics are those who live amongst us. They are the Trojan Horse skeptics of our community. As is discussed in "Skepticism is Relative" (Page 89) of Essay

6: *Paranormalist Community* (Page 83), a surprisingly high percentage of paranormalists discount any possibility of survival. Certainly, more parapsychologists accept the possibility of psi phenomena. Many amongst those who do, also accept a form of survival sometimes referred to as residual memory. (See: "A Divided Community" (Page 88) in Essay 6: *Paranormalist Community* (Page 85).)

The Trojan Horse skeptics also mislead our less well-informed seekers by posing as experts who are trying to help us understand these phenomena. With that said, I suspect one of the reasons we have such an *everyone is an expert* community is that, while seekers may not know the science, many trust their sense of truth more than scientific guidance. We may not know why people are wrong when they tell us the phenomena we experience is illusion, but for sure we know they are not right. The lack of respected leadership has produced a vacuum of learned guidance necessary for a healthy cooperative community.

So, while this and some of the following essays are more about community than metaphysical concepts, it is important that you are aware of the issues. Let them guide your seeking. Most of all, remember that it is the community that provides the information which populates your worldview.

Abstract

Perhaps the most important trait for anyone involved in the study of frontier subjects is the ability to maintain an open mind while practicing discernment. This attitude is sometimes referred to as skepticism; however, when a person is skeptical of something without rational reason, that person is known as a skeptic. Skepticism in itself is a healthy attitude so long as it is accompanied by open-minded investigation.

Rather than practicing discernment, skeptics actively campaign to teach the public to see such new thought as a danger to society. As is already occurring in some governments, including the USA, this vilification of frontier subjects has the potential to cause a social and governmental reaction that could at the least prevent further study and possibly provoke action harmful to people studying these subjects. Because of this, it is no longer realistic to ignore skeptics or their efforts.

Introduction

The word *skeptic* is based on a Greek term meaning thoughtful. According to TheFreeDictionary.com, a **skeptic** is:

1. **One who instinctively or habitually doubts, questions, or disagrees with assertions or generally accepted conclusions.**
2. **One inclined to skepticism in religious matters.**

From the Etheric Studies perspective, people who describe themselves as skeptics have the common interest of suppressing any idea or concept they believe is not supported by mainstream science. This perspective works for parapsychologists, as well. If you read their literature, you will see that they tend to obsess about being seen by mainstream thinkers as practicing good science. That *mainstream envy* is probably one of the reasons parapsychologists are so quick to denounce survival related phenomena.

A number of friends have urged me to use *pseudoskeptic*, as in false skeptic, to identify people I describe here; however, the mainstream people causing all the trouble are found with Internet searches for *skeptic*, and for the most part, remain invisible if searched for with *pseudoskeptic*. (86)

Because my intention is to inform our community about the problems skeptics are causing, I feel it is necessary to call them what they call themselves. The term has been co-opted by habitual detractors. Healthy skeptics, people who open-mindedly question to learn and understand or to assure the speaker knows what he or she is talking about, will need to find a better term for themselves. Perhaps *discerning*.

It seems that virtually every parapsychologist and psychical researcher has written articles complaining about skeptics. It is common to find laypeople in our community complaining about being belittled for believing what the skeptics consider nonsense.

The motto for my personal website, ethericstudies.org, is: *"Believe what you wish but understand the implications what you believe."* To *understand the implications of what you believe*, it is important to examine how you develop your assumptions about the world around you. The *Point of View* essay on ethericstudies.org addresses the relationship between what we come into this lifetime with, how we approach new learning situations and how we develop our point of view. (87) In this essay, I have addressed what skepticism is and what skeptics are from the point of view of the Implicit Cosmology. (12)

Skepticism and Scientism

Contrary to the objectives of healthy skepticism, skeptics tend to condemn the exploration of new ideas, thereby protecting the status quo. Their often-stated position is that something cannot be if it is not explicitly supported by existing, mainstream science. That is a form of **Scientism** which means **the ideological belief that science—mainstream science—is the only authority on the nature of reality**.

In his essay, "Sagan and Scientism," Greg Koukl defined scientism thus:

Scientism states this: only that which can be proved by science is true. Science can only prove things about the physical world; therefore if it doesn't prove something about the nonphysical world, which it's really not equipped to do, then the only rational belief is that only physical things exist and non-physical things like the mind or the soul don't exist. That is the doctrine of scientism…. (Paragraph 15) (88)

Tells of a Skeptic

Skeptics attempt to show the unacceptability of an idea by belittling it and associating it with obviously silly ideas, rather than relying on facts, evidence and sound logic to prove their point. They describe the idea and people associated with it in terms that would usually cause a fight if spoken face-to-face.

As I explain in the section below about *Wilhelm Reich* (Page 169), you should consider skeptics more than just a nuisance. They seem to assume that if *believers* are seen as second-class citizens by the mainstream public, they will not enjoy the protection of social norms afforded mainstream citizens. So, if you read something that calls someone a whacko, fraud or describes a practice as fraudulent or woo-woo, you can know that the material was written by a skeptic. Their intention is clearly to find a way to make your subject go away.

Name-calling is especially true of skeptics who focus on alternative or complementary health practices. They commonly refer to these with the derogatory term of *quackery* and practitioners as *quacks*. Of course, any such practice that is not specifically approved by the government is considered pseudoscience, and even some that are approved such as chiropractic feel the wrath of those who think *complementary* is just another word for *fraud*.

Without research to support their accusations, skeptics seldom add knowledge to the subjects they attack. They are only able to destroy knowledge. Here, the Latin term, *a priori* has special meaning. They routinely make statements about subjects for which they have no knowledge other than that their peer group is against it. In this context, the term means *without prior knowledge* and is used to say that the person is judging without having become informed about the subject. The practical result of this book burning mentality is that new ideas are suppressed and examination of new ideas by academically trained researchers has become probable professional suicide.

Comparing the View of Science with the View of Skeptics
Wikipedia editor Ludwigs2 expressed one of the better descriptions of the skeptic view:

> Science and skepticism are entirely different projects; they share the word skepticism, but it has different meanings for each group. For a scientist, skepticism means (roughly) *"I choose not to have any beliefs about a subject in the absence of evidence."* It's a philosophically conservative position designed to keep people from making a priori assertions about the world (except those dictated by logic or math). For skeptics, by contrast, skepticism means (roughly) *"I choose to believe that non-conventional ideas are wrong until they have met some burden of evidence."* This is an ideological position designed to advocate against certain kinds of viewpoints. See how these differ on (for example) acupuncture:
>
> Looking at something like acupuncture scientifically one would be forced to admit that there really isn't much evidence either way—there is no scientific reason to recommend its use, but no obvious reason to say that it's wrong, either. That is, acupuncture is morally neutral, like drinking tea with honey and lemon when you have a cold.
>
> Looking at something like acupuncture as a skeptic one would find oneself saying that acupuncture hasn't met the needed burden of evidence, and so acupuncture is wrong—and this will lead to ideological claims that people who take acupuncture are stupid, that people who do acupuncture are charlatans, and etc. That is acupuncture is morally bereft, like selling sugar pills as a cure for cancer.

Scientists and skeptics overlap in the assertion that one should use practices that have been borne out by systematic experience. But that's where the similarity ends: skeptics go on to make moral judgments about practices that science can never make, and to engage in advocacy with respect to those moral claims. Consider the vast range of skeptical literature, almost none of which contains any actual research (aside from literature reviews of other people's published work), and which is almost entirely dedicated to critical declamations against one or another questionable activity. Skepticism is (frankly) scientific punditry, and while I won't deny its value in that consumer advocate sort of way, one needs to be cautious with it as an intellectual enterprise."

Organized Harm to Society

Paranormalists represent a small and mostly unorganized community. The skeptical community, on the other hand, is relatively well organized and fast growing. In *Skepticism: The New Religion*, Roy Stemman notes that Spiritualism's public outreach is contracting while the skeptical community is becoming more organized and much more effective in influencing the media. (60)

A search of the Internet for *skeptic* will produce dozens of pages full of skeptical websites. The Association TransCommunication website (ATransC.org) and Etheric Studies (EthericStudies.org), which contains my personal writing, are clearly homegrown. By comparison, many of the skeptical websites are slick, professionally designed and maintained, and I am sure, well-funded. It is clear that paranormalists are losing the fight to gain public respect and support. That should be a concern if you enjoy the right to publicly study these subjects.

Some governments, including the US Government, have adopted the viewpoints espoused by organized skeptical groups and routinely label the study of paranormal phenomena as pseudoscience and cite harm to the best interest of the public caused by belief in pseudoscience. In many such claims, supporting references are skeptical sources, which in turn cite these government reports for support of their viewpoint. This is, in actuality, a form of circular referencing in which truth is invented as a means of vilifying ideas that do not agree with the prevailing scientific ideologies.

A Case Study: Government Acting on Skeptical Views

I am not well informed about how governments have acted against people who have been accused of activity deemed by skeptics as pseudoscience. Please be sure to examine this for yourself.

The first point I would make is that the study of frontier subjects is not protected by the law. Once it is socially okay to say that, what we study contradicts science and may be harmful to the public, it becomes possible for governments and organizations to make examples of individuals by suppressing their freedom. We have seen this reenacted many times in our history with everything from witch burning to internment during the Second World War.

Wilhelm Reich is a more recent example. The short story about Reich is that he was put in jail in 1956 for making claims about a hypothetical form of subtle energy deemed by skeptics to be unsupported by accepted science. He also developed devices that might put the energy to work and claimed he could heal people of some diseases with the energy.

In fact, he was jailed because he ignored the government charges (Food and Drug Administration (FDA)), and apparently because his partner transported some of their experimental material across state lines against government orders. That act was treated as criminal contempt of court. From the Wilhelm Reich Museum website (84):

> *While Reich appealed his sentence, the government carried out the destruction of orgone accumulators and literature. In Maine, several boxes of literature were burned, and accumulators and accumulator materials either destroyed or dismantled. In New York City, on August 23, 1956, the FDA supervised the burning of several tons of Reich's publications in one of the city's garbage incinerators, ... This destruction of literature constitutes one of the most heinous examples of censorship in United States history.*

Reich died of heart failure while in prison, and I understand his research partner committed suicide shortly thereafter.

Reviewing *What is Orgone Energy?* by Charles R. Kelley, Ph.D. will give you a sense of the nature of this subtle energy which Reich called *Orgone*. (89) You will also see that Reich's discovery is likely just one of many rediscoveries of the same energy. Today, it is being studied as psi or biofield.

Because of its apparent effect on living tissue, the influence intentionality has on this energy to heal a person is a primary means of studying the energy (see *An Unusual Form of Radiation has a Reproducible Effect in the Laboratory* by Robert A. Charman (90)). It also appears that meditation and group intention can reduce the randomness of random event generators. This effect may be the result of a change in the biofield and may also help explain how EVP are formed.

> Be aware that, like the psi field, Orgone is not a proper energy as energy is defined in mainstream physics. It appears to be more correct to refer to it as an etheric field which is influenced by intentionality, rather than such physical characteristics as gravity and difference of electrical potential. It is evidently not propagated as an electromagnetic field. Rather than vibration, it exhibits the characteristic of potential to manifest as an aspect of reality; a conceptual thing to which we may attribute physicality.

A Case Study: Skeptical Control of the Media

It is obvious from a simple search of the Internet that the skeptics dominate the media when it comes to public outreach about frontier subjects. Yes, there are thousands of ghost hunting club websites and websites promoting the many forms of complementary medicine, but if you look for substantive support for the concepts, you run into a wall of skeptical websites supported by skeptic clubs, universities and mainstream science organizations. One of the most dominant of the skeptic's media is Wikipedia. An article in Wikipedia is the first result for many search subjects.

If you are surprised to hear that Wikipedia is counted at the top of the skeptic's media, I recommend that you take some time to read the **Talk** Page associated with your favorite frontier subject on Wikipedia. There is usually an ongoing discussion amongst editors about the struggle to balance the point of view of the article–a cardinal rule of the online encyclopedia. The problem is that the rules favor the majority group of editors, which are skeptics and nearly all of the editors who seek a true balance have been permanently blocked from editing or simply run off. Subject-matter specialists are not allowed to edit subjects in which they may have a conflict of interest. See Essay 12: *Concerns with Wikipedia* (Page 193) for more on this.

The associated article **Talk** Pages is often a battleground in which naive new editors are attacked and eventually driven off by the dominating skeptical editors. An example is the biography of a living person for Rupert Sheldrake, (91) an article I am banned from editing for life. Especially, look at the **Archive** Pages (upper-right corner of the Talk Page). Being sensitive to prying eyes of the public, the skeptics quickly archive embarrassing exchanges to make them harder to find.

In a nutshell, a group of determined people hiding behind screen names have managed to gain control of what is thought by the public to be a respected online encyclopedia. Rather than writing the articles as *"This is what the subject is about, and here are the various viewpoints about the subject,"* articles about what they call *fringe* subjects are written in a tone that subtly gives the sense that the subject is nonsense and a danger to society. The articles may have a lot of information, but it is always couched in terms of believers, proponents and how it is pseudoscience or quackery.

The Internet has given skeptics considerably more access to the public so that people with strong opinions and too much time on their hands can substantially influence the opinions of many people. A little time spent reviewing the International Skeptics Forum (92) might shock you as to the strong opinions against frontier subjects spoken by ill-informed people hiding behind fictitious screen names.

Healthy Skepticism

There is a balance between a priori skepticism and open-minded gullibility. Some claims are not reasonable, even for frontier subjects. Many reported experiences are clearly delusion or the ordinary mistaken as unusual. Every bump in the night is not a ghost and not every instance of improved health is because of healing intention. Often, the difference between a report of a genuine paranormal experience and an ordinary one mistaken as paranormal is education of the witness.

On the other hand, some of these reported experiences are not explained by current principles of science and may point to new understanding of nature. It is not reasonable to accept some of the extreme explanations without substantiating research, but it is also not reasonable to discount the reports because they are not currently part of known science.

"Understand the implications of what you believe," from my motto, is based on the idea that we should practice critical thinking leading to discernment. That means we take the time to examine the evidence before adopting a conditional opinion. I say *conditional opinion* because the rest of the story is that whatever is decided should be routinely reexamined to see

if it still makes sense. If possible, the opinion should be tested. If the opinion cannot be reasonably based on evidence and good understanding of involved technology, then we should remain undecided. If there is not sufficient information to arrive at an informed opinion, no opinion is the only answer. To do otherwise is to base the opinion on faith derived from popular wisdom, superstition and/or the opinion of others who may be acting on an undisclosed agenda.

> The idea of suspended judgment is based on the understanding that our mostly unconscious mind tends to quickly integrate a decision into Worldview. The mindful approach to new ideas is to resist deciding until more information is available. A discerning person is one who does not assume he or she has all of the information necessary to decide.

A fair portion of academia believes all major principles in nature have been discovered and that all we are doing now is filling in the details. That is to say that there is only the physical universe ... period! Anything outside of that, such as a psi field, etheric personality and survival of personality after bodily death, is not included in these major principles of nature and therefore cannot be. As an a priori assumption, this is Scientism at its worst.

Each of us has a responsibility to practice discernment about what we believe. At the same time, anyone who makes a claim about these phenomena has the responsibility to clearly distinguish between what can be experimentally proven and what is believed as a matter of faith. Essay 11: *Pseudoscience* (Page 175) is a discussion about what these terms mean.

In essence, if study is not based on ideas developed from a clearly stated hypothesis (theory about the subject) following a predefined protocol (methodology of how to conduct the study) with the intention of publishing a report that will be vetted by subject-matter specialists, then resulting opinions cannot be claimed as experimentally proven. Fruits of the study can be described as an ongoing study but be careful not to claim science unless reasonably well-considered methodologies have been applied. And publication has been attempted.

A second, equally important consideration when claiming science is the qualifications of the people who conduct the research. Evan a person with a doctorate in parapsychology must establish credibility to study the particular subject. An example is a person trained as a psychologist conducting research concerning transcommunication. If it is about how a person experiences the phenomena, then the researcher should be considered qualified. If it is about how the phenomena are formed, then there is no reason to assume the person is any more qualified than an experienced paranormal field investigator who is able to put together and report on a sound protocol based on a well-conceived research question.

Our field is especially vulnerable to skeptical attacks based on qualifications of our specialists because it is so routine for retired professors to publish reports under cloak of academic authority about things paranormal about which they have no real training. Their peers are clearly not willing to police their practices, so it is up to the rest of us to make our concerns heard.

Your point of view need not be based on your research alone. EthericStudies.org and ATransC.org represent efforts to provide a growing body of material you can refer to. There are other association journals and judicious use of Internet material can help you develop a supportable statement about your understanding of these phenomena.

Be careful of the *Trojan Horse Effect*, however. As noted above, skeptics are sometimes members of organizations that are involved in paranormal research. In some cases, these organizations effectively function as debunkers for some concepts, especially survival related phenomena. There is a hierarchy of approval for concepts, so that while the mainstream objects to things paranormal, many paranormal organizations (or at least members of these organizations) regard the idea of survived personality as pure nonsense. Use discernment.

Skeptical About Skeptics

On the other side of the debate are people who are genuinely open-minded skeptics. The Skeptical About Skeptics website is a good example. (93) From the website:

Pseudoskeptics Revealed

Many self-proclaimed skeptics are committed to upholding the authority of established science by maintaining conventional taboos. They are intolerant of those who transgress the boundaries of scientific orthodoxy. These self-appointed gatekeepers of the dominant paradigm proudly call themselves skeptics but reveal themselves as fundamentalists who dismiss any evidence that challenges their belief system.

Skeptical About Skeptics examines the ill-informed attacks leveled by these pseudoskeptics. With articles by well-known scientists and thinkers, we reveal their faulty critiques and the underhanded methods they employ. We highlight controversies in specific fields of research and shine a light on prominent skeptics and skeptical organizations.

We are pro-science, and we are in favor of open-minded inquiry.

Each of us is a representative of our field. The skeptics have influence because they are zealous, not because they are right but mostly because we do not represent ourselves in a defensible way. Yes, there is still the problem that things paranormal are outside of known and therefore acceptable science, but that can never be addressed so long as the skeptics are able to make their ridicule of us so believable by using our own words and actions. It is for us to show the world they are wrong.

References and Alternative Sources
Listed at the end of the book beginning on Page 357.

Essay 11
Pseudoscience
2014

About This Essay

The *Glossary of Terms* on EthericStudies.org, (9) was written to support the Implicit Cosmology. In it, **concepts** are defined as ***fundamental ideas; root thoughtforms from which systems of thought can be derived.*** As fundamental elements of thought, they might be considered a building block of reality, but more importantly, concepts are the building block of understanding. Comprehension of concepts is fundamental to understanding the nature of reality.

The pseudoscience concept was brought to me by skeptics in Wikipedia. Before that, I assumed the term was a reference to really silly ideas such as proofs that the earth is flat. But then I was schooled by the dominant group of editors in Wikipedia. They are usually pretty smart people, but they are also typically adherents of Scientism. If their mainstream science god has not acknowledged the possibility of something, then it cannot be. Anyone claiming that it can is committing sacrilege in the form of pseudoscience … false science.

As I have said, the reason I am no longer allowed to edit the Rupert Sheldrake biography article (91) in Wikipedia is that I promoted his ideas as actually worth considering. The outcome of the Paranormal (94) and Pseudoscience (95) Arbitration Cases is now understood to mean *thou shalt not speak of pseudoscience as if it is real science*. Doing so can get you permanently blocked from editing Wikipedia if you get in the way of a skeptic editor.

As a lesson in human nature and the way people develop their personal reality, Wikipedia has given me a most valuable education. This essay is just one of the resulting studies I have used to develop my perspective on the nature of reality.

This essay is included in this book because it addresses an important part of the environment in which we must seek understanding. In fact, an accusation of pseudoscience is potentially a means by which mainstream critics might be able to deny us of the ability to openly study these phenomena.

Merriam-Webster Dictionary Definition of Pseudoscience:

A system of theories, assumptions, and methods erroneously regarded as scientific. pseu-do-sci-en-tif-ic, an adjective

Skeptic's Definition:

A belief or process which masquerades as science in an attempt to claim a legitimacy which it would not otherwise be able to achieve on its own terms; it is often known as fringe- or alternative science. The most important of its defects is usually the lack of the carefully controlled and thoughtfully interpreted experiments which provide the foundation of the natural sciences and which contribute to their advancement. (Stephen Lower, *Chem1 Virtual Textbook*) (96)

US Government:

Pseudoscience has been defined as *"claims presented so that they appear [to be] scientific even though they lack supporting evidence and plausibility"* (Shermer (97) 1997, p. 33). In contrast, science is *"a set of methods designed to describe and interpret observed and inferred phenomena, past or present, and aimed at building a testable body of knowledge open to rejection or confirmation"* (Shermer (97) 1997, p. 17). Science and Engineering Indicators 2006 National Science Board. (85)

Please note that Michael Shermer is an opinion setter for the skeptic community. In a practical way, the Skeptic's and US Government definitions are from the same opinion setter. The fact that skeptics want to make pseudoscience illegal to protect the country, and that the government echoes the skeptic's position, should give paranormalists reason for concern that their freedom to study these subjects might be in jeopardy.

Practical Definition:

A derogatory term coined by skeptics to label subjects with which they disagree. This disagreement is seldom based on the presence of bad science, but rather because, in the skeptic's view, the subject is not supported by orthodox science. This term is virtually always used in conjunction with efforts to convince an audience to dislike, mistrust or even fear the subject. Use of the term is often indicative of scientism. (Tom Butler See Essay 10: Skeptic (Page 163))

Scientism

The belief that science, the scientific method and work product is the only way to validate reality. In practical terms, scientism holds that, if something is not recognized by mainstream science, it is not real and is, therefore, impossible. When people under the cloak of authority of science advise the public about any subject without first becoming informed about its nature, for instance calling the subject pseudoscience, they are effectively practicing scientism. (Tom Butler)

Introduction

March 2014, I was notified by a Wikipedia Administrator that (in part):

> The following sanction now applies to you:
>
> Topic banned from Rupert Sheldrake in accordance with the terms at WP:TBAN
>
> You have been sanctioned per this arbitration enforcement request
>
> This sanction is imposed in my capacity as an uninvolved administrator under the authority of the Arbitration Committee's decision at Wikipedia:Requests for arbitration/Pseudoscience#Final decision and, if applicable, the procedure described at Wikipedia:Arbitration Committee/Discretionary sanctions.
>
> This sanction has been recorded in the log of sanctions for that decision. If the sanction includes a ban, please read the banning policy to ensure you understand what this means. If you do not comply with this sanction, you may be blocked for an extended period, by way of enforcement of this sanction—and you may also be made subject to further sanctions. (98)

In effect, I was banned forever from arguing that the work of Rupert Sheldrake was valid science; not pseudoscience. The complaint was stated in terms of *"Downplaying rejection by the scientific community"* and *"Further fringe promotion, rewording beliefs into 'hypothesis'"* It was brought to the sanctions enforcement court by *User:Second Quantization*, with the conclusion that *"This editor has been problematic over a prolonged period in the topic area of pseudoscience and fringe science."*

My *"prolonged period in the topic area of pseudoscience and fringe science"* began in 2006 with my attempts to balance the Electronic Voice Phenomena (EVP) page. (99)

Second Quantization is a pseudonym, which means I was charged in Wikipedia court by a *pseudoperson*. (humor) The complaint notes that *"User:76.107.171.90 did much of the legwork for the diffs"* (evidence). This editor is only identified with his IP address because the person has not bothered to establish even a screen name. It is in red because the editor has not added text to his/her personal page.

Implications

It is well-known that editing the articles about paranormal subjects in Wikipedia has been taken over by skeptics. While I was actively trying to balance the articles, I witnessed virtually all of the moderate editors driven off, banned for life, or like me, followed by skeptic trolls wanting to revert my every edit. (Yes, *Troll* is an official Wikipedia term.)

Consider the impact Wikipedia has on the general public. It is an important opinion setter, and typically the first result in searches for most subjects. The skeptic community is much better organized than the paranormalist community, and with the perception of mainstream truth, it has the ear of many governments, including the US Government. (60)

The objective of the skeptic community is to eliminate claims of truth that do not comply with established mainstream science. The term they use for such fringe subjects is pseudoscience. They literally want to make paranormal subjects illegal.

Fact of Paranormal Phenomena is the Issue

For nearly 60 years, people around the world have recorded EVP. Careful, well-educated people have devised ways to test EVP in an effort to determine what causes the phenomenal voices. In many cases, good science has been conducted, leading to peer reviewed reports that reinforce the one important fact that no known physical principle has been found to explain the existence of EVP. If a physical explanation cannot be found, then it is sensible to look for nonphysical explanations.

In a different forum, researchers have discovered that it is possible to influence the environment at a distance with intentionality. Substantial research has been conducted on what is commonly referred to as *psi functioning*. **Psi functioning is a term used to denote mental influence on a hypothetical psi field which is thought to permeate the physical.**

Since current instruments of science do not directly detect the psi field, it is studied by detecting how it affects physical or biological processes, or by examining the validity of apparent psi (psychic) information access. For instance, random event generators are known to become less random when near a group of meditating people. (100) Similar changes have been detected during successful remote viewing sessions. (101) If a physical explanation cannot be found to explain these effects, then it is sensible to look for nonphysical explanations.

By all reasonable standards, the scientific method is often followed in these studies, making the pseudoscience accusation technically baseless. Apparently, the real reason for the pseudoscience branding is that skeptics, acting as apologists for science, believe that mainstream science does not allow for the existence of these phenomena, and therefore, they cannot be real. The net result has been that the possible benefit to humankind brought by these and similar phenomena has only partially developed. If the skeptics prove successful, the benefit will never manifest.

This article about pseudoscience addresses this issue and explores ways our community might respond.

The Scientific Method

Science is basically the organized inquiry into the nature of reality. In its simplest form, it is observation of nature leading to a hypothesis describing what is observed. This, in turn, leads to predictions about the behavior of what has been observed. For science to be practiced, these predictions must be able to be tested, and test results must be able to be used to modify the hypotheses so that it can better describe the behavior of nature.

Anyone can conduct science; however, three very important elements are considered necessary if *real science* is to be conducted. The most important is that there must be a well-considered protocol describing how the predictions are to be tested. This protocol should be designed to assure that unnoticed artifacts of the experimental process do not contaminate the results or lead to misleading conclusions. The protocol must also allow for the collection of results that might confirm or disprove the theory.

The second element is development of a research report and some form of media for publication that allows for vetting of the report by a community of subject-matter specialists. Conducting science requires that qualified people are able to review the results and agree that the hypothesis has been tested and the results have been analyzed to produce a reasonable conclusion. Here, *reasonable* will generally be determined by best practices for that field of study. For instance, trans-etheric influences are experienced or detected differently than are physical phenomena such as apples falling from a tree. It may be unreasonable to arrive at a firm conclusion about the meaning of an EVP and there should be many more indeterminate results in etheric studies than there are in physics.

The third element is a history of prior research. This is a body of knowledge developed over time which will help to provide a foundation for development and evolution of the hypotheses. In principle, science is conducted in a continuum so that the present inherits some characteristics of the past and contributes to the future. Prior art is very important in Science.

In mainstream society, the practice of science is conducted by a well-established community supported by universities and professional organizations. It is funded by an established network of government and private funding. The community has evolved a culture of professionalism and peer pressure with many, often active lay-supporters.

Inappropriate Science

As a frontier field of study, the number of people studying paranormal phenomena is relatively small. It has only been recently that some of the New Age and religious beliefs have given way to research-based understanding. The net result is that the scientific history is very short, and the number of studies for any single aspect may be small if there are any at all.

Studying the effectiveness of alternative medicine is probably best done using the same techniques used for mainstream medical studies. But in some frontier subject, the usual methodology of mainstream science may be inappropriate. The etheric is hypothesized to be a mostly conceptual environment in which intention may be an equivalent to force in the physical. Any protocol that does not consider our current inability to shield from psi influences is simply inappropriate.

Statistical analysis has its place, but one should remember that some phenomena are very rare. Yes, a Class A example can be dependably recorded by a confident practitioner given enough sessions, but statistical analysis can be expected to reject the rare Class A example simply because it is considered a statistical outlier. (77)

Another example of possibly inappropriate science is the approach to testing experimental repeatability. This is discussed in my critique of two *failure to replicate*-kinds of articles published by the *Journal of Scientific Exploration*. (102) The study's protocol called for the use of untrained college students as practitioners. Recording EVP is repeatable to a point, but like many mundane practices, it is very difficult to conduct research if one does not have a skilled practitioner. The conclusion that the protocol had effectively tested the subject is not supported by best practices.

While it is fine for a layperson to conduct studies of paranormal phenomena, it is inappropriate to report the results as good science if the person is not trained in an applicable discipline. For instance, EVP needs to be studied from an electronic technology (physics) perspective first and perceptual (psychology) second. A degree in psychology alone is not sufficient unless the study is restricted to questions of perception.

A common reason for the accusation of bad science is the idea that only successful results might be reported while unsuccessful results *remain unreported in a file drawer*. However, as a point of order, the file drawer effect cannot apply to research if it is conducted by a person who is working out of his or her discipline. Research that fails this test should not pass peer review.

Pseudoscience

Pseudoscience is a term adopted by skeptics to describe everything that does not conform to their sense of proper science. It is a very effective term because one of the main characteristics listed for pseudoscience is that people who practice pseudoscience will naturally argue that they conduct real science, thus confirming the prediction.

To be clear, there are fields of study to which these characteristics apply. The problem is that skeptics likely do not know enough about the fields to simply write them off with a derogatory name. At the very most, one must consider that a questionable field is possibly emergent science, but remains theoretical, awaiting better research.

For instance, I have studied transcommunication for many years and still do not think I know enough to say any particular theory is wrong. Radio-sweep is a good example. I have come out against using it for EVP because it makes no metaphysical sense and produces too many false-positives to be practical for use by people new to the field. Still, I am open to the idea that proper studies might produce information that changes my mind. I am not wise enough to think otherwise.

Commonly Cited Characteristics of Pseudoscience

Having an easy name for the subject of conversation is useful for communicating ideas. Except in rare confrontation, skeptics only talk about the study of paranormal phenomena from the perspective of explaining the subject to the public. They write articles for the public, their websites and magazines are targeted for the public or other skeptics and they have conferences to promote their point of view to the public. As you read this essay, keep in mind that the term is about things paranormal, but its use is directed by the skeptics to the general public as a warning sign, so that: *"This is what pseudoscience means and beware that it describes this subject."*

The skeptic's message is that pseudoscience is dangerous and harmful to the greater good of our country. Once a subject is established as pseudoscience, it is a small step to make it illegal. We have a taste of this with the way the Federal Government jailed Wilhelm Reich and burned his books. (84) See "Wilhelm Reich" (Page 169) in Essay 10: *Skeptic* (Page 163).

Listing characteristics is a common approach used by skeptics to explain how to recognize pseudoscience. As you might expect, they include everything skeptics don't like about all things paranormal. Below, I used a list that begins with *dogmatic* because that is a characteristic I have most often noted amongst paranormalists.

While you may encounter these characteristics amongst paranormalists, it is important to note that there are reasons for this that are apparently not considered by the skeptics. As you read this list, try to first view the characteristic from the perspective of a doubter and then from the perspective of an advocate.

The characteristic often listed as pseudoscience by skeptics include: (103)

Dogmatic; ignores contradicting facts

Here, *contradictory facts* are taken from mainstream theory. The reason they are ignored by paranormalist is that the so-called facts tend to ignore the possibility of a psi field and survival. An informed paranormalist will know that they are being told an untrue story.

Dogmatic comes in when we are told that psi is pseudoscience and we say it is not. It is like saying you are dogmatic for denying that you are purple.

Paraphrasing from the paranormalist's point of view:

Frustrated, seeking better guidance from learned scientists.

Subject to confirmation-bias by selectively reporting evidence and research results

Confirmation bias can be understood as the tendency of a person to think evidence supports beliefs, even though it may not. It also implies that a person will only report experiences that support beliefs.

In terms of First Sight Theory, (16) (38) [Page 23] confirmation bias would relate to Corollary 8: **Bidirectionality**: (paraphrasing) *In this summative process, the person may turn toward information (signed positively) to include it in the construction of experience, affect or action, or turn away from information (signed negatively) and exclude it.* This is a mostly unconscious process which helps to determine experiences of which we will become consciously aware. In other words, if we have previously established an interest in things paranormal, we are more apt to notice things possibly paranormal.

Read Essay 14: *Open Letter to Paranormalists* (Page 223). In it, I describe how some parapsychologists deliberately ignore studies related to survival, apparently to strengthen their original assumptions. This would be a clear case of confirmation bias.

In science, the tendency to report only supporting research is referred to as the *file drawer effect*, as if research that does not support the theory is hidden in a file drawer. An alternative version of the *file drawer effect* is the rejection of reports the journal judges do not agree with, thus converging the field toward a status quo. (104)

In parapsychology, Exceptional Experiences Psychology seeks to identify ways in which people who believe in things paranormal tend to report experiences defined by the researchers as ordinary experiences. Such explanations should probably be ignored if they do not consider the Bidirectionality Corollary.

In a fifteen-second video loop Instrumental TransCommunication (ITC) session, I record around 450 video frames (like single pictures), but only keep fifteen or so and only find six or seven *keepers* which I save and sometimes report as examples. You can argue that this selection of only the feature-producing frames is selective reporting.

The same sort of selection happens in EVP as we listen to many minutes of EVP before finding a Class A or B utterance ... if we do at all. Meanwhile, we usually hear many more Class C, which we ignore.

The fact is, we are looking for an effect which is produced by applying a theory. We predict an average rate of occurrence of the examples, and so, the presence of an example supports the theory. You can apply the same test and demonstrate this for yourself. Fewer than expected features may indicate the equipment was not set up correctly but may also be due to the native ability of the practitioner.

Lisa's avatar is a video-loop ITC image from our early studies. You should see a woman demurely looking toward her left shoulder. She wears a dress with a V-neck collar and possibly a flowered hat.

The flesh-color of the skin and possible colored ribbon are visible in the original version.

And so, looking at it from the paranormalist's point of view, repeatability is based on having a qualified practitioner correctly applying a procedure. Validation is based on agreement with a set of previously known characteristics. We would paraphrase this point as:

> A practitioner reports examples which agree with previously known characteristics and ignores all else while seeking a reason for the rate of occurrence of the examples.

Hypotheses cannot be tested

This argument may be true of some of the global questions such as the existence of a first cause, but the real subject skeptics are trying to make go away concern psi and survival phenomena. Paranormalist researchers are not saying that there is some godly intervention which cannot be tested. They are saying that *"If we do this, this happens."* That is a very testable hypothesis.

For instance, if a video loop is set up in a certain, repeatable way, a resulting recording will often contain human faces that are detected by others without prompting. Where those faces come from, and why, are separate issues. People speculate, and in some cases, that speculation can be tested.

From the paranormalist's point of view, this is better stated as:

> *Inability to attract more and better-funded researchers has hampered examination of theory.*

No evolution in understanding or theory

This is a typical problem of frontier subjects, in that there is such a small population of people actually studying the phenomena. In fact, understanding does evolve depending on the time people are able to study the subject and available funding.

Consider the number of people involved, funding and mainstream popularity of such science projects as the Large Hadron Collider and the Hubble telescope. All of the people studying psi and survival phenomena would probably not match the number of people on the human resource recruiting staff for either project.

The funding and number of scientists have a direct effect on the progress in a field. Compared with mainstream science, there is hardly any noticeable progress. But there is progress. We have all the ATransC NewsJournals on ATransC.org. If you examine the first few and last few, you will see some progress in understanding. Then read *Your Immortal Self* or examine the Concepts section of EthericStudies.org. We are making considerable progress in survival research, and much of that is only because of the progress made in psi studies.

> Take time to read *A Model for EVP* at ATransC.org/model-for-evp/. (105) That essay includes a solid, testable model for the nature of EVP that did not exist just a few years ago.

From the paranormalist point of view, this can be stated as:

> *An observer informed about available resources in the field will note progress in theory and understanding.*

An appeal to recognized authority is used to support claims

There are two sides to this. In terms of major ideas such as quantum mechanics, psi phenomena behave a little as if they may be governed by quantum principles. This is a hypothetical link. Yet, it is becoming increasingly popular to claim some aspect of a favorite phenomenon is governed by quantum principles. Skeptics sometimes refer to this appeal to authority as *Quantum Mysticism*.

With that said, as we study these phenomena, it is important that we first attempt to apply known physical principles. It is expected that some will apply, but experience has shown that none explain the core characteristics indicating a psi field or survival. Researchers attempt to incorporate those that do seem to apply into their models. For instance, I speculate that stochastic amplification is involved in the expression of intended order. That, at least, gives me something to test.

Right or wrong, trying to apply existing theory is not proof of pseudoscience. It is proof of inadequate education or lack of resources to test a related hypothesis. In some cases, referencing existing authorities indicates that the researcher has conducted the necessary literature surveys to determine if existing work applies.

A paranormalist might paraphrase this point with:

> *Researchers are expected to recognize and test the applicability of known physical principles.*

Metaphorical/analogy driven thinking

When a person reports an experience for which there seems to be no normal explanation, it is natural to look about for alternative explanations. When the person finds that others have a similar experience and the community of experiencers generally agree that the experience is paranormal, then it is human for the experiencer to begin thinking the experience is also paranormal.

It is also normal to compare the strangeness of possibly paranormal experiences to the strangeness of quantum phenomena. For instance, the apparent nonlocality of quantum entanglement is comparable to the apparent nonlocality of psi functioning. The two may exhibit this characteristic for very different reasons, but researchers would be remiss if they did not at least attempt to integrate the two models.

Experiencer's acceptance of a belief-based explanation is likely what the skeptics are referring to. However, lacking learned guidance from science, it is natural for experiencers to develop a belief-based explanation for paranormal experiences. Metaphors are useful as a tool for understanding experiences and sharing ideas but are not intended to be science. In the end, it is up to the scientists in the community to provide constructive guidance.

This characteristic might better be stated as:

> *Observers of the paranormalist community must be sufficiently informed to distinguish between experiencer's belief and researcher's theoretical understanding.*

Anecdotes as evidence

It is true that, as with the early naturalists in mainstream science, paranormalist field research often involves observing and reporting on what was observed. This is especially true today for the study of survival-related phenomena. However, psi functioning studies have been conducted under very controlled conditions and managed following scientific methodologies.

The paranormalist might say:

> *Assumption of ignorance is the first tell of a skeptic.*

Lack of explicit mechanisms

My early education did not include the theory of Continental Drift. That came later as a mechanism for it was finally accepted. Looking back, many really good geologists were convinced of the theory before that. As I look at it now, their theory was seen as a hypothesis looking for a proper model. That was seen as good science.

Early efforts to explain some paranormal phenomena were actually efforts to develop a reasonable hypothesis. It has been only recently that useful models have begun to emerge. The Trans-Survival Hypothesis and resulting Implicit Cosmology are my efforts to define such a model. (10) (12)

The reason paranormalists prefer terms such as frontier science or emergent science is that they recognize the study is very new and that it is unreasonable to expect it to come into being with a full-fledged model. Skeptic's expectation that it should is further sign of their determination to protect the status quo, rather than to embrace new ideas.

A paranormalist would counter with:

Insistence on recognition of a mechanism before research has been conducted indicates poor understanding of science.

Special pleading (elusive evidence)

EVP have typical characteristics which make them difficult to understand. My studies indicate an average 25% correct word recognition of Class A examples by inexperienced listeners. (39) Anyone can expect to replicate the process of collecting an EVP, but Class A examples are relatively rare, and experience is sometimes necessary to collect one. As in most human endeavors, there appears to be a natural distribution of ability to record EVP, so that the combination of rarity and limits in natural ability makes it difficult for a casual observer to test the hypothesis. (102)

Trans-etheric influences can be described as a conceptual influence causing an objective effect. A major problem in the study of paranormal phenomena is that this influence appears to depend on the observer as a conduit, and cultural beliefs appear to influence the manifestation of the effect. Thus, we see Worldview play a large role in the way these effects manifest. In this way, we see that a person who accepts the possibility of paranormal phenomena is more apt to experience them.

To the uninformed, attempts to explain these limitations can appear to be special pleading; however, the ability to experience the phenomena can be taught to a willing person with objective results. In that way, the argument would appear to a lazy investigator as special pleading.

A better way to approach this characteristic would be to say:

Investigators should expect to do the work to become properly informed about the nature of the phenomena to be investigated.

Conspiracy theory

I was often assured by skeptic Wikipedia editors that there was no conspiracy by science to suppress our study. This is a good example of the tyranny of the majority. They can think they are doing us a favor by protecting us from delusion.

In a real sense, each person seems to assume that people further out on the frontier of thought is wrong. With that belief comes an apparent cultural norm that it is okay to ignore the more frontier person's argument. Even in the paranormalist community, the majority of parapsychologists do not accept survival hypothesis while a smaller number do not even accept the psi hypothesis.

Feedback I have received on Essay 14: *Open Letter to Paranormalists* (Page 223) makes it clear that many people think I am paranoid. This bothers me, but it is difficult to ignore the evidence. But which is it? Am I paranoid, or are people misinformed who think I am simply ignoring the facts?

This is one of those *time will tell* situations. I cannot prove something to people who refuse to consider the evidence. Conversely, it is illogical to think we who study these phenomena are going to simply give up because of lack of informed validation. I am confident that the accumulation of evidence and emerging understanding in mainstream science about how we think will change the discussion from condemnation of us by the mainstream to questions from people eager to learn more. If not now, soon, for many of us are diligently working to make it happen. Will you help?

I would answer this one with:

> *Mainstream and frontier scientists are expected to do a better job of explaining the implications of their assumptions to the public.*

Concept is described for the public rather than scientists

I write for the public but am mindful that academics might one day measure what I say. There are expectations scientists have that can only be met by other scientists. For instance, academics depends on their work being cited for credibility. In turn, more credibility results in more citations. In a real sense, that is how scientific truth is established. I am aware of no similar system of collaboration amongst paranormalist laypeople.

Without a doctorate, it is unlikely any of my work will be cited. So, while I attempt to be a good technical writer, I have no delusion of credibility amongst the paranormalists holding a Ph.D. Being ignored by academics is depressing, but not as depressing as knowing that the kind of shunning I experience from some Ph.Ds. is exactly the kind of shunning they experience from mainstream scientists. The resulting inability for parapsychologists to access research funding hurts all of our community.

Parapsychologists appear to be very aware of this accusation from mainstream scientists and respond by overcompensating with ultra-scientific writing styles full of statistical analysis and obscure terminology. This approach has not accomplished its intended objective but has made it difficult for laypeople to follow their work. When there is overcompensation, writing for scientists tends to defeat the purpose of writing.

The obvious answer is for all of us who are in this community to work together. I, for one, have exhausted my ideas for making that happen. See Essay 14: *Open Letter to Paranormalists* (Page 223).

I can think of no good counter statement for this, as many of us are, indeed, guilty of directing our work toward the greater community.

Alternative Terms for Pseudoscience

Other than *pseudoscience*, skeptics will sometimes refer to science they disagree with as *junk science*. This is often used in political and legal context to brand science as spurious and is commonly claimed by the skeptical community to be a form of fraud, or at the very least, ignorance.

A second common derogative term is *pathological science*, which is a reference to science which involves barely detectable phenomena that are then reported as being carefully studied.

It is interesting that this term is an example of circular referencing. Irving Langmuir coined the term. (106) He has the 1932 Nobel Prize in Chemistry and probably ran into a good deal of bad science. According to Langmuir, symptoms of pathological science are:

1. The maximum effect that is observed is produced by a causative agent of barely detectable intensity, and the magnitude of the effect is substantially independent of the intensity of the cause.
2. The effect is of a magnitude that remains close to the limit of detectability. Or, many measurements are necessary because of the very low statistical significance of the results.
3. Claims of great accuracy.
4. Fantastic theories contrary to experience.
5. Criticisms are met by ad hoc excuses thought up on the spur of the moment.
6. Ratio of supporters to critics rises up to somewhere near 50% and then falls gradually to oblivion.

These characteristics are very similar to other such lists that can be found around the Internet and rather similar to the one above for pseudoscience.

Scientism

If skeptics are associated with an ideology which amounts to a faith-based view, it would be *scientism*, (88) which is the ideological belief that science—mainstream science—is the only authority on the nature of reality. It is helpful to understand this. When confronted by skeptics, it is important to determine if they are concerned with the validity of your point of view because they have sincere questions or simply refuse to consider your proposition because it is contrary to their worldview. If it is the later, then you may as well change the subject.

It has been my experience that skeptics might be evasive about the reason for their interest. They will feign interest but disagree with you in the end no matter what you say. There have been times I have finished a conversation with a person, thinking I succeeded in making my point, only to learn in later days that the person was faking agreement and lacked the intellectual integrity to say as much at the time. More likely, the person felt no need to engage with someone who was on the wrong side of the academic-layperson partition.

Relative Scientism

A surprisingly common form of scientism in the paranormalist community shows up in the *bait-and-switch study*. Rather than thinking mainstream science has all the answers, it is a modified form of scientism in which mainstream science as it is enhanced by a favored theory has all of the answers.

To study paranormal phenomena under controlled conditions, it is necessary to have competent practitioners. In most examples I have seen of this form of scientism, the researcher ultimately does not accept the possibility of the phenomena, and so, seeks to use reportedly successful practitioners to prove their point. Of course, it is necessary to mislead practitioners into thinking the researcher has the practitioner's best interest in mind.

Community Response

If you are actively seeking understanding about possible survival of personality beyond physical death (transition), trans-etheric communication (transcommunication including audio ITC (EVP) and visual ITC), reported hauntings phenomena (trans-etheric influences) and the nature of subtle energy involved in such human abilities as remote viewing and healing intention (psi functioning), then you are a member of the paranormalist community.

Take a little time to search the Internet to find our community. If you search for *skeptic*, you will find page after page of listings for pro-skeptical websites. That community is clearly branded. In comparison, the paranormalist community has no such clear identity. For every apparently serious study group such as the Parapsychological Association, there are hundreds of groups talking about ghosts or trying to sell classes, and lately, selling ghost hunting hardware. The objective is lost in the clutter of the fantasy.

Did you know that the religion known as Spiritualism is more properly a member of this community than it is a religion? Did you know that parapsychology includes Ph.Ds. with interest ranging from anti-psi field and anti-survival phenomena to just a few who consider survival a possibility? Which amongst paranormalists groups, found with Internet searches, practice objective examination and which base their understanding on belief? If you cannot tell, don't expect mainstream society to know.

Our first task is to learn how to look like a community. We can begin to do that by learning to talk with a common vocabulary; one that does not feed the monster skeptic or make what we think is true sound like religious dogma. But before we begin, it is important that we know who we are.

Take a very close look at the way skeptics, and now the US Government, use pseudoscience in their literature. Pay attention to the fact that I was blocked from defending a parapsychologist in Wikipedia because the skeptics were able to argue that I was supporting pseudoscience. It did not matter what I said, just that I was openly in favor of something they have successfully identified as pseudoscience. Note also that they are the ones who defined the term.

Our freedom to study these phenomena is not assured. It is arguable that the primary motivation to attack us is not for the good of the country, but in defense of the skeptic's religion ... either scientism or one of the main religions. It may be illogical for them to attack us, but then, belief-based thought is seldom logical.

References and Alternative Sources
Listed at the end of the book beginning on Page 357.

Essay 12
Concerns with Wikipedia
2014

About This Essay
What would you do if you wanted to learn about your etheric nature but there were no books available and no Internet? That was about how it was when I was growing up. About the only books available to me were religious text and science fiction. As the years went on, a few more books became available. Many of them were useful, but some were outright misleading.

Looking back, it was the dominance of channeled material that cultivated my pragmatic approach to metaphysics. Virtually all of the channels and their unseen teachers eventually turned toward doom and gloom. Anyone with a good imagination could weave a story based on popular wisdom and claim it was from a *higher spirit*. If you believe the channeled material, humanity has been catastrophically destroyed a thousand times by now. Is that a window on our collective worldview?

We are still today, suffering under the legacy of Friedrich Jürgenson published examples of EVP. As a historical figure, it is important that he helped introduce the world to EVP. But, his experimental work was very early in the development of techniques and theory and has little educational and no technical value today. Yet, skeptics routinely use his really poor-quality examples as proof that we are all delusional. (107) (108) Judging by some of the people who still think he is the top authority in this field, the skeptics are right about at least some of us.

My point is that information is our most important tool for self-education. Today, we have more information than ever, but we have no real way of telling what of it is useful and what is misleading. We can't trust our learned parapsychologists because so many are trying to prove we are delusional. We can't trust the books we read or the thousands of paranormalist websites. Some are excellent, but some are not. Anyone can slap together a website or book and self-publish ... many do.

In effect, Wikipedia is one of those websites. It has been created by a collective *Joe Editor* who selectively quotes and freely translates sources according to its mainstream point of view.

Some kind of consortium created to establish standards in published material might help, but that is essentially what Wikipedia was supposed to be. Now, we see with the introduction of the Society for Psychical Research's Psi Encyclopedia that, when a respected parapsychological organization tries to standardize truth, they are almost as prejudiced toward survival as Wikipedia. (109)

Your best bet is to inform yourself so that you at least have a sense of the path you wish to follow; its nature and basic tenets. You must be informed enough to be able to vet a teacher ... and then find one to vet.

The value of teachers is that a good one can help you navigate the thicket of the sense and nonsense in the information world we have today. None of us self-proclaimed teachers know it all. Some of us are clearly on the wrong path for you. Even if we are on a good path, our approach may not suit your learning style.

Even the ancient teachers advised that we find a teacher in order to gain understanding. The advice stands today. Just as we need to have others act as our witness panel for phenomena we produce, so do we need a witness panel to give us a sanity check on the ideas we express based on our understanding.

If you do not have a teacher, find an online group, discussion board or local society and ask questions with an open mind.

This essay was written to warn others about the danger of misinformation posed by Wikipedia. The real lesson is in how the online encyclopedia was able to become such a potent tool for disinformation. That lesson is all about personal responsibility, cooperative communities and the way of a teacher.

Above all, know your detractors, as they often see more clearly than your supporters.

General

Wikipedia is hugely important as a means of documenting society and making that knowledgebase available to anyone with access to the Internet. For the most part, it has been successful. The point of this article is that Wikipedia has become a means for special interest groups to exert undue influence on public opinion. This is most obvious for subjects that are not part of mainstream thought, what is referred to here as a **Frontier Subject**: *The study, practice or experience of a phenomenon which has not been academically established as an accepted part of mainstream culture.*

The reason this is important is that a search of the Internet for almost any subject presents a Wikipedia article as first or nearly first choice. Citizens, and especially children, will often learn about a subject from Wikipedia first, and that means they may learn from a small group of editors pushing their particular view of the subject.

Treatment of Subjects

Wikipedia rules governing the content of articles are intended to assure a balanced disclosure of each subject in the detached style of traditional encyclopedias. Because the volunteer editors are seldom subject matter experts, everything in articles must be referenced. Consequently, the main rules used to control the tone of an article concerns the acceptability of the information source and neutral point of view of wording in the article. Original research is not allowed and material from frontier subjects journals is virtually not acceptable. The problem is that it is the dominant group of editors that decides what is acceptable and that determine how a subject is characterized.

Because of the rules, material in Wikipedia articles is at least second hand and often based on very outdated material. Because books are believed to be more authoritative than websites, they are preferred as references. Because it often takes years to publish a book, it is common to find book references that have long-since been outdated by new research published in journals and on websites. More importantly, references are often used that are unavailable to the reader, making it nearly impossible to verify that the included information is actually supported by the reference. Too often, it is not.

Original research means that what must be used is an article written by someone else about that research, other than the researcher. Small or non-mainstream publications are considered *fringe*, (110) and are therefore easily discounted by an editor determined not to allow its use. In most frontier subjects, there are only small publications because of the immaturity of the field. Mainstream publications will not venture to publish a positive report about a nonmainstream subject. Also, book publishers will not invest the resources to publish a book intended as a serious research report unless there is a large audience. All of this means that collaboration in frontier subjects is accomplished via newsletters, self-published books and websites. The most current research information is too often on the same search engine page as hobbyists speculating about the subject from the point of view of *"how it helped me today,"* rather than if it has any validity in fact.

Wikipedia Editing Rules

Wikipedia has rules governing the interaction of editors such as the need to assume good faith and the need to be civil toward other editors. There are also procedural rules, such as how often and why an editor might change another person's edits. Perhaps most important are the rules governing what may be included in articles. For instance, Articles are required to be written from a neutral point of view, and everything in articles must be based on verifiable references. The references themselves must not be out of the mainstream or self-serving to the author, and so there is also a conflict of interest rule.

An editor's failure to follow the rules is usually addressed by other editors, but if that does not work, then it is possible to bring an editor before a tribunal that has the power to ban an editor from making further contributions.

Who Can Edit Articles

The policy of Wikipedia, or at least the dominant culture's policy, is that subject matter experts are discouraged from editing articles within their area of expertise. In fact, it is common for subject matter experts to be so abused that they soon stop attempting to contribute content. One of the Wikipedia founders, Larry Sanger, has even written the article, *Why Wikipedia Must Jettison Its Anti-Elitism*, (111) explaining the pitfalls of editors not using their real name (no accountability) and not being knowledgeable about the subjects they edit. His response was to begin Citizendium, (112) which is intended to be a kinder and more dependable online encyclopedia.

In fact, anyone can edit Wikipedia articles because anyone can register under an assumed name. However, if an editor is found out as a person who might in some way benefit from what is said in an article, that person is considered to have a conflict of interest and is strongly discouraged from editing associated articles. This is an important rule because, in the case of frontier subjects, virtually all the people who are knowledgeable about the subject are the same people who are leading study groups, have websites, have or might write a book or give talks on the subject.

The Skeptical Community

By *special interest groups*, I am referring to the members of *Wikiproject:Rational Skepticism* (113) and those who are sympathetic to them. Based on my encounters with this group, they appear to be mostly James Randi (114) and Robert Carroll (115) adherents. People involved with frontier subjects often document their dismay at how unreasonably closed the skeptical community is to new thought and how ruthless its adherents are in their efforts to make sure the general public understands that the frontier subject is impossible, and therefore, cannot be. People believing such things are branded as delusional or possibly fraudulent.

This tradition of pathological skepticism is now an integral part of Wikipedia. This is a problem for all of us because the online encyclopedia has given skeptics inordinate access to students of the world looking for material to write a term paper. The skeptical theme is that anything that is not explicitly defined by mainstream science must not be shown in Wikipedia to have any form of possibility. Review of any article in Wikipedia will show that the subject is carefully characterized as fringe (116) and pseudoscience (117).

It is essential to remember that the skeptical community believes that it is executing the will of mainstream science to protect the community from being deceived. They edit from this perspective even though they seldom actually know anything about the subject. The inescapable conclusion is that the most aggressive skeptics have adopted a faith-based viewpoint and their argument is an emotional one cloaked with the authority of science.

Personal Attacks

It is common for skeptic editors to denounce anyone who studies frontier subjects as morons, idiots, deluded, or even more libelous, charlatans and frauds. There are administrative-level editors and procedures to request help from such abuse, but in many instances, complaints are answered by a barrage of comments agreeing with the original insult and adding many more disparaging words to the list. In the end, it is an inescapable conclusion that Wikipedia intends to maintain a civil work environment but is unable to apply existing rules to protect editors from other editors. The common term is *poisoned atmosphere*.

Why This Is Important

Electronic Voice Phenomena (EVP) is a good example of why you should be concerned about the influence of Wikipedia. The Association TransCommunication (ATransC) has gathered substantial information from members and the general public about the characteristics of EVP, including the circumstances under which they tend to occur, their typical characteristics and considerations about how to record, locate and listen to them.

The voices of EVP sometimes appear to be spoken by deceased people or at least include things they might say, in their voice and in response to specific questions. The ATransC has conducted well-organized studies and funded research by academically trained scientists. Careful analysis of this anecdotal and experimental data seems to indicate that no known physical principle accounts for their existence. (118)

> If you look through the ATransC NewsJournal Archive on ATransC.org, (32) you will see that over the years, other groups and individuals have also conducted excellent work. Alexander MacRae (119) is an important example, as is Keith Clark, (120) Sonia Rinaldi (121) and Anabela Cardoso. Anabela is one of the few remaining people publishing a journal dedicated to Instrumental TransCommunication (ITC) (itcjournal.org).

By any rational standard, EVP are a form of paranormal phenomena which clearly requires further investigation. If they are paranormal, the least benefit is that they represent an experimental tool with which other forms of apparent psi phenomena can be studied in controlled conditions.

The most important possible benefit of studying EVP is what they might tell us about our etheric nature and the well-being of our discarnate loved ones. But having been branded a pseudoscience by skeptics, mainstream society is cautioned to think of EVP as simply ordinary experiences mistaken as paranormal; at best illusion or fraud. (99)

With skeptic's success in branding EVP a pseudoscience, and their success in convincing governments that pseudoscience is a danger to society, funding for EVP research is virtually nonexistent compared to that available for mainstream subjects. Research funded is ultimately determined by the popular wisdom of mainstream culture. That opinion is shaped by mainstream science and opinion setters, such as the late Carl Sagan and presumably authoritative sources of information such as Wikipedia.

> **Abdication of responsibility in the scientific community:** In society, the responsibility of scientists is to explain Nature. If members of the scientific community ignore aspects of Nature being experienced by the citizens, the scientific community is effectively abdicating its responsibility. Conversely, it is intellectually arrogant for scientists to ridicule laypeople who take it upon themselves to seek explanations for the experiences. It is also unethical for scientists to then comment on the work done by laypeople without becoming familiar with protocols used and what has been learned.

Peer pressure and popular wisdom are other important factors. Scientists dare not associate themselves with a subject that is characterized as fringe or pseudoscience for fear of ruining their career. History will surely show that the skeptical community is delaying discovery of many new ideas because of the peer pressure they bring to mainstream science. Frontier subjects will not be studied by mainstream science until opinion setters become more accepting of new ideas. When the public demands to know more, funding will follow, and scientists will follow the funding.

The Internet provides extraordinary access to the public making it possible for a determined, but minority group of people to have extraordinary influence on what people believe. It is not realistic to think that private websites can be controlled, but Wikipedia is publicly funded and affords public access to virtually everything that is written in its pages.

What can be Done?

The best way for paranormalists to counter Wikipedia is to tell our story in a levelheaded manner. Whatever your subject, learn to talk about it in terms that a person new to the subject will understand. Giving talks, writing articles and being on talk shows give opportunities to learn what works and what does not. Much of the criticism of the frontier subjects is due to the failure of people who study them to communicate what they are and why they are important.

Maintain the point of view that the subject is an observed phenomenon and that there is a need to study what it is. Avoid a single conclusion by clearly explaining the working hypothesis that best explains the evidence **at this time** and always try to leave the discussion open for alternative explanations. Do not appear to be determined to prove anything. Let the evidence determine the next step.

Establish a presence on the Internet with an informative website. It need not be slick or even pretty so long as it conveys a sense that you are levelheaded and that you know what you are talking about. Keep it current. If there is empirical evidence for some parts of what you want to say, then clearly explain that. If parts of it are based on assumption or belief, then clearly explain those parts and clearly distinguish the two.

Make the difference between demonstrably objective and theoretical assumption clear if you wish to be seen as a researcher with integrity. Probably most important is that the articles that others might link to for citations are stable. Credible articles that can be referenced in other work have become an important replacement for scholarly books in frontier subjects.

> Mainstream scientists live and die with citations of their work by others. It is a way that researchers can gain a sense of how credible the person's work is seen to be by others. We in the paranormalist community are no different. If you refer to an idea posed by another, include some kind of a reference so that your readers can find the source and learn more. Links are very useful for quick access. It is the citizenship thing to do in a cooperative community.

Seek critique and feedback from friends, or even better, from webmasters related to other frontier subjects. Mainstream science has a system of societies, universities and publications that enable collaboration and archiving of the knowledgebase. This is missing for most frontier subjects, so it is important to establish a culture of cooperation amongst interested people. This essay is about an issue that is common to all frontier subjects and it is not necessary for us to have common subject matter interests for this issue to be addressed.

Write an article about your subject that is suitable for an encyclopedia and include it on the website. For all of its faults, Wikipedia is a good place to see what formats work best. The article you would write is not to prove your point in any way. It should be a clear explanation of what your subject is without too much emphasis on proving your point. Let your reader decide. Perhaps people outside your circle of experts should help draft the article because it needs to be a serious *"What I would like to see in Wikipedia"* article that is written from a neutral point of view with good, solid references. The good should be shown with the bad. Seek and include viable alternative explanations. This article may also make an excellent *white paper* to be used as a handout at conferences.

Place this link logo on your website and help others become informed about Wikipedia. (ethericstudies.org/concerns-with-wikipedia/)

Alternatively, write an article yourself warning people about Wikipedia. You can also use this logo to link to the *Concerns with Wikipedia* article on EthericStudies.org. It is worth noting that the content of ethericstudies.org is *copyright free* (Creative Commons (122)), and available for you to use as you see fit with appropriate attribution. But remember, the more links there are to a web page, the higher it will be in the search engines and the more people will read it. This is all about public education, so often include links to other paranormalist pages on your website or writing to help counter Wikipedia's influence.

We do not recommend that anyone becomes an editor at Wikipedia. Until the environment has become more civil, we feel that the anger you will certainly come to know will do more harm than good. If you do want to contribute to an online encyclopedia, we recommend the Paranormal Subgroup of Citizendium (112). Wherever you edit online, always use your real name. Reading the essay about the *Lord of the Flies* (83) will tell you why we feel Wikipedia is able to sustain such a gangland-like atmosphere amongst editors.

Navigation Guide for Wikipedia

The most important thing anyone can do to help attract serious research is to become personally educated about this field of study and how people react to it. What are the arguments used to discount the work? Insight into this can be found by reading the **Talk** Page associated with **Article** Pages in Wikipedia--in effect, by looking behind the curtain. Always look at the archives listed on the Talk page, as the skeptic editors seek to hide bad things.

Read the History of both the article and talk pages. For instance, on the **History** page for the *Rupert Sheldrake* Article, (91) you can see in **Archive 19** (123) that I was trying to talk the skeptic editors out of calling Sheldrake's work *pseudoscience*. But on 5 March 2014, the skeptics were able to permanently ban me from editing the Rupert Sheldrake article because I was promoting pseudoscience.

> (See the ban notice at User_talk:Tom_Butler. (98) and read *Wikipedia Under Threat* on Rupert Sheldrake's website (124)

A very important point is that the skeptical community has even managed to establish an article category of pseudoscience. (125) The term is usually used to reduce the credibility of frontier subjects and is generally considered a derogatory term. Likely your subject is on the list or will soon be added. Qualifying terms or phrases that imply unscientific thinking or practices are routinely used to suggest through innuendo that the subject is not credible. Something as subtle as relating a subject to religious or spiritual subjects effectively changes a frontier subject from an effort to apply good science to understand something, to a belief system that is not to be taken seriously.

Editors are supposed to sign their posts in the discussion pages. By clicking on the editor's name at the end of the post, for instance, ScienceApologist (Talk) (now signing as 9SGjOSfyHJaQVsEmy9NS ... I will call him User:SA), you will go to the editor's personal page. Every editor has one, and it is considered off-limits for others to post anything there.

You will first see that User:SA does not much like what he prefers to as *fringe subject* or *believers*, and that he is dedicated to protecting the status quo as he understands it to be defined by mainstream science. User:SA has apparently received his Ph.D. and is now a teacher, so as 9SGjOSfyHJaQVsEmy9NS, he appears to be trying to look more professional on his personal page. However, take a look at one of his old personal pages as ScienceApologist. That is the User:SA I dealt with. (126)

If you click on the *(Talk)* after the editor's name, you will go directly to its discussion page. It is there that Administrators (Admins) alert editors about formal complaints, warnings and advice. It is educational to see who is commenting there and why.

> **As an aside,** consider the influence User:SA has had on Wikipedia, and by extension, on Wikipedia readers. He has been instrumental in the skeptic's takeover of Wikipedia. As a young astronomy student, there is virtually zero reason to think he knows more than the average person about things paranormal. Yet, he has pronounced over and over again about the pseudoscience nature of all things paranormal ... often in very demeaning terms. You can thank the rules of Wikipedia for his success, and the silence of other paranormalists as a few lonely editors like me failed to stop the skeptic takeover.
>
> One more point. User:SA is likely teaching young people under cloak of academic authority the same scientism he practices.

There is a virtual labyrinth of administrative and policy pages in Wikipedia. One sure way to navigate the maze is to follow comments from interesting editors. For instance, an editor might make a complaint about incivility at *Wikipedia:Wikiquette Alerts*. Someone will warn another about civility with, *"You have violated WP:CIVIL,"* which is a link to the *Wikipedia:Civility* Page containing the policy. (Please note that the policy is also edited and has been occasionally diluted by editors wanting to be allowed more leeway in how rude they are to others.

Another good place to look is the Request for Arbitration Pages. Two important ones are Paranormal (94) and Martinphi-ScienceApologist (User:SA). (127) Both have several associated pages for evidence and such, and although painful to read, they offer an important lesson for all of us. There are others, such as the *Incident notice board* (116) and the *Arbitration enforcement board*.

Conclusion

It is important to keep in mind that Wikipedia is not the evil empire, it is a very important tool that needs a few changes to keep it from being a platform for social engineering. It really is not realistic to say that one person is at fault for the harmful social engineering a few skeptics are able to accomplish with its content.

Looking behind the curtain, it is evident that the skeptical community of Wikipedia is out of control, and that as long as people can insult people with impunity and ignore consensus and balanced reporting, it is essential that the public be told that the online encyclopedia cannot be trusted as a knowledge base.

References and Alternative Sources

Listed at the end of the book beginning on Page 357.

Essay 13
Arrogance of Scientific Authority
2015

About This Essay

In the ancient past, the public only heard rumors of esoteric schools in which wise men taught about the wonders of hidden realities and magical practices. The reason they call them esoteric schools is that they were secret. They taught hidden wisdom. The wise men were masters of the occult.

The schools were hidden from the public for several important reasons. They were not about religion or faith. The lessons were taught as a form of natural science. A sequence of lessons and initiations was designed to transmute the seeker from a common, dull-sensed citizen to a master of the natural principles of nature. You will recognize this as the transmitting of lead into gold, as taught in the Hermetic Wisdom Schools.

Even today, most people do not understand the difference between belief and understanding. Religions are all about completely trusting the priest and accepting faith-based dogma. The wisdom schools were all about understanding the nature of reality as it pertained to personal wisdom. In wisdom schools, seekers are intended to test their teachers.

The problem is that esoteric teachings and religious subjects are essentially the same. A wisdom school lesson about our relationship with the infinite sounds a lot like a church lesson in our relationship with God. Thus, there was a conflict of authority between the priests and the wisdom masters, and resulting clash posed a danger to the unaffiliated wisdom schools.

The Law of Silence applies here as well. Imagine that you had a most wonderful experience while on vacation, about which you eagerly tell your co-workers. If they rejoice with you, the sharing reinforces the joy of your experience. If they pretty much ignore your story while telling you about their work, their reaction tends to subtract from your joy. See Essay 20: *Law of Silence* (Page 331).

Fun vacations are relatively objective, so imagine you had a life-changing experience while in deep meditation. That is a most conceptual experience. The influence of the memory can be tenuous and easily dissipated with doubt. Telling others about such an experience is risky because, if they do not react in a reinforcing way, it is likely the warm memory will quickly fade. Doubt is a most insidious enemy of conceptual lessons.

The third reason for secrecy is the most applicable to our work today. Seekers are on a difficult path that frequently tests their determination to gain understanding. It is so much easier to have a beer and enjoy the Sunday game. Like a very young athlete training for the Olympics. Seeking requires dedication and sacrifice. Unlike the athlete, a seeker must be mindful throughout life.

Remember the Zen Buddhist saying, *"Before enlightenment chop wood–carry water, after enlightenment chop wood–carry water."* One must do the work to learn. One must do the work to test. One must do the work to understand. Understanding is relative and its pursuit if forever.

For the average citizen, seekers must seem like peculiar people with strange ideas about life. It is easy for people to feel a degree of jealousy if they learn of the seeker's objectives. After all, the seeker is working to become an enlightened being while the average observer is not.

Secrecy is a practical matter of self-preservation. It protects the school from the church, it helps protect the fragile emerging understanding of the seeker and it protects the seeker from local jealousy.

I think mostly with the influence of the Internet, the veil is being lifted from the secret wisdom schools. From my experience, this is not a good thing for the seeker, but it is very good for potential seekers who wish to find a path to personal enlightenment. Our task as a community is to learn how to help potential seekers while protecting our ability to seek.

As a practical matter, the paranormalist community is teetering between a return to secrecy and even further disclosure. The determining factor is the ability and willingness of the paranormalist community to protect its seekers and practitioners.

On the one side of the scale is a New Age-like community future in which paranormalists seek to sell to paranormalists and everyone believes whatever feels good. Such populist behavior is no threat to mainstream society. Accusing citizens of that community of practicing pseudoscience would be silly because such shallow beliefs are not likely to be mistaken as science.

The other side of the scale is as it has been in the past. Religions would dominate the teaching of the nature of reality, secret wisdom schools would disappear into the shadows, and a gray zone of counter-culture ghost hunters would flourish somewhere between.

Parapsychologists are the determining factor. They are divided by mainstream skeptic posing as open-minded parapsychologists, psi phenomena advocates suffering from mainstream envy and a few who seriously consider survival.

The average paranormalist seeks to understand these phenomena for personal reasons. They may simply be fascinated by the unknown, perhaps they seek to know their transitioned loved one is okay or perhaps they fear death and are looking for reassurance. Many are in it for the social scene of ghost hunters.

Whatever the reason, paranormalists depend on scientists to tell them the true nature of their interest. Are the phenomena real? If so, how do we work with them? What about them is verified and what is just popular wisdom. None of us want to be delusional.

As I discuss in Essay 14: *Open Letter to Paranormalists* (Page 223), our parapsychologists are mostly interested in furthering their personal beliefs. While they depend on the rest of us as experiencers, and practitioners, they tend to mislead us to gain our cooperation so that they can prove we are delusional. In doing so, they are driving our serious seekers back into the shadows.

This essay is about the way one team of parapsychologists have attacked their physical medium, research subject. I have become outspoken about the abuse of our practitioners because it has so obviously interfered with my efforts on behalf of the ATransC. It is easy for me as a practitioner to imagine being treated the same way. There is no reason you cannot be treated the same way.

The problem goes beyond this offending team of parapsychologists, making me think it is a cultural byproduct of the Academic-Layperson Partition. The way it is organized, the parapsychological community encourages its people to produce study reports about subjects for which they have no training. It is common to read anomalistic psychology reports in the journals that are about how *believers* foolishly attribute paranormality to mundane things based on gotcha-like studies.

The average paranormalist is not well-informed about the more detailed aspect of these phenomena. This is not a criticism. All of us are seekers at one stage or another of our education. But, parapsychologists' criticism tends to drive serious practitioners and seekers into the shadows. I hear them saying

"I will not share my understanding with people who might doubt my experiences, rather than help me better understand them."

This essay was written because of my disgust about how a good man was so unfairly and needlessly attacked by people who are supposed to know better. The simple fact is that the researchers do not know enough about physical mediumship to be so judgmental. It is recognition of that lack of understanding which is supposed to be guiding their research. Having read books and sat in séances does not make one sufficiently knowledgeable to make eye-witness field study reports as experts. That alone is intellectually unethical.

None of us know enough about these phenomena to say all the ways it will or will not manifest. Possibly excepting a few theories, such as the Implicit Cosmology, we lack the necessary models on which we might base our characterization of these phenomena. We are pretty much still in the naturalist phase of this emerging science.

It is tradition in Spiritualism that mediumship is in service to the individual and only publicly demonstrated to the public as a means of showing our immortality. The Felix Experimental Group is a development circle that demonstrates to the public as a service. The medium has accepted direction from his etheric helpers to devote himself to public demonstration of an ability that has taken many years to develop. If he wanted to earn a decent living, he would need to find a real job.

If we wish to continue having the opportunity to witness such demonstrations, it is necessary that we find ways to protect our practitioners, lest they protect themselves by once again retreating into the shadows.

> An aside about retreating into the shadows. As I proofread this book, there is increasing pressure on the government here in the USA to figure out a way to keep guns away from mentally ill people. While that is a good idea, one of the measures being discussed for how to recognize mentally ill people is if they hear voices.
>
> Hearing voices is one way to describe mental mediumship. For all of us who seek to develop our mediumship ability, it is a good idea to find a way to clearly distinguish ourselves from mentally ill people. I am not qualified to make suggestions for this. For myself, I will speak in terms of depending on external verification of sensed information as a way to know it is objective.

Suggestions are welcome. I will try to generate a discussion about this on the ATransC Idea Exchange.

Background

With an invitation from the Felix Experimental Group (FEG) circle leader, Stephen Braude assembled a team to study the phenomena demonstrated by the FEG medium. The team included Michael Nahm and Peter Mulacz. After the multi-year study, it is my understanding that Braude was about to publish a generally positive report confirming the mediumship phenomena, but then Nahm indicated he would submit his own report detailing his suspicions of fraud. As I understand from Braude, he rewrote his report to agree with Nahm's and both were published in the *Journal of Scientific Exploration*, issue 28-2, Summer 2014. (128) (129)

My essay on EthericStudies.org, *Debunking Survival Under Cover of False Academic Authority*, (70) also includes a brief discussion of the original attack. Here are the links to the two articles that began what I think of as an unethical attack on a research subject.

- A brief of the two articles can be found on history buff and strong Braude supporter, Carlos Alvarado's blog under *Questions about the Physical Phenomena of the Felix Circle*. (130)

- Braude holds a doctorate in philosophy and apparently specializes in the philosophy of science. He is Editor-in-Chief of the *Journal of Scientific Exploration*, a Society for Scientific Exploration publication. His article, "Investigations of the Felix Experimental Group: 2010–2013," can be read on *Academia.edu*. (It is free to become a member of Academia.edu) (128)

- Nahm holds a doctorate in biology with a specialty of Forest Science. His article is "The Development and Phenomena of a Circle for Physical Mediumship." (129)

The suspicions expressed by Nahm involve his belief that, in the early years of his work, the medium used several tricks to fake phenomena. I stress here that Nahm based his suspicions on a number of sources, which from my experience in sitting with the medium, depend on circumstantial evidence that could as easily be interpreted in other ways. For his part, Braude took the mediums reticence to respond to direct questions about the accusations as an implicit admission of guilt.

> It is important to note that, even though the medium speaks pretty good English, it is his second language. We often confused him during conversations with our American references. Braude's supposed *gotcha* occurred during a Skype call in which he confronted the medium. I expect the medium hesitated to digest what he was being asked ... a sure sign of guilt, according to Braude.

Apparently, there were no indications of trickery in any of the séances on which the research team was supposed to base their study. The medium did produce phenomena and it was witnessed under at least some level of investigator control (reportedly including an invasive cavity search for cheesecloth).

I asked Braude why he had not included tests for Nahm's suspicions in the protocol, to which he replied something to the effect that *"Things come up."* So, in effect, the team has ignored their findings and focused on Nahm's suspicions. Very scientific indeed!

After the publication of the original report, both Braude and Nahm repeated their accusations in every speaking forum they could find. For instance "Fall of the House of Felix" by team member Peter Mulacz (Ph.D. psychology) is the leading article in the Spring 2015 Society for Psychical Research *Paranormal Review* magazine. (131) (132) The issue also includes articles about Stefan Braude (Ph.D. philosophy) and one by Michael Nahm (Ph.D. biology). The issue is generally negative toward the Felix Experimental Group's (FEG) medium while celebrating the Myers Memorial Medal the SPR has given Braude for his "contribution to the subject of Psychical Research."

Here, I will say that I have sat with the medium at least seven times in three venues. We (ATransC) hosted him for demonstrations on two, multi-session occasions. For the five or so sessions we hosted, I had complete control of the room and inspected it before each session. Sitters were searched, and I inspected the medium's near-naked body before each séance. After those strip inspections and my watching him drink a liter of colored juice, I escorted him to the séance room. There, he was constantly under the control of two experienced sitters. They held his hands and arms, sometimes making contact with one another to assure everyone was doing their job. I inspected the room after each session. In each séance I witnessed, phenomena were displayed which could not be explained in mundane terms; some in pretty good red-light elimination.

As an engineer, I am trained to look for how things work. Braude is trained in philosophy, Nahm in biology and Mulacz in psychology. Trickery of the kind claimed by Nahm requires devices and sleight of hand. The training best suited for detecting fraud is more in how things work than how people think. As such, I will argue that, skill for skill, my training is more to the point for judging the authenticity of the physical phenomena.

My objective here is not to say that I am a better observer or more knowledgeable about how the phenomena are formed. As a certified medium and healer, and experienced ITC practitioner, I have produced some of the phenomena myself. After more than fifteen years examining thousands of examples of photographic and audio evidence of the paranormal, I have come to understand how even expert witnesses can be confused by photographs in poor light.

Our mind is not trained to process some of the phenomena of darkroom séances such as impossible lights and unexpected faces in ectoplasm. I have written before about the perceptual process. For many, the only valid perspective is suspended judgment. Certainly, deciding there is fraud based on very unusual witness conditions and then acting on that decision to ruin a person's reputation is not the best approach to understanding. Certainly, no rational observer would think of such behavior as scientific.

Science in the Paranormalist Community

It took me a long time to figure out why I have been so indignant about how the three parapsychologists have treated the FEG medium. The common factor of our community is our interest in paranormal phenomena. As is true of society in general, either directly or via our opinion setters, we all look to our scientists for guidance. It is that guidance that is becoming a problem for the paranormalist community.

In Your Immortal Self, I begin the section on science by stating: "Hands down the most disruptive influence for the paranormalist community is science." (1) Consider these points:

- Three professional/academic organizations we have all come to trust are the Society for Psychical Research (SPR), (53) the Parapsychological Association (PA) (52) and the Society for Scientific Exploration (SSE). (133) They are ostensibly seeking to understand all paranormal phenomena ... and presumably report their findings to the community so that we might better understand and work with these phenomena. That is how science is supposed to work.

- In fact, all three organizations focus on understanding psi phenomena and/or understanding why people believe in it (aka delusional). With the exception of mental mediumship, survival-related phenomena are seldom studied. Yes, reincarnation, near-death and out-of-body experiences are studied, but those groups are pretty insular and play only a small part in the paranormalist community.
- Parapsychologists are obsessive about proof and have seemingly turned inward to battle skeptics rather than reaching out to the paranormalist community. Even without bad behavior such as the subject of this essay, we (ATransC) have determined that there is no sense in looking to parapsychology for our science.
- The SSE has been especially hard on transcommunication by publishing *failure to replicate*-types of research reports written by a clearly unqualified doctorate in psychology. (102) Braude is the chief editor for the peer reviewed SSE journal. Peer review is in secret, and Braude has been ... unfriendly toward comments about articles from the non-doctorate community.
- The slanderous articles about the FEG medium were poorly considered, and in most communities, such treatment of the research subject would be considered unethical. (134) Nahm's article is a study in pure *what if*, *could be* and *might be*. There is no way it could be construed as a research report. Neither article should have passed peer review unless Braude abused his authority to demand that they be accepted.
- Now the SPR has allowed Braude and Nahm's unchallenged comments in their Paranormal Review, which is intended for a more general audience. In fact, the SPR has recently given Braude the Myers Memorial Medal.
- The researchers obtained grants money to conduct the research from the Parapsychological Association (PA). Also, the PA cooperates with the SPR, making all three of our more important organizations solidly on the wrong side of our best interest. As we watch, the parapsychological community has, in effect, *circled the wagons* in support of Braude and his team.
- The SPR has included an article titled "Felix Experimental Group" by Braude in their new *Psi Encyclopedia*. That is an online information

source intended as an outreach to the public, meaning we must assume Braude's repeated accusations about the FEG medium in his contribution are official SPR policy. (135)

Human Research Subject

Take a little time to look over the *Research Ethics* guidelines published by the University of Washington School of Medicine. (134) It is loosely based on the *Belmont Report* published by the U.S. Department of Health and Human Services. (51) The *Belmont Report* appears to be the golden standard for research ethics. Major points from the report that apply to this essay are:

Respect for persons: *"To show lack of respect for an autonomous agent is to repudiate that person's considered judgments, to deny an individual the freedom to act on those considered judgments, or to withhold information necessary to make a considered judgment when there are no compelling reasons to do so."*

> It is clear from conversations with the medium prior, during and immediately after the study that he had an expectation of fair reporting of the data. The medium was excited! There was no expectation that possible information outside of the protocol would be used. In research, the protocol is all there is. Part of the funding process for research is review of ethics-related portions of the research protocol. This does not appear to be the case for parapsychology.

Beneficence: *"Persons are treated in an ethical manner not only by respecting their decisions and protecting them from harm, but also by making efforts to secure their well-being."*

> The research team operated under the cloak of scientific authority. This is at least one point psychologist Mulacz should have understood. While the medium is clearly more knowledgeable about the phenomena than the researchers, he and his circle members have been culturally trained to respect perceived scientific authority. The medium and his circle opened their home to the investigators with the expectation of fair treatment. As such, the investigators had a moral obligation to protect their research subject.

Justice: *"Who ought to receive the benefits of research and bear its burdens? This is a question of justice, in the sense of 'fairness in distribution' or 'what is* *deserved.' An injustice occurs when some benefit to which a person is entitled is denied without good reason or when some burden is imposed unduly."*

> To be clear, the medium is not making a fortune by causing himself, his wife and his circle to take their time for the séances the investigators requested. In fact, most people would not give so freely of their time. The medium also risked a lot by agreeing to be the test subject. It took him years of great personal cost to develop his ability. Each session demands that he pay a high penalty in physical effort. There are very few physical mediums able to demonstrate even the basic phenomena. For a scientist seeking to win the approval of his academic community, the FEG medium represented a once-in-a-lifetime opportunity.
>
> At the same time, the resulting research reports have cast a shadow over the medium's reputation. The resulting costs to the medium are incalculable.

Libel and Slander

According to The Free Dictionary by Farlex: ***Collectively known as defamation, libel and slander are civil wrongs that harm a reputation; decrease respect, regard, or confidence; or induce disparaging, hostile, or disagreeable opinions or feelings against an individual or entity. The injury to one's good name or reputation is affected through written or spoken words or visual images.*** (136)

> **From the abstract of Braude's report:** *"...Regrettably, recent indications of fraud (explored also by Michael Nahm in this issue) have tarnished the case as a whole. ..."* (128)
>
> **From Nahm's report Abstract:** *"...and explain why I finally arrived at the conclusion that considerable parts of the phenomena were produced by fraudulent means."* (129)
>
> **The last statement in of Mulacz's article:** *"Where there is conscious and deliberate fraud it is absurd to speculate whether some of the ostensible phenomena might perhaps be genuinely paranormal (although the desire of some disappointed observers 'to save what can be saved' is psychologically understandable) and it is a waste of resources—time as well as money—to continue investigations of such pseudo-mediums. From my observations, I conclude that (the medium)'s 'physical mediumship' is a deliberate deception from beginning to end."* (132)

These accusations of fraud are made under cover of academic authority. That gives their word considerably more weight than if they were spoken by a person hiding behind a screen name on a discussion board. So not only are the accusations of fraud hurtful and unnecessarily damaging to the medium and our community, they are spoken by people others are culturally conditioned to believe without question. That undue authority makes their words especially harmful.

If you read the reports that provide a context for the above quotes, you will see that they spring from accusations made by Nahm concerning suspicions related to séances that were conducted years before the actual study began. There is no evidence based on scientific inquiry provided by the supposed scientists to support the accusations. Certainly, no effort is made to account for the apparent phenomena produced during the study, making it clear that discrediting the medium is of a higher priority than discovery of possibly new principles of nature.

Opinion About Arrogance of Scientific Authority

Here are my concerns:

- As with many parapsychologists, it seems that the three investigators come to the subject more as a hobby than as qualified observers. Braude is a philosophy major and Nahm is a biologist. Both are clearly unfamiliar with the methodologies of science. In the context of studying mediumship phenomena, they have no more standing than the average, well-educated observer and certainly no standing as doctorates in the field of mediumistic phenomena.

- Some of their comments are naive. Braude has a history of assuming authority that he does not have. For instance, he has written at length about how Rupert Sheldrake, a cellular biologist, is wrong about his theories concerning cellular morphogenesis. (137) In that, and his comments about séance phenomena, he clearly lacks understanding about the metaphysics.

- One of Nahm's accusations is that the medium purchased Halloween cobweb with the intention of faking ectoplasm during a séance. This is one of those circumstantial situations that all of us need to be aware of. Having been accused of regurgitating cheesecloth or the fake cobweb, it is natural that the medium experimented to see if those materials made convincing ectoplasm. In fact, they do not;

never did! Any even halfway decent observer would see that the medium's ectoplasm is not so contrived.

We have never found residue of the ectoplasm in the room after seances. The large quantity of colored liquid the medium drinks just before going into the séance room cabinet would stain cheesecloth or Halloween cobweb. The body juices should also make the regurgitated material smell of bile. I have had the ectoplasm touch my face, yet there was no odor, it was very wet, but then it was gone.

- The parapsychological community is standing behind the three researchers by enabling the widest possible distribution of their libelous articles. The Academic-Layperson Partition problem is encountered in a lot of ways, but the most damning is the attitude doctorates—qualified parapsychologists and academic hobbyists alike—have toward the layperson practitioner. This is further illustrated in the next point.

In the *"They eat their own, don't they"* department, the negative effect of the criticism is pretty well summed up by one post on the Spiritualismlink.com concerning the original articles: *"Another one bites the dust..."*

- Mulacz conveys a sense of disdain for the medium, FEG and Braude in his comments ending with: "Attending such séances at home circles has but little value; scientific séances must occur under laboratory conditions where the investigators are in control." (132) This is an expression of academic arrogance, as he clearly thinks the medium should submit to every demand made by the investigators.

In the model I have been working with, collecting examples of trans-etheric phenomena involves a particular and extraordinary mindset and/or the presence of a contact field which can enable trans-etheric influences. Much of the activity in a physical séance is a process of developing such a contact field to which the sitters must contribute. Too sharp of a focus on the process or the distraction of external controls are known to defeat this process. (138) The medium expressed too me great relief that he was able to produce any phenomena at all for the researchers.

- Like it or not, parapsychologists are bound to layperson practitioners by virtue of what they study. But they seldom behave as if they are, choosing instead to treat those they depend on for the production of

- Having spent time as an editor in Wikipedia, I know it will be just a matter of time before these articles are used to justify negative comments in paranormal articles. Even as parapsychologists rail against skeptics, they give the skeptics ammunition to demonstrate the social danger of belief in the paranormal.

Again, I need to stress that this is all my opinion. Obviously, the doctorates who govern the parapsychological organizations think the articles they publish represent good science. If you agree with my assessment that the pseudo-researchers effectively threw the FEG medium under the bus, then you should also be very cautious about possible dealings with the supposed scientists who might wish to study your work.

Establishing a research capability as part of an organization dedicated to supporting mediums, such as the Forever Family Foundation (139) is probably the only way to assure that our phenomena are studied by people who actually understand the metaphysics. As it stands now, the Academic-Layperson Partition assures that cooperation is a one-way street.

While I trust the FEG medium, I have not personally witnessed some of the séances Nahm referred to as instances of possible cheating. The concern I have is that the three investigators had the moral obligation in the researcher-test subject relationship and in the parapsychologist-paranormal community relationship, to stick to their protocol. As far as I can tell, every instance of possible fraud they have complained about falls into the category of maybe so, but no hard evidence ... certainly no evidence during execution of the protocol.

From My Experience

2010, my wife Lisa and I attended a séance in the FEG medium's home, (140) I believe well before the supposed trickery Nahm complains about. There were a number of phenomena demonstrated that were beyond the reasonable scope of trickery. For instance, raps on the wall directly behind my head, almost instantly followed by others on the ceiling, all well away from the medium. In the dark, with no one aware of my intention, I dared the communicators to place the rapidly moving spirit light in the palm of my hand. A moment later, the light hit the palm of my hand.

At another time, I dared the communicators to make an OK sign with the spirit hand that was occluding a luminous plate on the floor. Moments later, I saw the hand momentarily form the universal OK sign.

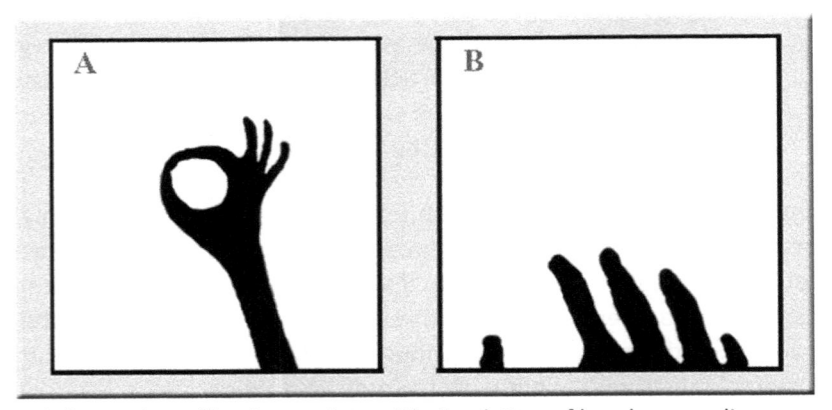

A) Illustration of luminous plate with simulation of hand responding request to show an "OK" sign.
B) Same plate illustrating relative size of other simulated hands that we saw.

The séance was amongst friends and there was little done to assure no trickery, but there was also good light from time to time that enabled reasonably careful observation. As an engineer, I was almost manically looking for alternative explanations for the astounding phenomena demonstrated that day. I cannot speak to all of his demonstrations, but then and in many subsequent séances in which I did have complete access for inspections, I have no rational reason to think the medium is other than genuine.

By the way, the spirit light, hand and raps were moving about far out of reach of the medium or any likely tricky device postulated by Nahm. There was a lot of ectoplasm and a hand did form which rose up a little and seemed to wave about. Good red light, two observers holding the cabinet curtain open sat very close to him and the rest of us were trying to see … everything. String to hold the ectoplasmic hand up, and make it move about as it did, would have been quickly detected. As always, the ectoplasm came and went without residue, smell or even a hint of trickery, save the cloth-like appearance.

13. Arrogance of Scientific Authority

Cameo Faces in Ectoplasm: FEG Medium Kai Mügge separating ectoplasm to reveal three cameo-like faces. The image below shows the three faces in closeup. Reportedly, all three were recognized by sitters.

In later séances, the cameo-like faces that formed in the ectoplasm were much like historical examples I have seen in the literature. *Cameo Faces in Ectoplasm* photograph shows the FEG medium holding ectoplasm during a séance. We did not attend that séance, but the cameo-like faces shown in the photograph are similar to other faces we witnessed.

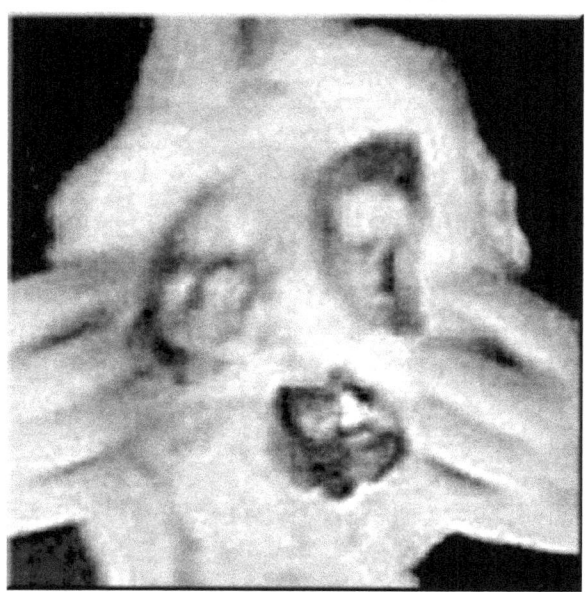

In one of those seances, I clearly recognized Hans Bender as one face. A member of our local Spiritualist Society recognized a loved one. The faces appeared to be formed as a precipitated simulation of a face, rather than as a photograph and had a three-dimensional quality. There is quite a lot of precedence in Instrumental TransCommunication for such features.

When the medium stood and circled the room in red light with a handful of ectoplasm, I could see that the substance was dripping water. When he came to me, he unceremoniously flopped the ectoplasm against my forehead. It made my head very wet. Yet, when I inspected the unpainted plywood panel we had placed under his chair to cover the rough brick floor, there was no hint of moisture residue from where the ectoplasm had been. By the way, at another time, the substance glowed bioluminescent green as it came out of his mouth!

I do not know if the medium has tried ways of cheating. He is constantly looking for ways to assure sitters that there is no fraud, so it is possible he has tried to cheat to learn how to guard against it. I would!

The glaring double standard here is that parapsychologists discount many examples of our phenomena because there were no scientific controls when the examples were collected. At the same time, I am sure no one actually trained in the use of the scientific method would propose a finding in science based on the level of evidence being used to smear the FEG medium. How was an article smearing the medium's good name, based on lame theories, innuendo and magical thinking, published in what is supposed to be a peer reviewed science journal? Perhaps the answer is that that part of our community is just a toy for retired professors.

From the perspective of a director of an organization *"Founded in 1982 by Sarah Estep to Provide Objective Evidence That We Survive Death in an Individual Conscious State"* I believe the treatment received at the hand of these supposed scientists has set our work back in ways we are still learning.

Beyond the grief imposed on the medium, the most damning of all of this is that there are so few people speaking out to make sure this will not happen again. Certainly, the parapsychological community seems just fine with the treatment.

Update

Instances in which the parapsychological team has publicly attacked the FEG medium:

Coast-To-Coast: February 22, 2016 - From the Coast-To-Coast AM website. (141) An example of continuing public accusation well beyond any research or scientific value:

Braude was able to investigate the German physical medium Kai Muegge of the Felix Circle under somewhat controlled conditions, and searched his body before the videotaped experiment. During the séance, there were some table levitations, and Kai produced large quantities of ectoplasm from his mouth, in a seemingly unexplained manner. Yet, Braude learned later that there was conclusive evidence that Kai had cheated at some other seances he hadn't supervised. "Even so, just because a medium uses deception on occasion, it doesn't mean they don't have genuine abilities," Braude pointed out, adding that "these so-called 'mixed mediums' sometime use trickery out of convenience or necessity when they make their living from producing spirit contacts.

Journal of Scientific Exploration, Vol. 30, No. 1, pp. 5–9, 2016: (142)

Follow-Up Investigation of the Felix Circle by Stephen Braude in which he essentially rehashes previous reports. Remembering that Braude is Chief Editor of the Journal, I note with much dismay, the arrogant conclusion of his editorial in the same issue *"For reasons I discuss, it seems unlikely that Kai will again submit himself to examination by me or any other careful researcher. It appears, instead, that he would prefer to continue shooting himself in the foot."*

Further Comments about Kai Mügge's Alleged Mediumship and Recent Developments by Michael Nahm in which he endlessly elaborates on his accusations. These are still based mostly on an *"it seems to be" kind of attack*. (143)

Psi Encyclopedia owned by the Society for Psychical Research

This new online encyclopedia is supposed to provide true information to the public to counter Wikipedia. However, they asked Stephen Braude to write the mediumship article. He simply wrote more of the same to continue his efforts to defame the FEG medium. Not only must we contend with the very biased article about paranormal subjects in Wikipedia, but we must also deal with attacks from the SPR. (135)

References and Alternative Sources

Listed at the end of the book beginning on Page 357.

Intentional Blank Page

Video-loop ITC Image collected by Tom and Lisa Butler.

This is one of our first captures. It appears to be a man standing in front of a round window; perhaps a large porthole. Notice the way the head seems to be oddly attached to the neck.

We have collected many faces that appear to be from a different line of evolution than our own.

Essay 14
Open Letter to Paranormalists
2017

About This Essay

The full title of this essay is Open Letter to Paranormalists: Limits of science, trust and responsibility. Writing it was a most difficult task.

As I learned more about ITC, my writing has become more focused on community, ethics and mindfulness. While the phenomena are paramount in showing us the way, they are just that, a beacon shining a light on truths revealed to us in every generation from Hermes of ancient Egypt to our contemporary theorists.

It is telling that I am not citing contemporary teachers. I feel one has yet to emerge in our community. But, I know what the new teachers will say. Each will in some way tell you that the one lesson you are to learn is that you are immortal self experiencing this lifetime to gain understanding. Virtually all of the phenomena we study, especially communication from our transitioned loved ones, exists to show us the reality of that one thing.

The fact of our immortality is in itself a simple lesson. But as they say, the devil is in the detail. There are implications brought by the fact of our immortality, and those have consequences.

The Implicit Cosmology (12) is my version of the nature of those implications and consequences. The new teachers will have different, probably more easily understood cosmologies. I know it is important to speak to the heart in these matters, rather than just the brain. Speaking to the heart is something for which I am not particularly well equipped. But, their cosmologies will undoubtedly look a lot like the Implicit Cosmology. It is a good place to begin while you await a teacher.

Knowing that I think this way, can you see why I might be concerned about supposed scientists attacking our practitioners without due science? In doing so, they put our emerging understanding at risk.

Essay 13: *Arrogance of Scientific Authority* (Page 205) explains one of my complaints. That was a difficult essay to write because I used it to call out parapsychologists by name. Doing so puts me even further on the fringe of our community, certainly outcast from parapsychology. However, while highlighting conflicts within our community is divisive, saying nothing is the greater offense.

This essay began as an open letter to parapsychologists. My point was that, to mislead the rest of the community, even to the point of condoning unethical treatment of research subjects, is an abuse of academic authority. My intention was to post grievances, in response to which they might reconsider their ways.

One grievance is that the offending researchers focused their report on unproven accusations about behavior rumored to have occurred before execution of the research protocol. Rather than challenging the report on grounds of bad science, the parapsychological community gave the researchers every opportunity to publicly repeat their accusations. They even invited the team leader to repeat the claim in the Society for Psychical Research's (53) new Psi Encyclopedia. (109)

I spell out other grievances in the essay. My sense is that the offenses are only possible because of the Academic-Layperson Partition. If the self-superior academically trained people feel above those of us who simply produce the phenomena, it becomes permissible amongst their ranks to treat us as they wish. Without an outcry from the greater paranormalist community over such abuse, there is no penalty for such treatment and no likelihood of change.

As you might think, every attempt I made to write this essay turned into a rant, which served no good for any of us. Finally, I began composing it in terms of a warning to other practitioners. In effect, I have given up on parapsychology. The message to you now is to beware of the Trojan Horse researcher and make darn sure your claims of paranormal phenomena are rational and well-supported.

Keep Essay 13 in mind as you read this essay. Learning about these phenomena is important because of the light it shines on the Mindful Way. A good percentage of the parapsychological community will seek ways to prove you are ... well, to prove you are delusional, if nothing else works, so be informed.

Introduction

Science is good. The *science* practiced by parapsychologists is not necessarily good. Much of it is done to prove paranormalists are delusional. You and I know that to prove we are delusional, they must ignore or falsely represent our evidence.

Scientists are supposed to be our friends. Some are, but the majority consider the average paranormalist inferior in many ways ... as second-class citizens that are not as smart, as well educated and as wise as people with a Ph.D. Most people calling themselves paranormalists are retired from unrelated careers and are using the study of paranormal phenomena as a hobby. They are likely less informed about the actual nature of these phenomena than you, the practitioners and experiencers.

If you come away from reading this with nothing else, I pray that you remember these points. If you want to see these phenomena properly studied, if you want informed scientists to help you understand your experiences, if you want to see this field of study evolve into a well-understood science, then it is important that you know who to trust, who to believe and with whom it is safe to trust your phenomena.

It is important that you encourage the pretenders to go away so that real scientists will feel free to help. It is most important that you speak up!

About This Letter

While many of us look forward to the day a scientist will want to study our work, few of us realize the potential problems that can come from being studied. In simple terms, science is great; scientists are not always so great. It is for you to be aware of the differences, because many of those who have not been aware of the difference, have regretted ever volunteering to be research subjects.

To understand the science of paranormal phenomena, it is necessary to understand the phenomena. You may not agree with the model I use for this letter but take time to think about it. If I did the work correctly, the model should be reasonably close to reality, if not in detail, at least in principle.

The model is based on currently understood mainstream and parapsychological science. Unlike more widely accepted models, it is greatly informed by lessons learned from mediumship and Instrumental TransCommunication (ITC), especially Electronic Voice Phenomena (EVP). It will evolve over time, but for now, I am sticking with it because I need a target of sorts toward which I can develop arguments and collect supporting data. I suggest you also pick a story about which you can learn to talk.

The Paranormalist Community

In the context of this letter, paranormalist are people who experience, study or have a more than casual interest in psychic ability (psi functioning, remote viewing, healing intention), healing intention (biofield healing, distant healing, healing prayer) and the phenomena related to survival of consciousness (mediumship, visual and audible ITC, hauntings).

The paranormalist community is actually several communities of interest that are in the same boat, so to speak, because they all seek to understand the same phenomena but with different intentions. A useful way to think of the paranormalist community is as an aggregate of communities of interest such as ghost hunters, experiencers seeking understanding of often disturbing experiences, grief management, investigators and academics working as parapsychologists.

Theories of Reality

Because it is important that you are clear about who believes what, the three dominant theories designed to explain the nature of reality are provided here. This is taken from the book, *Your Immortal Self*: (1)

The Physical Universe Hypothesis

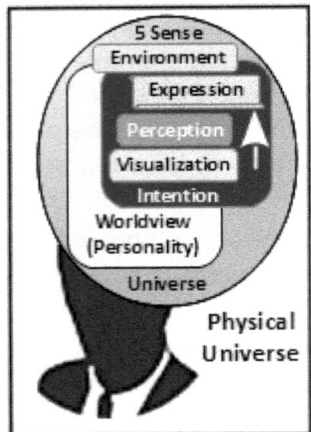

- All that exists is the physical universe.
- The universe has evolved from a singularity into what it is today.
- Life has evolved on earth into what it is today.
- Mind has evolved as a product of brain which is a product of evolution.
- Memory is an artifact of mind.
- When the brain dies, mind and memory cease to exist.

- People have five senses: smell, sight, hearing, touch and taste.

To simplify conversation, people who think the Physical Universe Hypothesis is correct are described here as **Normalists**. Parapsychologists who lean toward the Physical Universe Hypothesis often work under the banner of Anomalistic Psychology, (58) which holds that reported paranormal experiences are actually ordinary-world experiences mistaken as paranormal. In effect, these parapsychologists are paranormal phenomena debunkers determined to find a normal explanation for all psi and survival-related phenomena. They try to do so by ignoring existing research seemingly supporting the existence of a psi field and survival.

The Super-Psi Hypothesis

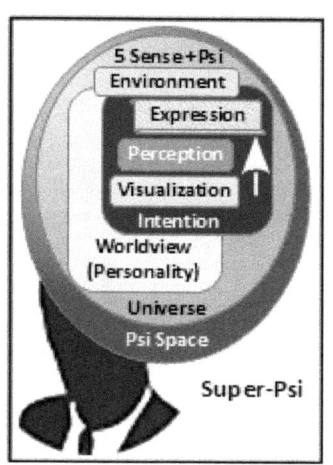

- All that exists is the physical universe.
- The universe may have evolved from a singularity into what it is today.
- An as yet unidentified form of space called psi (psi field) permeates all of physical reality.
- Life has evolved on earth into what it is today.
- Mind exists in the psi field and continues beyond death of the brain as differentiated, residual energy.
- Brain is a transmitter/receiver for mind.
- Thought, memory and emotions are retained in the psi field.

- People experience reality via five bodily senses that are informed by impressions from the psi field.

People who think the Super-Psi Hypothesis is correct are described here as **Psi+ Normalists**. Parapsychologists who lean toward the Super-Psi Hypothesis are increasingly working under the banner of Exceptional Experiences Psychology, (57) which holds that reported paranormal experiences may be ordinary-world experiences mistaken as paranormal but may also be evidence of psi. In effect, these parapsychologists are survival phenomena debunkers determined to find a normal or psi explanation for all psi and survival-related phenomena.

The Survival Hypothesis

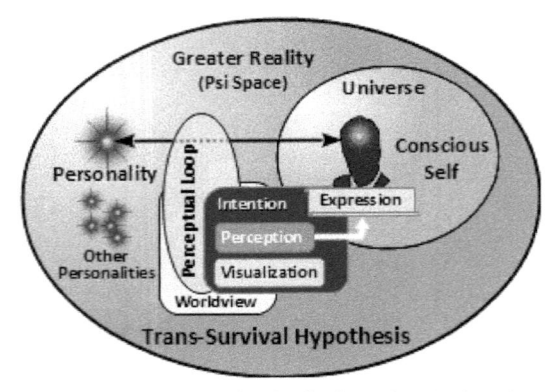

- There is a greater reality of which the physical universe is an aspect.
- An as yet unidentified form of space called psi permeates all of reality.
- The psi field is an aspect of the greater reality.
- Mind, with its thoughts, memories and emotions, has evolved in the greater reality and continues to exist beyond death of the brain.
- For a lifetime, mind and brain are entangled to produce a physical-etheric link: a person.
- During a lifetime, mind is expressed as consciousness (*I think I am this*) and a mostly unconscious etheric personality (*I am this*).
- Mostly unconscious mind is informed by the person's five physical senses and psi signatures from the environment.

- Mostly unconscious mind expresses to conscious self an understanding of the environment as it is informed by Worldview (memory, experience, human and personality instincts).

People who think the Survival Hypothesis is correct are described here as **Dualists**. Some Psi+ Normalists accept that mind is different from body. The distinction is that Dualists think mind preceded body and continues after the body in a sentient form. Psi+ Normalists think mind is a product of body and sentience ceases when the body dies. For them, all evidence of survival is just evidence of survived memory.

Experiencing Phenomena

It is true that paranormalists, mainly experiencers and practitioners, are occasionally guilty of making claims that simply do not make sense when the larger field of study is considered. From my experience, it is clear that this reporting error can be corrected with better training.

Here are a few points that are true or as good as true, and which will help you speak intelligently about the phenomena:

1. **We, or an interested observer, provide the channel through which paranormal phenomena are produced.**

   The avatar model (80) seems to best describe what we know about our etheric-physical nature. The idea is that a person is the conscious self of an immortal etheric personality which becomes entangled with a human body at the moment of the human's birth. The conscious self experiences a lifetime from the perspective of the human.

   Personality and body consciousness share the mostly unconscious mental processes, including Worldview. At first, conscious self is mostly influenced by the human's instincts, but with maturity, gains control as it acquires new understanding. In this model, it is the understanding that is returned to the immortal, intelligent core.

   There is an extensive explanation of this model in *Your Immortal Self*, (1) and early essays under the Concepts tab of EthericStudies.org. Here, I will explain that all the functional areas in the *Life Field with Avatar* Diagram (Page 230), except the human body itself, are etheric. If you take a little time to contemplate the implications of this point, I think you will see that conscious self has experiences, and by convention, assigns physicality to experiences encountered from the avatar perspective.

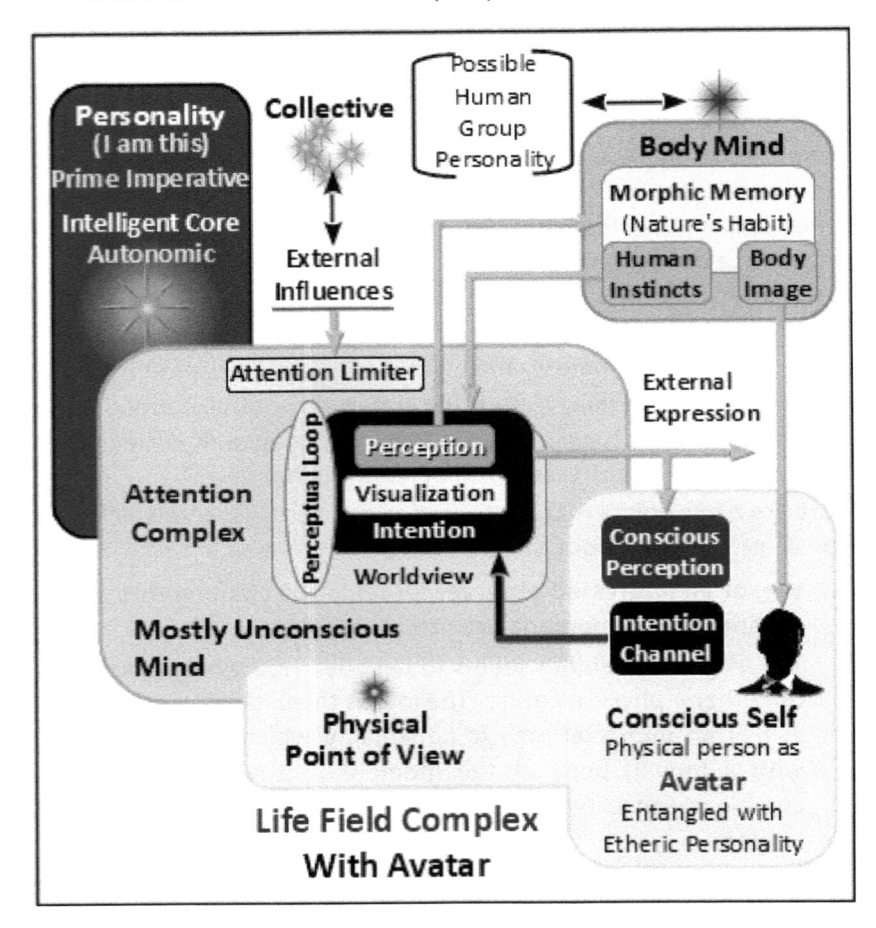

Personality (I am this) Prime Imperative — Intelligent Core Autonomic

Collective — External Influences

Possible Human Group Personality

Body Mind — Morphic Memory (Nature's Habit) — Human Instincts | Body Image

Attention Limiter

Attention Complex

Perceptual Loop: Perception, Visualization, Intention, Worldview

External Expression

Conscious Perception

Intention Channel

Mostly Unconscious Mind

Physical Point of View

Life Field Complex With Avatar

Conscious Self — Physical person as Avatar — Entangled with Etheric Personality

In a very real sense, we create our world. It only exists as our mostly unconscious perceptual processes assign meaning to sensed environmental psi signals. Those psi signals come from our loved ones, our collective of fellow personalities, thoughtforms, our body's self-image and our physical body's five physical senses.

A model that is useful and may as well be correct is that a person is necessary for a psi influence to manifest in the physical. Put differently, we, or an interested observer, provide the channel for trans-etheric influences such as EVP, remote viewing, precipitation and haunting phenomena. This is especially meaningful in view of the next item.

2. **All mental activity is psi functioning.**
   If Item 1 is correct, we need to think of such phenomena as mental mediumship, remote viewing and precipitation as essentially the same process, just with different intent. If we are not our body, then we should think that we are our mind.

   The greater reality (etheric) is a conceptual environment, meaning that, instead of physical objects, it is made of fields of influence. The field of influence of our life field is anything we turn our attention to such as a loved one or work. From the parapsychological perspective, our life field, decision process and influence are all psi.

3. **What we experience is always modified in our unconscious mind before we become aware of the information.**
   Always! Our experience in the ATransC has shown that a person tends to record EVP that tend to confirm unconsciously held beliefs. It is important to note that information in the psi signal which is the kernel for EVP formation is understood to be real. But since etheric space is conceptual, psi information is necessarily conceptual as well and must be embodied into objective form. It is how that initial conceptual signal, say from a discarnate loved one, is embodied through our mental channel that determines the content of the message.

   For instance, Lisa and another person recorded for EVP in the same dark, reportedly haunted place. As a pragmatic researcher, Lisa is neither afraid of the dark nor does she believe in demons. She recorded helpful information and names. The other person, who liked being afraid, recorded mostly scary EVP such as *"Get out!"* and *"I hate you."* (Again, it is important to note that the phenomena are real. It is just that the intended meaning can be changed by Worldview as they are manifest into the physical.)

   In a similar example, a Korean television crew came to our home to make a segment on EVP for their Korean audience. The reporter and crew became very excited during the on-camera playback of an EVP recording session. We had no idea why until they explained that we had recorded an appropriate response in their language. In that case, we believe the message had come through the trans-etheric channel provided by one of the Korean crew, and it was in that person's language as expected since it passed through his worldview.

Mainstream and parapsychological research is showing that our mostly unconscious mind acts as a filter to present only information to our conscious self that conforms to our expectations held in Worldview. (38) (67) (66) We will sometimes not even notice information if it is something for which we have little interest because it will be discarded by our mind.

This science is still evolving, and these points are only indicators. The message to us is that what we experience tends to agree with what we believe. We tend to more often have possibly genuine paranormal experience if we believe in the paranormal. Conversely, if we do not believe in the paranormal, we might not even notice such experiences.

4. **Cultural expectations tend to contaminate perception and subsequent effects.**

One of the ways to recognize that a person's worldview is involved in formation of physical phenomena is if the result sounds like something the person would normally say. For instance, if many individuals in a group of hauntings investigators are recording for EVP, the resulting messages can be expected to be couched in terms which are consistent with the way the individual practitioners think of the world. As such, more religious practitioners will tend to record more religiously influenced messages while people who believe in etheric stations will tend to record more references to those stations.

We must also be mindful of erroneous perception. I refer to the complex of ways in which we become delusional about paranormal phenomena as hyperlucidity. (29) *Lucidity* is used here to indicate the degree to which conscious self has become aware of mostly unconscious mind. That is the objective of mindfulness. *Hyperlucidity* is used here to identify situations in which the person thinks they are very lucid, but are in fact, fooling themselves by imagining they are aware of their unconscious mind. This can become evident when people mistakenly think they are hearing words from discarnate loved ones in EVP when there is only noise. (75)

5. **Some of the experiences you think are paranormal can be explained in ordinary terms.**

   My experience is with Buck Fever: It is true that many a cow standing in front of a tree has been mistaken as a buck.

   Research has shown that people who believe in paranormal phenomena are more apt to think a picture is paranormal then those who do not believe. This is true of all forms of phenomena. It does not mean there are no paranormal phenomena, only that some of us are not as discerning as we need to be.

   This tendency to error on the side of paranormal is used by Anomalistic Psychologists to prove that all reported phenomena are errors in perception. Since established evidence of the existence of things paranormal is deliberately ignored, it can be difficult to know when reading Anomalistic Psychology research reports that the intention is to debunk rather than to understand.

6. Etheric space is nonlocal, meaning that everywhere is here.

   Parapsychologists argue that the psi field is nonlocal, meaning that a psi influence is experienced everywhere at the same time. To my knowledge, parapsychologists retain the understanding that there is physical distance. The difference is important. This item makes essentially the same argument, except to say that a psi influence is experienced everywhere at the same time because everywhere is the same place. There is no distance in the etheric.

   The effect is that people in a different part of the physical world are in the same part of etheric space. When we record an EVP, the etheric personality producing the voice can be anywhere relative to our physical location. We can be in San Francisco, speaking over the phone with a person in New York while both of us are operating a recorder. Either one of us can provide the channel for an EVP on either recorder ... at the same instant. The EVP could be initiated by the thoughts of anyone living or dead, anywhere in physical or etheric space. It is who we intend to contact that determines who we contact.

   The same goes for haunted locations. The evidence thus far is that an EVP apparently initiated by a personality associated with a place does not mean that the personality is in the physical space. It only means we, *or an interested observer,* have provided a channel for a message from the intended personality.

This can also be said for faces in visual ITC phenomena. For instance, a face in a photographed orb only means that light, probably reflected from dust, produced sufficiently chaotic noise for transform phenomena to occur. Still in the flesh or not, the personality initiating the face is in etheric space in which everywhere is here.

To make this clear, current understanding is that we contact the local ghost because we expect to. We see this effect when a mental medium brings information from a loved one. When this happens, there is no reason to think the loved one has rushed to the scene to make contact.

7. **Time is a practical necessity.**

Process is sequential because the nature of the next event in a process is dependent on the nature of the preceding event. Each event is a concept which is created by the thinker/observer. I cannot argue if time exists, but I can argue that time may as well exist.

In the Implicit Cosmology, (12) there is a delay in perception between reception of a psi signal and presentation to conscious mind. Sensing a psi signal can produce a physiological change before the person becomes conscious of the event.

In psi research showing the presentiment effect, the apparent precognition of the nature of the next picture in a study is possibly explained as the person responding to this psi signal before the information comes to conscious awareness. (144)

Processes are modeled in the Implicit Cosmology as related fields which change in character as the process develops. Since these are psi signals and people are normally unconsciously aware of such signals, developing potential events will also be sensed by people. Many possible events might never develop into an actual event and potential events might merge as the moment for the event comes closer. If this model is correct, precognition might actually be the sensing of these potential events. That would help explain why some predictions never come to pass.

8. **Physical principles might resemble paranormal phenomena, but not necessarily account for those phenomena.**

Electromagnetism, vibration, spatial direction, physical energy, holography and quantum mechanics are often used to describe, even explain paranormal phenomena. Using these physical world concepts is convenient as an analogy to describe phenomena, but it is important not to assume they actually explain phenomena. Making this mistake, especially references to energy and electromagnetism, has caused many researchers to investigate dead-end research questions.

Etheric space is conceptual in nature, and the principles, forces and processes of physical space are considered objective. (We assign the characteristic of physicality to things we consider objective.) At best, they are a derivative of a psi influence. To my knowledge, the only etheric to physical influence that has objective support is the effect of stochastic amplification on small signals. Even there, the causative relationship has only been marginally empirically established.

Many of these points are the consequence of duality. If we are not our body, the implied consequence of our mind not being our body is that there must be a flow of information from our mostly unconscious mind to our conscious self. Put another way, if we accept that we existed before this lifetime and will continue to exist after, then we must also accept the idea that our intelligence creates our awareness (conscious self), but that our awareness is only a part of who we really are.

Our usual perspective is from inside of our head and looking at the world via our body's eyes (body-centric). Think of that perspective as a video camera for our mind. When we sleep, that video camera is aimed at other aspects of our personal reality (more immortal self-centric). In the end, we are not our body. We are in an avatar relationship with our body.

To speak intelligently about the phenomena, it is necessary to understand that there are some well-established ... if not facts, at least points which are arguably true, given the evidence. The most common red flag comment I hear is *"We will never know for sure."* The reason we study these phenomena is because we think it is possible to know for sure. The study is producing understanding that, if not truth, may as well be truth, as it provides a solid foundation for further study.

What You Need to Know About Science

The natural order of things in a mature society is mutually beneficial exchange of informed guidance from scientists to the general public in return for financial support of higher education and research. The public is conditioned to trust scientists without question. In turn, scientists are trained to follow the scientific method, ethically and truthfully. They are also expected to assure *useful* information about research is made available to the public. (Here, I emphasize *useful*.)

A hypothesis is a statement of theory. For instance: *Objects have an invisible force that attracts other objects*. Corollaries of the hypothesis explore the *so what* of the hypothesis, for instance: *The earth* (object 1) *has gravity* (invisible force) *that attracts objects* (object 2). Research questions are designed to test the corollary, for instance: *Will an object fall to the ground?*

Once the question has been established, a series of events must take place for the work to be respected as actual scientific research:

1. A literature review should be conducted, and subject matter specialists consulted to determine what is already known about the question.

   For paranormalist interested in survival-related phenomena, a review of the literature must include published studies about survival subjects such as EVP. A point I will make here is that much of the lay-literature is not considered in a proper literature review because it is neither peer reviewed nor produced by a Ph.D.

2. Assumptions should be stated which lead to the development of a step-by-step test procedure (the protocol). The protocol is designed to direct actions intended to test the assumptions related to the question.

3. The protocol should be closely followed while the results are recorded. Deviations from the protocol may possibly discredit the research.

4. Concluding remarks should be limited to findings resulting from following the protocol.

5. For the work to be considered research, a final report must be readied for peer review with the intention that it will eventually be

published in a respected journal. At the least, it must be made available for others to review *in the literature*.

Research reports should include a brief discussion of the literature and prior study supporting protocol design. This should include the foundation theory and important contending theories. The conclusions should also address if and how well research results support the foundation theory. This is where the contending theories should be considered with an explanation as to whether they are supported by the study, and if so, in what way.

Paranormalists may be academics, but most are laypeople, meaning they do not have an advanced college degree. Lay-paranormalists seldom have access to the library systems used by Ph.Ds. It has been my experience that few laypeople have access to parapsychological journals. (We subscribe to four journals at the cost of about $400 a year. Even so, I am not allowed full membership because I do not have a Ph.D.) To be available to the larger community, research reports should be written in terms understandable to a wide audience and made accessible via an Internet search.

The report may be published behind a paywall. If so, to fulfill the scientist-public support social contract, the essence of the report should be accessible to the public at no charge, and via a commonly used access method.

6. Based on research results, the hypothesis is reviewed for possible redefinition, probably leading to a new and improved research question.

The scientific method is really a system involving people, institutions and media. In simplistic terms, scientists develop and test theory, the results of which are passed to engineers who produce usable products for the public based on that research.

The scientific method must be adapted to each situation, but the common factor must be the intention to organize findings into an understandable form for the intended audience and making that information available to the public. In mainstream society, the population of scientists is great enough that it is reasonable for scientists to write just for other scientists without writing a version of the report for public reading. But in fact, they are failing an implicit contract requiring them to find a way to communicate with the public.

In many areas, media reporters acting as science writers have assumed the role of disseminating research results to the public. However, the paranormalist community is relatively small, and while some of us have assumed the science writer role, an effective path of communication between scientists and lay people has not yet developed.

A final point about proper science concerns ethical treatment of research subjects. I address this later in this letter.

Qualified to Practice Science

It is understood that scientists hold a Ph.D. in the field to which they apply the scientific method. Yes, anyone can conduct science, but the system is designed to filter out all but academically trained people. I hold a BSEE and it is acceptable for me to say that I study a subject but saying that I am researching a subject is not technically acceptable. The culture only allows people without a Ph.D. to be citizen scientists, presumably in service to a Ph.D. Mainstream culture is that well organized.

For Normalists, a degree in psychology is assumed to be sufficient to study paranormal phenomena because the study is seen as actually a study of mental problems. If you are a Psi+ Normalist, then a degree in psychology may apply to some study. For instance, a person's point of view appears to have a lot to do with whether a person experiences phenomena, and if so, in what way.

If the nature of psi phenomena is being considered, a degree in physics might be most appropriate. There is also a need for substantial understanding of the applicable technology. For instance, an experiment to determine if psi produces an objective effect is a question of physical principles, psi interaction with the test equipment and the nature of the psi field. A psychologist is no more qualified to study those than is a plumber.

EVP has a physical effect but appears to have a mental component. It is thought to be related to survived personality, which itself, requires an unidentified form of space in which to survive. In the case of instrumental forms of transcommunication such as EVP, both a psychology and a physical science degree is needed. In every case, an understanding of currently popular cosmologies is essential. Of course, studying EVP without understanding the idiosyncrasies of the audio devices is simply naive.

> If a psychologist is involved in research of transcommunication, the way people develop perception and the way people respond to

cultural influences must be considered as two different factors. It may be that a Cognitive Psychology specialty is required.

The question of who is qualified to study what is fundamental to the issues addressed in this letter. In fact, I cannot name a Ph.D. in our community that I know to be qualified to study EVP without a multidisciplinary team. For the study of EVP, a retired professor of psychology, biology or philosophy has little academic standing. Yes, the retired professor may have superior reasoning skills and a good understanding of the scientific method. But, without well-qualified practitioners, technologists and metaphysicians on the team, research results must be reported in terms of a hobby or a personal study. According to the preferences of the academic community, such qualifications alone are certainly not sufficient for research.

The reason this is important is that parapsychologists are rather obsessive about conducting research that is seen as good science by mainstream scientists. Two of the problems they avoid are known as the *File Drawer Effect* and *Selective Reporting*. To avoid these, all research should be candidly reported. My contention is that anyone working outside of their area of expertise is not conducting science and the file drawer problem does not apply. In fact, their work should never pass peer review if it is reported as science.

General Rules for Qualifications

Most parapsychologists seem to assume a Ph.D. in any subject is all that is required to study paranormal phenomena. In fact, more specific skill sets are required to study them under cloak of academic authority, that is, as a person trained in college to study the particular subject. In every case, an understanding of contemporary metaphysics and an understanding of the methods of working with the phenomena is essential.

Here is a summary of qualifications I think are necessary if researchers are to have the authority to speak under cloak of a Ph.D.:

Type A: Physical Trans-Etheric Influence: The study of phenomena that involves an effect anyone might physically witness.

This class includes such phenomena as Instrumental TransCommunication (ITC) (both visual and audio (EVP)), apparitions if witnessed by many, phantom lights, movement of objects, direct voice and healing intention.

This does not include apparent phenomena that require direct human involvement to manifest such as movement of a planchette, mental mediumship and automatic writing.

Expected Qualifications: Requires understanding of cognitive psychology, physics, applicable technology and prior art for that form of phenomena.

Type B: Mental Trans-Etheric Influence: The study of phenomena that is only experienced by one person, or that must be reported by one person for others to be aware of the nature of the event.

This includes mental mediumship, any physical manifestation that requires physical contact or interpretation of the practitioner such as Ouija Board-like devices, automatic writing, mental mediumship and remote viewing, near-death and out of body experiences.

Expected Qualifications: Requires understanding of prior art for that form of phenomena. It is a psychological study if verifiable information is not produced. It is a psychological and applicable technology study if verifiable information is reported. Data mining is often required to establish verifiability of information.

Study of Paranormal Phenomena with Instruments

Whether or not the study involves a physical effect, if instruments are used, it is necessary for the study team to include a person who is knowledgeable about electronics and trained in ITC studies.

The need for this comes from the study of EVP. Transform EVP is believed to be formed via stochastic amplification in a nonlinear, active region of a device. This means that any electronic instrument with such a component is apt to be influenced by psi. At the same time, the instrument may only be able to display that influence as an unrelated change according to its design.

An example is a magnetic field detector which is designed to report changes in the magnetic environment of the detector as a change in magnetic field. A psi influence on the detector would display as a change in magnetism, when in fact, there might not have been a change in magnetism, but just the influence of intention on the detector. A person versed in the technology and the phenomena should be able to identify this possible ambiguity.

A second aspect of this is precedence. Most of the literature for many of these phenomena is in the form of anecdotal reports. While a Ph.D. might not find applicable information in a proper literature review, a well-informed practitioner should know that there is no precedence for changes in magnetic field in response to psi influence. While this does not mean there is none, an instrument signaling such a change should raise a red flag for researchers.

Citizen Scientists as Naturalists

In mainstream society, citizen scientists are thought of as people who help Ph.D. scientists conducting research. For instance, linking a home personal computer to a network so that a central computer can use available computing time from each linked computer to conduct data analysis, or so that the volunteer can help examine photographs.

As I have observed, research in the paranormalist community is not so well funded or organized, and there is little activity for citizen scientists. Yes, a person does act as a citizen scientist when volunteering to take a survey. However, being a test subject for psi function tests such as guessing next events is not being a citizen scientist. It is being a test subject. Expected ethical conduct, I often discuss, apply. When a practitioner volunteers to be part of a study, as the FEG medium did, it is as a human test subject and not as a citizen scientist.

As discussed in Essay 6: *Paranormalist Community* (Page 83), people who seek to further understanding of one or more aspect of the paranormal can be thought of as citizen scientists. The methodology may even be the same as used by trained scientists, but because they either do not hold a Ph.D. or have one in an unrelated discipline, it is best if they do not claim science. This is especially true if their final report is not published in a peer reviewed journal.

From my experience, the most common methodology used by citizen scientists is observational studies typical of the naturalist's approach to science. The major element of this include:

- **Education** – After intention to gain understanding, the first characteristic distinguishing an accidental witness and a naturalist is education. Simply witnessing phenomena does not further our understanding. Of the many unexplained events in our life, few, if any, have a nonphysical origin. Being informed means being knowledgeable about how phenomena have occurred in the past, current theories explaining them, idiosyncrasies of human cognition and the ways in which naturally occurring artifacts might seem to be paranormal.

- **Observations** – Record keeping based on observation of the phenomena as it occurs in usually uncontrolled field conditions, as opposed to more controlled laboratory conditions. This is typically in the form of first-person accounts, such as accumulating observation records from hauntings investigations or séances.

- **Modeling** – This is an important step in any form of field studies. The naturalist should be aware of existing models (theories, hypotheses) and be able to relate field observations in best-fit comparisons. In other words, does what has been observed agree with commonly held theories about what the experience represents? If different, in what way? This comparison may be to the naturalist's own models or to popular wisdom.

- **Reporting** – Communication skills are vital for anyone seriously studying phenomena of any form. A milestone-triggered, periodic report should be submitted to one of the layperson-oriented publications, posted on a website or included in a formal report on academia.edu or similar. The process is not complete until the public has been informed.

The protocol in the naturalist methodology is the observation of events as they unfold, analysis and reporting. If it is your intention to conduct a study involving induced phenomena following a specific protocol, the methodology described in *What You Need to Know About Science* (Page 236) of this essay would apply.

Develop the habit of documenting experience (perhaps journaling) and take advantage of opportunities to discuss observations and theories in an informed way. The value of any scientific methodology is not in the degree held by the researcher, but how well informed the person has become and how effective the person is in communicating acquired understanding. This is true no matter your level of education.

Pseudoscience

The study of the phenomena we are interested in is branded by skeptics and the mainstream academic community as pseudoscience, meaning false science. They argue that the existence of paranormal phenomena is not supported by known physical principles. Thus, arguing that such phenomena are being scientifically studied leads the average observer to think there is more to science than is possible.

Belief in pseudoscience is said to be a danger to society because it promotes science illiteracy. Such belief is thought to diminish the ability of citizens to distinguish between *good science* and *bad science*, as measured by mainstream academia.

Studying the phenomena can cost a mainstream scientist his or her career. Perhaps that is why it is mostly retired professors who are brave enough to study these phenomena. To be sure, it has historically been professional suicide for an academic to be associated with paranormal phenomena. We all should acknowledge that possible sacrifice.

As a mirror of mainstream science, skeptic Wikipedia editors have managed to make the online encyclopedia's official policy that pseudoscience is a danger to society. The arbitration cases that set the classification standard for Pseudoscience, Fringe and Paranormal can be accessed via the *Wikipedia Arbitration* Essay EthericStudies.org/wikipedia-arbitration/). (145) As I experienced, it is now possible for skeptic editors to cause an editor to be banned for claiming a paranormal subject is not pseudoscience. An example is how I was banned from editing the Rupert Sheldrake article in Wikipedia. (98)

The idea that the study of paranormal phenomena is pseudoscience, and that pseudoscience is a danger to society, has made it more acceptable for the Federal Government to actively suppress unpopular science. (85) The Federal Government has already adopted the skeptic's lead and officially maintains that belief in pseudoscience is a hazard to the public welfare. This is important to all of us because it demonstrates that our ability to work with and study these phenomena is not guaranteed.

The case of Wilhelm Reich is a good example of what can happen when skeptics apply pressure on the government to act against someone promoting *pseudoscience*. (84) The short story is that Reich was put in jail for making claims about subtle energy which were deemed by the skeptics to be unsupportable and therefore pseudoscience. He also developed devices that might put the energy to work and claimed he could heal people of some diseases with the energy.

In fact, the actual jail time was because he ignored the charges, and apparently, his partner transported some of their material across state lines, which was seen as criminal contempt of court. The following is from the Wilhelm Reich Museum website:

> *While Reich appealed his sentence, the government carried out the destruction of orgone accumulators and literature. In Maine, several*

boxes of literature were burned, and accumulators and accumulator materials either destroyed or dismantled. In New York City, on August 23, 1956, the FDA supervised the burning of several tons of Reich's publications in one of the city's garbage incinerators.... This destruction of literature constitutes one of the most heinous examples of censorship in United States history. (84)

Reich died of heart failure while in prison, and I understand his research partner committed suicide shortly thereafter.

The only real protection we have from skeptics and skeptic-promoted suppression of our study is proper application of the scientific method. Assumed belief without rational reason is why pseudoscience is seen as a public danger. In the view of mainstream culture, rational explanations about the nature of reality must only come from scientists. The point I will make below is that our parapsychologists are actually defeating our efforts to develop rational explanations about the nature of psi and survival-related phenomena.

Science and the Paranormalist Community

Most paranormalists see parapsychologists as *our scientists*. Yet, as I have explained here, parapsychologists are not a homogenous group diligently working to help us understand the nature of the phenomena we experience. In fact, most are actually anti-paranormal or at least anti-survival.

It should also be obvious that I speak as a Dualist. *Your Immortal Self* is essentially a book about survival, the nature of survival, available proof, and how it affects our life. This letter also applies to you if you consider yourself a Psi+ Normalist, as our best science indicates that the media, by which etheric concepts interface with the physical world, is psi, the psi field and the way in which a person expresses intention.

But old habits are hard to break. After most of a lifetime learning to see the world as a person, which is an etheric personality entangled with a human body, it is almost impossible for people to imagine anything else. But consider this, if we are a nonphysical mind first and a human body second, then the proper way to interact with reality is from the perspective of that nonphysical mind. To be true to Psi+ Normalist or Dualist understanding, our point of view must be as an immortal self-centric and not as a person or body-centric.

Practicing Science

It is common to see paranormalist organizations announce that they are a scientific organization or that they plan to conduct research. Of the several dozen times I have seen this, to my knowledge, not one has actually done so. It is fair to say some have conducted studies, but most have only made a record of what was done without the expected protocol designed with controls. Conducting science can be a lot of work.

The parapsychological organizations are little help. They won't even let people have full membership in their organizations unless they have a doctorate and a recommendation from a current member. The literature is mostly not available to non-academics and the literature that is, is too often written in terms intended for other academics familiar with the terminology of statistical analysis.

The Academic-Layperson Partition I describe below might help make sense of the fact that many of these subjects must be studied by laypeople because our academic friends will not, at least not with good science. It is ultimately up to us, but we need to be better informed to do so.

Academic-Layperson Partition

Besides division by interests, the paranormalist community is divided by what I refer to as the Academic-Layperson Partition. The partition has been created by those who claim intellectual authority based on their college degree and their subsequent reluctance to collaborate with people they (apparently) perceive to be of lesser intellect. The effect is that people who are trained in the scientific method too often try to conduct research without collaborating with people who are able to produce the phenomena and who have personal experience.

Trojan Horse Science

Paranormalists are all in the same boat. Parapsychologists seek credibility, probably all of us want to understand these phenomena and the urge to commune with spirit is strong in anyone who has realized they might be more than their physical body. Like the blind men and the elephant parable I explain below, the elephant-paranormal is not to be described by examining just one part.

At the same time, we live in that proverbial glass house. The rest of society is watching us. They think we may be crazy. Anything any one of us does to support that belief casts doubt on all of us. While you might think what you do concerning things paranormal is your business alone, it is not. If it is public, it is all of our business.

That is why it is so important that you are able to recognize the difference between a parapsychologist who wants to help you understand and one that wants to prove you are delusional. That is also why I think it is a good idea to stop calling them parapsychologists. Call them what they are, Anomalistic Psychologist (Normalists), Exceptional Experience Psychologist (Psi+ Normalists) or survival researcher (Dualist). After all, if survival is real, any discipline based on psychology is probably not the right field of study ... when practiced by itself.

I have come to think of people, organizations and websites, that represent themselves as pro-paranormal, even pro-survival, as *Trojan Horses* when they turn out to be anti-paranormal or anti-survival. The parapsychological journals are full of such Trojan Horse articles. Anti-psi or survival researchers routinely attract the cooperation of practitioners and witnesses by seeming to be pro-paranormal.

You may find it distasteful to call out Trojan Horses. I do. So, consider opening a discussion about them on a discussion board such as the ATransC Idea Exchange at atransc.org/forum/. At least there, you can get a second opinion and perhaps work with others to find a way to neutralize bad players without making a public spectacle or without leaving the problem for the next person to become a victim.

Good Science ... Bad Science

Yes, I am obsessing about this, but I want to make sure I only need to write this essay once.

Consider all the theories

Good science means considering all contending theories; at least the major ones. We are concerned with *"is it normal mistaken as paranormal?"* and *"If it is paranormal, is it evidence of survival?"*

If good research indicates a reported EVP is not paranormal, it is necessary to either report an indeterminate outcome or explain with good science and logic why it is not. Saying a reported EVP, which has been collected in a light/sound/EMF shielded chamber and correctly heard by several uncoached listeners, is just illusion or stray radio waves is only acceptable if a mechanism for radio signals to enter a shielded compartment is identified. Else, it is not science.

Witness reports are not science

Anyone can sit in on a séance or witness a paranormal event and tell others about it. Doing so is not being a scientist. It is only being a witness. A walkabout in a reportedly haunted house does not make science. Reporting about it does not make science.

In field studies, good science means having a protocol defining what will be done in the walkabout. There should be proper controls such as control recorders and cameras. If there is a walkabout, the space should be locked down. Following the protocol should produce a detailed report which considers the collected data and explore alternative explanations. At a minimum, a survey of historical documents, resident questionnaires and a record of environmental conditions are required. And then it is science only if the report is submitted to a review of subject matter specialists and made available to the public in an easily accessible way.

If part of the credibility is the fact that the report has been reviewed by subject-matter specialists (peers), the specialists need to be identified along with the qualifications that make them a specialist for that review. In my opinion, that is *vetting*, rather than *peer review*.

This is an important point. Reports about walkabouts or séances for the popular press or blogs is good and useful, but not if it is expressed as science. A Ph.D. sitting in on a few séances and writing a first-person account is not science. This is especially true if it is done under cloak of academic authority.

Be the adult

Please don't take this the wrong way. It is good to have fun with things paranormal. If you are not reverent toward your neighbors now, don't be reverent toward them on the other side. Have fun, but you still have to do the work. Sloppy work signals sloppy thinking. You will notice that Ph.Ds. try to act *academic*. They play the part of the learned ones, even if it is in a *fake it until you convince everyone* sort of way. That is actually a good way to be because it fosters exactly that behavior in the actors and those around them.

Mature behavior can be characterized as being pragmatic, discerning, practicing suspended judgment and mindfulness. Learn the meaning of these terms and seek to make them your own. See the Glossary at EthericStudies.org/glossary-of-terms/.

Peer review

The claim made by the parapsychological organizations, that they publish peer reviewed journals, is misleading. My apologies to certain editors, but secret peer review is just a way for good ol' boys and girls to give a wink and nod to poorly conceived articles. I have seen too many articles written by people who are not trained or even well informed about the subject published in supposedly peer reviewed journals. It is bad enough that such articles are written under inappropriate cloak of authority. It compounds the offense when the articles are supposed to have been judged by qualified peers who are judging under inappropriate cloak of authority.

Don't make that mistake. If you are presenting an article as a study or research report, make sure you ask people outside of your circle of influence to look over the work beforehand. Find a friend willing to irritate you.

Remember that there is a difference between a review of the quality of communication and a review of content source. Quality of communication, such as grammar and clarity of meaning, requires no disclosure of reviewer qualifications. Claims of content source, how they were formed and whether they are trans-etheric influences, do require review.

Claiming an article about the communication is peer reviewed is saying that you are not the only one making the point. Peer review gives the article additional intellectual authority, and in effect, the reviewers are co-claimants. As such, their qualifications are nearly as important as yours and need to be disclosed.

Note my comments above about vetting in *Witness reports are not science* (Page 247).

Ethics

Mainstream science conducting research involving living test subjects are obliged to have some kind of ethics board overview. The

Parapsychological Association (146) has published a good ethics guideline which addresses the need to inform human test subject about the study. Under the subheading of *Treatment of Participants*, it is written that *"Participants should be treated with respect, concern for their welfare, and recognition of their own needs which are being subserved by participation in a study."*

As with peer review, the parapsychological organizations have not shown the willingness to oversee ethical practices of its members. As a professional organization, membership should include submission to such oversight. That they are not, suggests unethical treatment of research subjects, and misinformation about the actual objective of studies, is a systemic problem. This leaves us with the need to manage the ethical considerations ourselves before we agree to be a test subject.

If a Normalist or a Psi+ Normalist asks you to participate as a practitioner in a research project or study, make sure you have a written agreement signed by the scientist and witnessed by a friend. It is best that you ask someone outside of your circle of influence to review the agreement. The agreement should include the research question, how the questions will be tested (protocol) and how the results will be reported.

> An effort to form a Practitioner Advocacy Panel for the purpose of representing the lay-person community to the academic community is stalled due to lack of interest. The ATransC is willing to support this effort, should there be more interest. (147)

Of course, you expect to be treated fairly and with respect, but your most important concern should be how the conclusions are written. Remember I said that Anomalistic Psychologists deliberately ignore evidence of paranormal phenomena. To make their point, they must find ways to conduct research with practitioners that will show the practitioner is delusional or cheating. From experience, there is a good possibility that, even if you produce phenomena under controlled conditions, the resulting report will be written to suggest that you did not or in some way may have been cheating. For instance, there might be ten words acknowledging the phenomena and a hundred words explaining it away.

In one example, a recent study of séances resulted in an aggressive, public character assassination of the practitioner based on rumors of prior bad acts, even though phenomena were recorded. The only protection you have against such treatment is group support, witnesses and a line item in the contract clearly stating that you have final approval of the report, or at the very least, that the report will only include reference to the actual data collected during the protocol. See Essay 13: *Arrogance of Scientific Authority* (Page 205).

The golden standard for research ethics is *The Belmont Report*. (51) The main points of the report are:

Respect for persons: *"To show lack of respect for an autonomous agent is to repudiate that person's considered judgments, to deny an individual the freedom to act on those considered judgments, or to withhold information necessary to make a considered judgment, when there are no compelling reasons to do so."*

Beneficence: *"Persons are treated in an ethical manner not only by respecting their decisions and protecting them from harm, but also by making efforts to secure their well-being."*

Justice: *"Who ought to receive the benefits of research and bear its burdens? This is a question of justice, in the sense of 'fairness in distribution' or 'what is deserved.' An injustice occurs when some benefit to which a person is entitled is denied without good reason or when some burden is imposed unduly."*

Practitioners also have ethical responsibilities. When making a public comment about another paranormalist (any person, really), it is important to remember that the consequences of the comments may have ethical and legal consequences. In Essay 5: *Ethics as a Personal Code for Mindfulness* (Page 75), I have recommended a personal code of ethics to help develop mindfulness. It is the least we can do as citizens of this community.

Personal responsibility is an issue in the paranormalist community. I have witnessed too many people and organizations relentlessly attack individual practitioners because they question the truthfulness of the demonstrated phenomena. In fact, the accusations were based on what I have come to think of as the trapdoor defense. In that, people who do not know enough about the phenomena, the situation or the metaphysics, make accusations such as *"There must have been a trap door"* about what is demonstrably valid phenomena. I have seen one Spiritualist group join into a sort of accusation feeding frenzy against a well-known medium, a grief management group attack another medium without proof and academics discount phenomena based on supposition.

Blind People and an Elephant

Do you remember the *Blind Men and an Elephant* Parable? In it, blind men were tasked with describing an elephant based on the body part near which they stood. As you might expect, the elephant was variously described as a tree, wall, spear, even a snake. The point of the parable is that one cannot know what is being studied by only studying a part.

In a very practical sense, Normalists, Psi+ Normalists and Dualists are as blind people examining the elephant-paranormal. Errors resulting from this lack of collaboration are exasperated by the Academic-Layperson Partition. It is further compounded by the tendency of laypeople to assume academics have subject matter understanding that does not exist.

Concluding Comments

This is a long letter, but I think you will see that we can all benefit by being aware of the many considerations listed here. It is mostly about science, but there is much here about our community and culture. It is ultimately about citizenship.

There is probably no source of information for how many paranormalists this letter applies to. For discussion, if there are 10,000 paranormalists, maybe twenty will read all of this letter. If you think this information is useful, consider mentioning it to your friends.

Comments from the Media

The *Skeptic's Boot* website has posted a pretty comprehensive review of this letter titled: "Science is not the enemy: A Response to 'Limits of science, trust and responsibility.'" (148) It has a derisive tone but is useful for learning an alternative point of view. Much of what author Brian Cox has to say reflects the mainstream's self-image, and some such as his denial that Anomalistic Psychology deliberately ignores the possibility of psi or survival, are simply knee-jerk denials.

The author of "What is Anomalistic Psychology?" (58) defines paranormal as:

> "Alleged phenomena that cannot be accounted for in terms of conventional scientific theories." He explains that "Anomalistic psychology may be defined as the study of extraordinary phenomena of behavior and experience, including (but not restricted to) those which are often labeled 'paranormal.' It is directed towards understanding bizarre experiences that many people have without assuming a priori that there

is anything paranormal involved. It entails attempting to explain paranormal and related beliefs and ostensibly paranormal experiences in terms of known psychological and physical factors."

The only way these phenomena can be explained "... *in terms of known psychological and physical factors"* is to discount, perhaps even ignore contemporary research. It is like proving the earth is flat while ignoring evidence of the ever-receding horizon.

References and Alternative Sources
Listed at the end of the book beginning on Page 357.

Essay 15
Let's Talk About God
2016

About This Essay
The saying, *Old habits are hard to break,* is especially true when it comes to our self-identity. For instance, it is common to hear Spiritualist say that we are spirit having a physical experience. Yet, few behave as if it is true. From my experience, the hardest habit of all to change is to stop thinking like a person and begin thinking as an immortal personality.

From years of trying to communicate these ideas to others, I know part of the problem is that we are taught from birth to think of ourselves as our body. Our cultural indoctrination begins soon after birth with local beliefs and religious dogma. Virtually every aspect of our early training is colored by belief in our relationship with God, death and morality.

According to religions, God lives in heaven. When we die, we hopefully go to heaven. Even though the idea of Etheric Studies is to develop an objective, evidence-based understanding of the nature of reality, it shares many of the same words used in religious dogma. The dominant cultural references are religious. If listeners are not prepared to change their mental references, they will always hear terms like death to mean going to heaven or hell. Immortal personality is likely understood as forever in heaven. The intentionally generic term of Source will be first understood as a father god unless the listener is mindful to look for the difference.

Religious training is a life-long experience while most paranormalists come to etheric studies later in life. That usually means that the first thing a person thinks of when god is mentioned, is the religious God and all of the associated dogma.

It works the other way, as well. Atheists reject the god concept, so it is natural for them to turn away from any talk of a metaphysical first cause or Source. Even though they are probably in agreement with the Implicit Cosmology, it is likely they do not get past the subject line.

The fact that cultural training tends to bias how people hear and understand the terms I use is why I so often restate or rephrase my point. *Your Immortal Self* (1) even has a comprehensive glossary of terms. The glossary is also available on EthericStudies.org. (9) I intentionally repeat explanations of concepts I think are essential for you to understand the nature of your immortal self.

To be clear, your mostly unconscious expectation of what I am trying to say is dominated by your cultural training. Even the way you allow paranormal experiences to unfold around you is dominated by prior training. If you do not consciously seek to become a more open-minded experiencer, you will finish this lifetime and go well into the rest of your eternity thinking what your church has told you is true.

Of the ideas most difficult to stop thinking about in religious terms, the god concept is probably the most difficult. I am not a religious person, and I go to extremes to avoid taking any of the metaphysical concepts I work with on faith. So, the first promise I can make to you is that there is nothing in this book I am asking you to believe as a matter of faith.

People say my writing is too complex. If it is, it is because I write to be specific and to show what is reasonably established as actual and what remains to be proven. Explaining what I think is actual reality is complicated by what I think you have been taught to think is true.

To say a concept is faith-based, is to say there is little or no objective evidence that can be experienced by others without prompting. It can only be real as a matter of faith or belief in the supposition that it is actual.

Since we fool ourselves into thinking the world is as we have been taught, objective evidence is not enough. There is also a need for experimental support which produces a rational, testable hypothesis.

The Implicit Cosmology, (12) which is explained in *Your Immortal Self*, has many explanations that depend on the validity of the Trans-Survival Hypothesis. That, in turn, is extensively supported in the book with evidence based on mainstream and parapsychological research and theory derived from the way Instrumental TransCommunication (ITC) and other survival-oriented phenomena are experienced.

I cannot prove Source (God) exists, the only reason I write about it is to bound the scope of the hypothesis and to model a consistent cosmology. The Source concept is based on the life filed concept. If we can say that our thinking mechanism works in a particular way, and that instances of life are quantitively different but qualitatively the same, it becomes reasonable to think of life as a building block of reality. At the same time, the expressions of life would be the stuff of which our experiential world is made.

We have good science indicating how we think. The same can be said for the existence of a subtle field of influence permeating the physical known as the psi field. The only real question that needs to be answered in order for us to reasonably speculate about Source is whether there is a purpose that unifies us.

If a common purpose for our existence can be identified, that would be the influence that binds reality into a single thoughtform. We could reasonable model reality as the expression of that unifying purpose.

If not, then it is enough to say the Source or God concept is just a logical mechanism which is useful for establishing the limits of the Implicit Cosmology. The Source concept would otherwise have no real significance.

With all of this said, I wrote this essay as an effort to turn your attention away from belief in a religious God and more toward understanding a more pragmatic, evidential Source.

> As an aside, I should say that belief in God gives us something greater than ourselves to turn to in times of emotional need. If you are an aspect of a purposeful Source, it is reasonable to think achieving that purpose is important to all life. When you pray for help. Pray to your loved ones and your collective, your teachers and guides on the other side. You are never alone. By praying for help, you give your friends on the other side permission to help you as you proceed to make it happen.

Introduction

This essay introduces a metaphysical model for God that satisfies current understanding of the nature of reality based on parapsychological and survival research with emphasis on the evidence of transcommunication.

It is common to hear people speak of God, usually in grand terms such as "God is the all-knowing yet unknowable punisher of the wicked and benign provider to the deserving." The usual proof of this god is "Look around. How else could all of this be possible?"

Spiritualists speak of God as a ubiquitous presence and the source of Natural Law. The National Spiritualist Society of Churches (NSAC), with which I am affiliated, rejects the *father god* characterization. But, individually, Spiritualists tend to stay with the old ways, characterizing God as a benefactor and giver of laws.

For many people, the need to look to a higher source is stronger than the desire to maintain a rational world perspective. A person may not believe in God but still look to a divine influence to make sense of life. We are good at compartmentalizing our need for divinity from our need for rationality.

Informed Perception of God

If God is a father god living in some distant part of reality like the mythological gods of Mount Olympus, then perhaps God will remain unknowable. However, ancient wisdom schools, especially Hermetic Traditions, tend to be knowledge-based, rather than faith-based. The assumption that reality, and by extension, God is knowable, is the foundation assumption of systems teaching the concept of Natural Law. So, let's talk about what we know, beginning with three assumptions:

1. **God is the source of reality:** The Source life field is the reality field.
2. **God is expressed as order in reality:** This is Natural Law or Organizing Principles.
3. **God is knowable:** This is fundamental in the Prime Imperative concept and seeking progression. See Essay 3: *Prime Imperative* (Page 49)
4. **Each instance of life is an aspect of God:** *Aspectation* is defined in the Implicit Cosmology as *"The creative process of intention acting on an imagined result produces aspects of reality which are a subset of personality's personal reality."* Consider The Creative Process, (19) and Life Fields. (13)

Taken as a logical argument, these assumptions imply a conclusion which might be worded as:

Therefore: By knowing about ourselves, we can know about God.

Building a Cosmology

A cosmology is a model of reality based on the metaphysical study of the nature of reality. A useful tool for developing a cosmology is black box analysis. The technique is often used for reverse engineering a competitor's device when only the purpose, input and output are known. The result is supposed to be a device, process, technique ... something that will satisfy the input and output to fulfill the same purpose.

Black box analysis is also a way to develop a metaphysical cosmology that may not be actually correct but is functionally correct and useful for further study. The theory presented here has evolved from such a study.

It is important to note that the functional areas in the *Functional Areas of a Life Field* Diagram (Page 259) are not necessarily accurate representations of reality. Psychologists probably will not recognize them or agree that they are useful representations. However, as a practical matter, they provide a useful guide for personal development.

Foundation Concepts

A number of concepts need to be part of this study. It is important that you understand and are comfortable with them, so please feel free to use the contact tool at the bottom of every page of EthericStudies.org and ATransC.org if you would like a better explanation, have questions or suggestions:

- **Reality is referred to here as the etheric, as opposed to the physical which is referred to as an aspect of the etheric.** The psi field described in parapsychology is referred to by me in this cosmology as the etheric. The confusion comes from the fact that psi field researchers speak of it as an aspect of the physical which is a body-centric perspective. From the perspective of immortal self, one must use the immortal self-centric explanation for the psi field. That is, that it is actually a local manifestation of the etheric. It is an aspect of reality, sensed signals of which we assign physicality.

  Trying to retain old references for clarity of communication, I do often say psi field rather than etheric.

- **A person is an immortal self entangled with a human in an avatar relationship.** The *Functional Areas of a Life Field* Diagram (Page 259) below illustrates many of these points. (13)

- **Our natural habitat is the etheric.** This includes our personality (who we really are, Core Intelligence as *I am this*), mostly unconscious mind, perception and expression processes and conscious self (*I think I am this*). Our natural perspective is etheric-centric. As a person, our conscious self experiences reality from the perspective of the human body. That is body-centric.

- **Conscious self's perception is normally dominated by Worldview.** Worldview is populated with a combination of human instincts, inherited urge to learn, inherited and acquired understanding from immortal self and beliefs acquired during the current lifetime (cultural contamination).

- **Sensed information from the human body is transformed via the brain into etheric form.** The brain transforms bioelectric signals from the body's five senses into psi signals mostly unconscious mind in the etheric is able to process.

Once it is in etheric form, the physically sensed information is processed in the perception and expression functional areas of our mostly unconscious mind (This is the External Influence signal shown in the *Functional Areas of a Life Field* Diagram (Page 259)). Information from the body and other etheric personalities is processed in the same way. The meaning of that information is assigned based on Worldview. Information intended to control the body is transformed from etheric to physical form in the brain.

- **The Prime Imperative.** We inherit an urge to gain understanding about the nature of reality. As such, we enter into a lifetime to experience reality as it is expressed in the physical. It is this acquired understanding we retain after transitioning out of a lifetime. Understanding determines perception, and therefore, the aspects of reality we are able to access. See Essay 3: *Prime Imperative* (Page 49).
- **Life fields are the basic building block of reality.** As objects of reality, fields consist of related elements which are attracted to an intelligent core (common or shared influence). Reality consists of life fields and expressions of life fields. (13)
- **All life fields have the same basic characteristics, but express them to a different degree, depending on perception.** The basic functional characteristics include are illustrated in the *Functional Areas of a Life Field* Diagram (Page 259). The behavioral characteristics of life fields include:
    - **Purpose motivated:** Spiritual instinct to gain understanding which is ultimately integrated with the collective understanding of other life fields.
    - **Perception:** Selectively sensing to environmental signals.
    - **Expression:** Selectively responding to environmental signals.

- ○ **Curiosity:** Urge to seek experiences that offer opportunities to gain specific understanding.
- ○ **Collective cooperation:** Shared influence on perception to create opportunities for specific understanding.
- **Source is the Core Intelligence of reality.** Source (God for this discussion) is the top Life field, and therefore, is the reality field from which the Prime Imperative is expressed. In the cosmology, life fields are arranged as a hierarchy of nested fields. Functional areas of each instance of a life field are more or less expressed. For instance, compare human life field to cell life field. (149)

Implicit Cosmology

The Implicit Cosmology (12) represents my effort to model what we think we know about reality, our immortal self and our relationship to reality. The functional areas in the *Functional Areas of a Life Field* Diagram provide an overview of the cosmology.

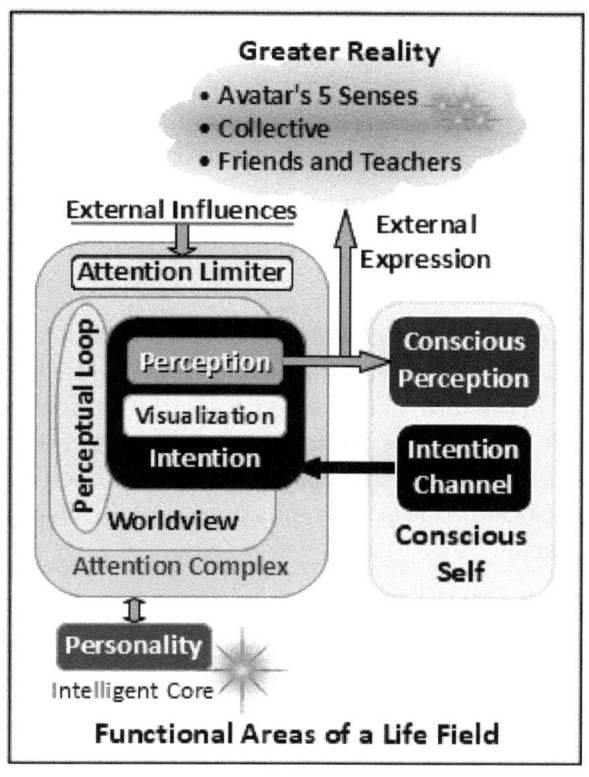

Functional Areas of a Life Field

Our perspective is always the Conscious Self functional area. The Attention Complex is our mostly unconscious mind and Personality functional area is our *higher self* and that which is immortal. The Intelligent Core is the autonomic system of our life field in the same way its body consciousness is the autonomic system for our human body (avatar).

Think of the Conscious Self functional area as a traveling perspective which is the experiencer part of who we are. It represents our etheric eyes and ears and typically thinks it is our physical body during the waking state. When free of our body during sleep or meditation, it is the experiencer of our inner consciousness. Always, what it experiences is moderated by the Attention Complex, specifically the contents of Worldview, which acts as our inner judge.

As Above, So Below

Another assumption is that reality is homogenous, in that knowing the nature of one part indicates the nature of the rest. Hermes is said to have taught this fundamental concept around 6,000 years ago via the *Emerald Tablet*. (28)

The first two lines of the *Emerald Tablet* read:

1. *It is true and no lie, certain and to be depended upon, that which is above is as that which is below; and that which is below is as that which is above, for the performance of the one truly great work.*

2. *And as all things are from only one thing, by will of the one God, so all things have their origin in this one power, by adaptation to their individual purposes.*

The *Great Work* is the process of transmuting a faith-based Worldview into one that is in accordance with the actual nature of reality. Of course, the *One Thing* is God in the form of expressed organizing principles, purpose and understanding. It is important to note that, while Hermes has told us all things are of God, he also said that each individual has been differentiated as a unique aspect of God. This echoes the concept of life as a fractal in which God is the top fractal.

Morphic Fields

In the Hypothesis of Formative Causation proposed by Rupert Sheldrake, (15) formation of a living organism is managed via what he refers to as a morphic field. The field is nonphysical, and formation is based on what he refers to as Nature's Habit. In a physical organism, the fields are arranged in a hierarchy of nested fields, meaning that there is a top field (Intelligent core is body consciousness) and many dependent fields such as skin, bones, organs and cells. There is a many-to-one relationship (nested) so that for instance, many cell morphic fields would be associated with the skin morphic field.

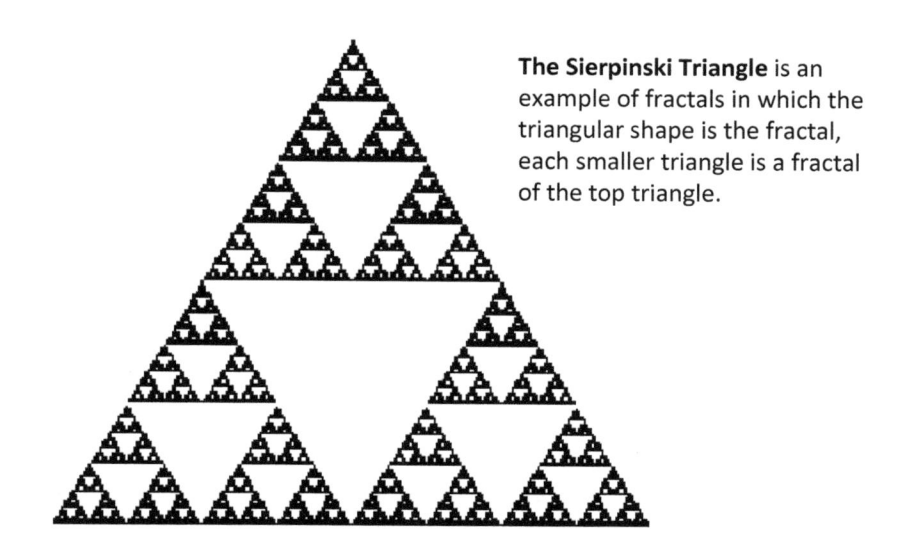

The Sierpinski Triangle is an example of fractals in which the triangular shape is the fractal, each smaller triangle is a fractal of the top triangle.

Speaking in terms of cosmology, morphic fields are related to the etheric as opposed to the physical. To work, they need the same functional areas shown in the *Functional Areas of a Life Field* Diagram (Page 259). Nature's Habit is like the Worldview functional area. The ability to evolve Nature's Habit is in the Perceptual Loop. External expression in the diagram represents the organizing influence a morphic field has on the physical processes of organism formation.

Sheldrake's Hypothesis of Formative Causation is by no means established science. In fact, it tends to be an updated variation of Lamarckian Evolution (150) which competed for a time with Darwinian Evolution but then was discarded for lack of a recognize physical mechanism for transmitting change to the next generation. The difference is that Sheldrake has given a mechanism for inheritance that makes sense considering current understanding of the etheric.

There is some analytical support for the inheritance of acquired traits, (151) but the real interest for this discussion is how Sheldrake applies Nature's Habit. Expression by a life field and a morphic field must follow the same principles. If morphic fields are validated as the building block of organic life, the same principles and research should apply to life fields. And, of course, Hermes told us so in line one of the *Emerald Tablet*.

Unconscious Perception

Who we are, is not who we think we are. (152) As the *Functional Areas of a Life Field* Diagram (Page 261) shows, the Attention Complex is a mostly unconscious part of our mind. We become consciously aware of the output of the Perceptual Loop. And so, what we become consciously aware of is based on our worldview. (67) (66) (68)

The importance of unconscious perception is that we use the same mechanisms that organize cell formation to develop an objective image of our physical world. This is an emerging realization of mainstream science (67) and you can expect to hear a lot more about it in the future.

Source as a Life Field

Considering what has been presented here, reality can be modeled as a hierarchy of nested life fields. Source is the top life field, and as such, it is also the reality field. In effect, everything in reality is in Source's life field by way of the thread of entanglement between first cause and subsequent aspectations. This influence of entanglement is a function of intention, attention and imagined outcomes.

A morphic field is modeled so that the organism it influences is in the field. This is because the morphic field imposes an influence on the biological process. *In the field* means within the field of influence expressed by the morphic field or life field. Since the etheric is conceptual space in which everywhere is *here*, this is a conceptual *In the field* and not something like a physical envelope around the organism.

This hypothetical model is important to explain that God (Source) is the reality field. Consider how we express our personal reality. What we imagine remains associated with us by way of a link of attention. What we expect to express via our creative process remains in our influence field. It is in effect, in our field of influence, just as we are in God's field of influence (reality field).

Personal Reality

Key to this discussion is the idea that we make our world. We have a personal reality based on what we believe to be true (Worldview). Since we only become consciously aware of what our mostly unconscious perceptual processes present to our conscious self, we literally experience objective reality as it is biased by our beliefs. This is becoming established science. (67) (66) (68) We manage our sense of reality by learning to change Worldview. One way to accomplish this is via mindful living. (48) In effect, that is what we do while learning to understand the nature of our local reality.

If life fields are in Source's reality, then it is arguable that Source also has a personal reality and is seeking to align it with its actual nature. As such, God is still learning.

Self-Organizing Reality

An important indication about the nature of God (Source) is that the Organizing Principles (Natural Law) enable reality to self-organize without an all-knowing god. If a cosmology begins with a sentient personality that is given the functional attributes of a life field, then all else follows without much intervention.

Notice in the *Functional Areas of a Life Field* Diagram (Page 259) that there is an Intention Channel between conscious self and the Attention Complex. That is the one conscious influence we have on our unconscious perceptual processes.

If Source is given the attribute of curiosity about its own nature, it is arguable that it will attempt to visualize itself and its environment as a means of satisfying that curiosity. Using ourselves as a model, our curiosity about something naturally initiates a mental exercise in which we imagine various aspects of it as it might relate to us.

A useful exercise is to imagine what it would be like to own a new sports car. It is likely that we would imagine a situation in which we would be driving the car. If we really want to know what it is like, we would give that *little me* self-determination. When we have explored the situation, our *little me* would return understanding of the experience to our worldview.

Such imaginings are usually only a moment in duration in our mostly unconscious mind and only the sense of the experience emerges into conscious awareness. However, the exercise will become part of our personal reality if we focus our attention on the question and allow our conscious exploration to run its course.

In terms of cosmology, the process of creating *little me* and the situation we want our *little me* to explore is referred to here as aspectation and the process is differentiation. It is our intention that gives the imagined *little me* purpose and attention which determines the duration of the exercise.

In this model, the physical aspect of reality is such an imagining done by creators of our venue for learning (local source). It is shared by many personalities via their collective of aspects.

The idea of a shared venue for learning is useful to explain the many references in religion to personalities holding the physical aspect of reality in their mind (angels, devas, nature spirits) and we who experience lifetimes in the physical.

As it is modeled in the Implicit Cosmology, (12) many personalities are using the physical venue to gain understanding about specific aspect of Source's nature. Their imaginings produce many aspect personalities which remain entangled with them so that they are the top life field for the collective of their imagined personalities. I am part of such a collective. You may be part of the same collective (Soulmate?), but more likely part of another.

Examples of Self-Organization

The functional areas of Life Fields naturally occur in response to the influence of curiosity (intention and perception) and the state of understanding (Worldview). This assumption is based on the expectation that such functional areas are necessary to produce known response patterns.

An organizing principle that naturally results from our perceptual processes is **Perceptual Agreement** which can be stated as: **Personality must be in perceptual agreement with the aspect of reality with which it will associate.** The implication is that we are not able to experience parts of reality which do not agree with our expectations (personal reality). There is no need for an ethereal being to say we cannot go to heaven. If our worldview will not allow our sensing of heaven to emerge to conscious self from the Perceptual Loop, we will simply not be able to experience that aspect of reality.

A second organizing principle is **Cooperative Communities** which may be stated as: **An effort to express understanding is necessary for progression.** This can be understood in simpler terms as: **Personalities are attracted to communities of like-minded people cooperating to facilitate personal progression.** The conscious decision to express an idea by one person must be coupled with the conscious intention to understand by others in the community. A continuing exchange of ideas can affect Worldview in the same way as a Maybe outcome of the Perceptual Loop. (A Maybe outcome can change Worldview).

A more pragmatic way to express the **Golden Rule: Do unto others as you would have them do unto you** is to say **Teach me as I teach you**. The objective of a lifetime is to gain understanding. As it is modeled, the collective may not fulfill its initial purpose until the intended understanding has been gained by every member of the collective. Cooperative communities are a natural response to that imperative. See Essay 3: *Prime Imperative* (Page 49).

God and Gods

In a sense, people speak as if there is an impersonal ubiquitous god of Natural Law and a personal god of our reason for being. We find meaning in communion with our loved ones and enjoy the comfort of knowing we have guides dedicated to our progression. Our inner space is populated with loved ones and guides, ... sometimes with nature spirits of one character or another.

Always implicit in our sense of inner community is the understanding that there are ethereal beings at the edge of our awareness; present but acting through our more accessible loved ones and guides. It is to those implied gods that I believe we direct our prayers; perhaps not as the deliberate act of a metaphysician, but the instinctive act of respect for those whom we sense care if we progress in our understanding.

In this cosmology, these ethereal beings are the top life field of our collective. Just as we have many *little me* aspects of ourselves populating our imagined inner worlds, created by us to experience an imagined situation, so we are *little mes* for these beings. It is those pinnacle life fields who must wait until we gain sufficient perceptual agreement to return our treasure of understanding so that they may have the perceptual awareness to move on toward their source. As I have come to understand order, it is through this progression that Source will eventually come to understand its nature. As expressions of Source, we are part of its nature.

References and Alternative Sources

Listed at the end of the book beginning on Page 357.

Intentional Blank Page

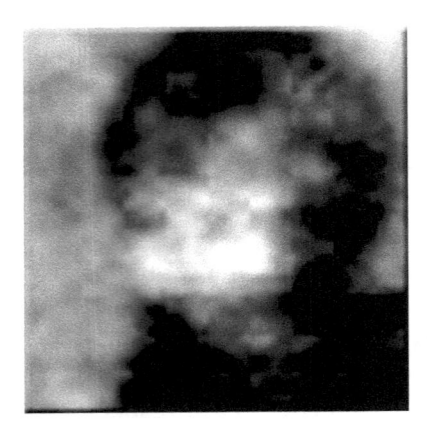

Video-loop ITC Image collected by Tom and Lisa Butler.

This appears to be a child with short hear ... perhaps a girl. We have captured many such children in our Visual ITC work.

Essay 16
What is it Like on the Other Side
2014

About This Essay

This essay is put together with bits and pieces of information about the other side I have gathered over the years. That information has been updated and organized here based on the Implicit Cosmology. That means it is based on the most current information I have been able to find about the other side.

To be sure, I am not an authority on dying. I am also not well studied in near-death or out-of-body experiences. Of course, I have spent quite a lot of time in trance and have had a few noteworthy out-of-body experiences. But, as I explain in the essay, I think that is not the same as the mind-state we experience during transition.

We say that we die when our human avatar is no longer able to support us in this lifetime. When that happens, and our worldview is free of our human's instincts, we move on to a different aspect of reality.

It is a given that we will die someday. Nothing more than that is certain. That is one of the reasons they refer to the separation between our awake life and the aspect of reality beyond death as the veil.

> The veil earns its name from the phrase, "Veil of forgetfulness." The idea is that we existed before this lifetime and memory of that prior existence becomes obscure at birth and quickly fades in our infancy. The veil is also a factor as we change from the disassociated mental states of sleep and trance back to an awake state.
>
> The veil represents the difference in perception between raw environmental information coming to mostly unconscious mind and that which is presented to conscious self. In effect, we expect to experience things in physical terms. That expectation is something our immortal self must learn when first becoming entangled with its human. This might explain why some infants seem to remember their pre-birth existence. As we learn on the Mindful Way to assume more control over Worldview, the veil is expected to thin. We gain in lucidity.

The fear of death is part of our psyche. It colors much of our thinking about aging. In fact, a loved one's transition and the sense of growing old are two of the primary reasons people take up a personal study of these phenomena. This essay is as much written to explain what is known as it is to help people feel more comfortable with the idea of their inevitable transition.

This essay is written in terms of our exit from this lifetime as a journey, rather than an event. I refer to this journey as transition. Separation from our human is an event.

Our process of going out of this lifetime and into another venue for learning is ended, we think, with something like a mini-transition as we exit the *in-between* and enter into the new experience. In-between is something of a place for rest and recuperation as we digest what we have learned and decide what we will seek to learn in a new venue. As a matter of speaking, we die out of that place of rest. It appears to be the way we move from perspective to perspective.

Introduction

This essay began as a response to a question from a website visitor about the nature of the other side. I answered the question from the perspective of the actual nature of reality, as opposed to the popular wisdom commonly held by many modern seekers. The difference is a point of view in which I am always striving for objective understanding of the actual nature of reality, as opposed to one accepting the ancient wisdoms and modern channeled material which is unavoidably colored by the ancient wisdoms.

The answer given the website visitor was written before I began formulating the Implicit Cosmology. (12) So naturally, it has undergone a number of revisions as I aligned it with my current understanding.

Old Beliefs Versus New Understanding

I began answering the question by explaining why the usual sources for information are not widely accepted by researchers who seek an objective view of reality. Having studied these concepts since the early 1950s, I can say that the old truths based on ancient wisdoms and channeled guidance are slowly giving way to more contemporary understanding based on well-considered research and better-informed models.

Most of what we are told about the other side is conveyed to us via a medium. That is, a physical person has made contact with an etheric personality and has in some way conveyed information from the etheric to the physical.

Having a person in the path of communication is seen as a problem because research is showing that the beliefs of the medium are more involved in the message than popular wisdom would have us believe. In mediumship, what they call *coloring* is the medium's own worldview contaminating the information coming from the etheric communicator. This is a known problem, and because of it, mediumistic messages are almost always taken with considerable reservation by researchers seeking objective evidence. (153)

For instance, mediums who are very religious might color a message originating as *"I am part of your group"* to be delivered as *"I am God."* People who are fond of conspiracy theories might color a message that originated as *"Change is a natural part of growth"* to be delivered as *"beware of change in your life."*

Careful reading of the Trans-Survival Hypothesis (10) will show that we think all transcommunication must pass through the worldview of a physical person. That is because, based on emerging mainstream and parapsychological research, all information first comes to our mostly unconscious mind where it is filtered according to Worldview before being presented to conscious awareness. See Essay 1: *Conditional Free Will* (Page 9) for a brief explanation of the Implicit Cosmology.

The filtering process based on Worldview is where the coloring occurs. For most of us, it is beyond conscious control. The only way we can control it is by managing what is allowed into Worldview beforehand. That is accomplished by habitually having the intention to examine our thoughts while trying to understand their consequences. I refer to developing this habit as the Mindful Way.

Lucidity is the conscious awareness of our perceptual and expressive processes. It ranges from unquestioning acceptance of what we have been taught and its implications to spontaneous examination and understanding of our worldview.

> The mostly unconscious mind is sometimes referred to as the judge because it decides what it thinks about sensed information before, and if, it is passed to conscious self. in involuntary coloring, the medium has little control over what emerges into awareness each moment. The effect of mostly unconscious cultural contamination can be minimized through habitually intending to sense reality as it is, rather than as the medium has been taught. That is described here as the Mindful Way.

Another form of coloring might best be described as profiling. The medium's message is developed around visual and behavioral cues from the sitter. Developing a message around questions and answers is a form of profiling.

The form of coloring that is usually very obvious and is certainly unethical is the *"Spirit tells me you should …"* kind of message sometimes referred to as social engineering.

The medium's competence as a clear channel is determined by how well he or she is able to manage coloring by developing lucidity and adhering to a well-considered personal code of ethics.

Tree of Life	Influence	Self	Plane
1 Crown	Cosmic Self		
2 Wisdom	Life Force	Causal Body	
3 Understanding	Intuition	Buddic Body	Divine Soul
0 Knowledge			Gnosis
4 Mercy	Memory		Higher Mental Plane Retrospective
5 Severity	Volition		Higher Mental Plane Prospective
6 Beauty	The Ego	Egoic Body	Ego Plane
7 Victory	Desire	Desire Body	Lower Mental Plane Retrospective
8 Splendor	Intellect	Mental Body	Lower Mental Plane Prospective
9 Foundation	Automatic Consciousness	Astral Body	Astral Plane Vital Soul
10 Kingdom	Sensation	Physical Body	Physical Plane
	Fire		Man
	Water		Animal Kingdom
	Air		Plant Kingdom
	Earth		Mineral Kingdom

From the *Handbook of Metaphysics*: The hierarchical cosmology as described in the Hermetic Qabalah (as if a layer cake) works in conversation as a means of indicating *here* and relatively *not here* as one *moves* into other aspects of reality. (2)

Old Models of Reality

As you can tell from my other essays, finding ancient wisdoms that seem to agree with our more contemporary models is especially interesting for me. As it turns out, what I tell you about mindfulness has been revealed to humanity many times over the millennia. However, ancient translators tended to hide the lessons in local religious beliefs or the social engineering preferred by the priesthood.

It is understandable that there would be some cultural contamination. For people who think they are their body, the nature of reality must be explained from that body-centric perspective. We stand on the ground and look up to heaven. We understand concepts like vibration and density. It seems obvious that humans are above animals who are above plants and minerals. But then, gods must always be above humans. Some humans are more spiritually advanced than others, so they must exist above the rest of us but not all the way to the gods.

Probably as an early attempt to explain concepts such as progression and spiritual maturity, ancient teachers developed cosmologies based on something of a layer-cake model. The models are not necessarily wrong from the human perspective, but they tend to cloud the ancient messages when considered from our actual etheric nature.

> If you should ever attempt to explain a metaphysical concept using the perspective of the Implicit Cosmology, you will learn that modern people still find it difficult to comprehend reality from the perspective of their immortal self. They might understand they are not their body but remain unable to shift their perspective. It has taken me years and remains a journey not yet finished.

My first effort to develop a model of reality was with the *Handbook of Metaphysics* (2) published by Christopher Publishing House in 1994. The book included a discussion about several of the more popular cosmologies.

> The *Handbook of Metaphysics* was published. But, it was my first effort and I did not understand publisher and author roles. The publisher took my manuscript and went right to publishing without any copyediting. It was not widely distributed and is now out of date. Until I find time to produce a second edition, it remains a rarity on the market, but otherwise just a title.

I have also studied mediumship and the information we receive via mediumship, both (mostly) conscious and deep trance. This study has included attending many classes and continued efforts to develop my mediumistic ability. While I am convinced some information brought via mediumship is meaningful, I am painfully aware of the problem of coloring.

This applies to all forms of information access that involves our mostly unconscious-to-conscious self as the channel.

The most common response I receive when trying to explain elements of the Implicit Cosmology are based on the other person's strong convictions that the layer-cake model is correct. For instance: *"... the astral-ethereal body is a substantive second body."* Or, *" ... It is not an illusionary construct; it's a very real body ...."*

"Very real body" is a direct rejection of the suggestion that we experience reality as an ideoplastic construct based on Worldview. It is true that the body is experienced as being very real, but the speaker is ignoring what is being learned about our perceptual process and how it is guided by what we have been taught.

> Ideoplastic: The Ideo- prefix means idea or image. Ideoplastic is used here to mean the nature of objects formed in the creative process as mind to object expression.
>
> **Ideoplasy** means **the process of formation**. The term was coined by Max Vorworm. While ectoplasmic forms in the séance room are ideoplastic, according to the Implicit Cosmology, all of the physical is an idioplasmic creation of minds.

The old models have some value, but they need to be reconsidered with emerging understanding about the nature of mind. A typical error is thinking being more or less spiritual is being of higher or lower vibration. That too easily leads the person to think of reality as different frequencies of electromagnetic radiation. The next step is to assume transcommunication must be a matter of changing electromagnetic vibration.

None of these models stand up to controlled study or logical consequences of the nonphysical concept. Thinking in these terms has sent many metaphysicians and experimenters down blind alleys.

Transition

Death or dying is a meaningful term when we think of ourselves as our physical body. However, we are not our physical body. Our body is just our host and avatar for this lifetime. When our avatar is no longer able to support us in this lifetime, we begin a process of transition out of this lifetime that is thought to eventually take us to a new venue for learning.

Think of our conscious self as a video camera. The lens of the camera is our human's eyes when our body is awake, and our perspective is associated with it.

It is normal for us to dissociate from our body. Daydreaming is a light form of dissociation. A light trance is a form of dissociation. We momentarily dissociate when we access information from our memory. Sleep is a more complete dissociation. However, while the body lives, we continue to share Worldview, so that our stream of consciousness is always colored by our human's instincts.

As we understand the process, when we die, this dissociation becomes complete and permanent. It appears the difference between a deep dissociative state and death is the release from the influence of our human instincts. This release begins the transition process and also marks the first time we have been without the burden of human instincts since birth.

Beginning Transition

We think all of us experience a short period of shock when we physically die, followed by a period of *getting well*. That is thought to be followed by a period of adjustment and finally, a second transition into a new venue for learning. The new venue could be in the physical, but it could as well be in some other imagined world.

During the initial shock of transition, we think it is relatively easy for the entity to remain perceptually close and communicate with people in the physical. Keep in mind that this is not physically close. Think of yourself as the center of reality. In fact, think of yourself as the Axis Mundi spoken of in many systems of thought. Axis Mundi is a reference to the center of the world or connection between heaven and earth. We are that. Being perceptually close means that we are able to clearly visualize our physical life as it was. By doing so, in effect, we are able to be there.

The early phase of transition seems to be when we are apt to experience the tunnel some people report in near-death experiences. I am not qualified to speak of the near-death experience; however, I believe the Implicit Cosmology is rather different from what most theoreticians have considered, so I am going to speculate. The Implicit Cosmology predicts the influence of our human's mind is separated from our worldview at the beginning of transition. That sudden freedom from controlling instincts must be expected to produce a period of confusion.

Since Worldview is thought to be the primary factor forming our perception of reality, the perception of passing through a tunnel may possibly be a resulting perceptual artifact. Our human instincts cause us to always be aware of our surroundings. Let us call that awareness, peripheral attention. When we are free of that urge to see everything, we are, probably for the first time in the lifetime, able to truly focus our attention. It may be that we experience this new focus literally as tunnel vision. This would only last while the Attention Complex of our mind adjusts its perceptual processes. Refer to the *Basic Functional Areas of Perception* Diagram.

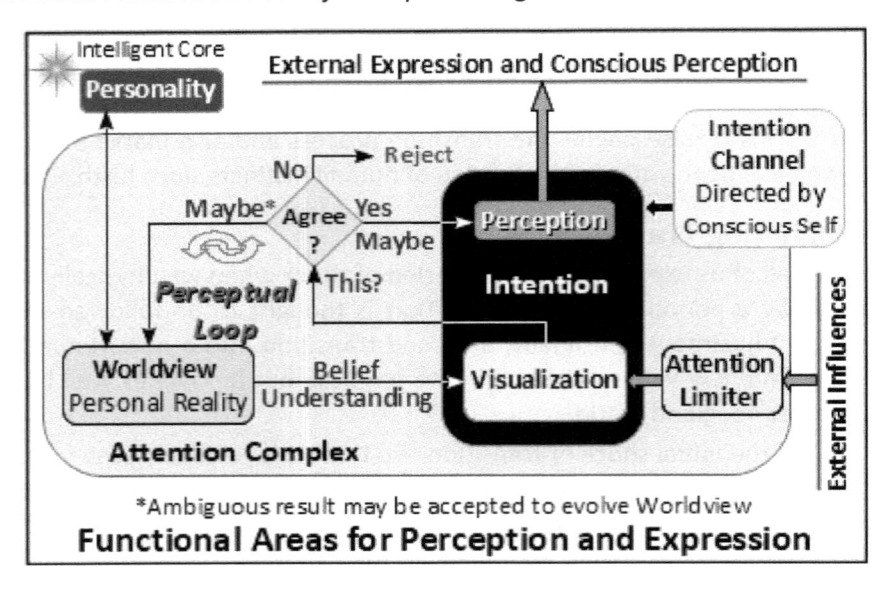

*Ambiguous result may be accepted to evolve Worldview

Functional Areas for Perception and Expression

Even though our natural habitat is the etheric, we spend a lifetime being convinced we are our body. We create our physical environment according to what we have learned. Based on reports from recently transitioned people, we will continue to create a version of our body during the transition period, and probably a version of our familiar world. Those of us who have been conditioned to think of the next world as a brighter, more colorful and magical place, are also conditioned to find their new world to be that way. In effect, that has become as real a New Ager's heaven as that promised by religions.

Imagine the shock of suddenly finding that we are no longer in our lifetime but are still in a body. Old rules would seem not to apply as we realized our intention was so influential that we could actually imagine ourselves somewhere and be there. In time, we would discover that our body was no longer crippled with age. We would feel young and strong again, and so, find that our body was also once again young and strong.

From the reports, we think memory of our physical life remains strong as a visualized reality. As with our body, at first, we probably continue to create physical reality more or less as we remember. During that time, we are more able to impress our influence on the physical and our thoughts on people close to us.

Our ability to influence the physical is a function of how well we can mentally associate ourselves with it. As we come to realize we are no longer part of that old reality, our perception will begin to shift toward our new situation. If for some reason, we do not allow ourselves to turn toward the new situation, it is possible we will remain perceptually close to the physical. It may be this refusal to accept change that causes some personalities to seem earthbound, possibly even to produce haunting-like effects in the physical.

> As a device in the novel, *Two Worlds, One Heart*, (154) I used the idea that it was possible to move between alternate timestreams by changing the person's perspective to agree with the intended reality. As a modern Californian, time shifting to join a pre-colonial Native American society was accomplished by remembering or clearly visualizing what it is to be immersed in that perspective.

Guidance

An important experience reported by near-death experiencers is the presence of loved ones and guides. Again, I am not qualified to speak with any authority about near-death experiences; however, the Implicit Cosmology does provide a little guidance.

The way we experience the perceptual presence of a loved one or interested personality is subject to what we have been taught to expect. While still more associated with the physical, we expect to experience the presence of a person as a physical person.

As a mental medium, I am occasionally aware of the presence of other people's discarnate loved ones. I am even sometimes able to describe them and their mannerisms. Still, there are times when I only sense their attention on me and the intended receiver of the message. I am never sure, but I think it is by following the sensed personality's thread of attention that I discover for whom the message is intended.

> On days in which my lucidity is not clear, my delivery becomes *"I have a message for someone in this part of the room."* In Spiritualism, this is bad form for a medium. Part of the demonstration of mediumship in Spiritualism is supposed to be the ability to know who the message is for. I should be calling out the intended sitter. For instance, *"Jeff, may I come to you?"*

So, as a practical matter, I know etheric personalities are at least occasionally close to us. This is not a *"Move over; you are standing on your uncle"* kind of thing. The personality comes perceptually close by turning its attention to us in the same way that we can see a loved one in our mind's eye by thinking of them.

We have quite a lot of friends on the other side. The ones I think are arguably true include:

Our collective

> In the Implicit Cosmology, each of us is an aspect of a local creator personality, even as it is an aspect of another, and so on, as one looks toward Source. We are all direct or indirect aspects of Source.
>
> When we create an aspect of ourselves to experience an imagined situation, that aspect and attendant memory of the experience remains with us. Think of the aspects you have created as a collective of personalities sharing you as their source. In the same sense, you are part of a collective.
>
> Think of where those aspects exist. They are not *out there* somewhere in space. They are where you are, in your life field. Everywhere is here.
>
> According to this model, aspects are created to gain understanding about some part of reality. In principle, the source personality has not gained the understanding it seeks until the collective of personalities it created have gained the portion of that understanding for which they were created. In the end, they are all one mind just as your aspects are all one mind ... as they are held in our worldview.
>
> The guide we have been taught to look for on the other side is likely part of our collective. Members of our collective are vested in your success and can be expected to help you cross. See "Cooperative Collective" (Page 53) in Essay 3: *Prime Imperative* (Page 49).

Loved ones

People in the physical with whom we have a close relationship are apt to become *more present* for us when we enter into the etheric. This is equally true of close acquaintances such as workmates and fellow military. These friends come to us when we think of them now, not in a physical sense, but perceptually near.

Ethical Understanding	Ethical Principles	Ethical Expressions
Do not violate *Seth*	• Respect • Kindness • Do no harm • Justice • Fairness • Suspended judgment • Courage • Discernment	• Just because I can, doesn't mean I should • Mindfulness is a way of life • Citizenship means cooperation • How will my actions affect me and others? • Is it a belief or a supportable understanding? • I will not impose my will on others • Lessons come from new experiences • Contemplation not meditation

Possible Mindfulness Personal Code of Ethics

In Essay 5: *Ethics as a Personal Code for Mindfulness* (Page 75), I began an example code of ethics with advice from the channeled entity known as Seth who advised: **Do not violate**. Common guidance for seeking help from our friends on the other side is that we must ask them. They cannot impose their help on us. There are also practical limits to the kind of help we can be given. For instance, they are not going to help us impose our will on someone or interfere with a lesson our core intelligence seeks. The point is that they will be present to help as they can, should we be open to the help.

Teachers

This is an interesting concept. In the Implicit Cosmology, (12) the **Cooperative Communities** Organizing Principle is defined as ***An effort to express understanding is necessary for progression.*** (27) The urge to teach is part of a cycle of seeking understanding. A seeker learns to integrate information into a logical form of knowledge, but it does not become understanding until the seeker is able to express it in a meaningful form. Teaching others is a means by which the seeker is able to consolidate understanding. For the student, being taught is the first step toward understanding.

While you read this book, I am your teacher. In effect, you have asked for help by taking the time to read and I have agreed to provide that help as I am able by having written the book. The urge to teach appears to be with each of us. Even after so many years trying to communicate these ideas, I still have no idea what motivates me. It remains an itch I can only scratch by teaching.

We know that some personalities purposely remain perceptually close to the physical so as to be better able to provide guidance. The control personalities known to help physical mediums are one example. Channeled entities are another. Jane Roberts' Seth, (47) for instance.

Following the Implicit Cosmology, we will have attracted a teacher personality if we have consciously expressed the intention to gain understanding. Presumably, the same intended understanding might be gained from many different kinds of experiences, such as baseball or hunting. Thus, it is reasonable to think a great warrior personality might attract a warrior-like personality as an etheric teacher.

If we have a conscious relationship with such a teacher, it is reasonable that we would be greeted by that teacher when we make our transition.

Devic personalities

This is a complex concept. Our culture is influenced by nature spirit myths, which when deeply integrated into our worldview, effectively makes their influence real. As such, they may as well be real, and as thoughtforms, probably are amongst our friends on the other side.

My early efforts to develop a cosmology included a source at the top of two parallel hierarchies of personalities. One hierarchy represented experiencers such as you and me. The other represented formative personalities generally known as devas, but also known as angels and nature spirits. The idea was that reality needed personalities who were not so concerned with gaining understanding as they were in helping others to have experiences to gain understanding.

The lord of the physical plane was actually a formative personality in that cosmology. One source for that thinking came from the work of Paul Twitchell. (155) He taught about a personality named Kal Niranjan who is said to be the lord of the lower planes. Kal is also known as the Devil because he is thought to be the instrument of our lessons, many of which can be learned via painful experiences.

In the Trimurti concept of Hinduism, the cosmic functions of creation, maintenance, and destruction are personified by Brahma the creator, Vishnu the maintainer or preserver and Shiva the destroyer or transformer. Kal Niranjan relates to the Shiva aspect, seemingly the destroyer or Devil, but actually the transformer or teacher.

In terms of the Bible, devic personalities are angels. In other systems, they are nature spirits. Every aspect of life is said to have a devic personality assigned to help in the life functions. For instance, the elementals of Western mythology and New Age lore.

As I evolved the Implicit Cosmology, it became clear that reality is self-regulated, so that the formative function is an integral part of every life field, as what I refer to as a core intelligence. For our human avatar, this is the body mind function which maintains the body image used for the ongoing formation of every cell. For our immortal self, it is associated with our etheric personality, which is the intelligence (*I am this*) for which we are conscious self (*I think I am this*).

We, as immortal self, create thoughtforms based on our beliefs held in Worldview. It is possible, even likely that, when we begin transition and are free of the blinding influence of our human body vision, we will become aware of the thoughtforms we have created. For instance, if we are very involved in systems of thought based on gardening with nature spirits, we can expect to encounter nature spirits in our new reality.

In the same way, a strong belief in Jesus will likely produce a Jesus thoughtform during your transition. I expect Tinkerbell will show up for me, as I so much wanted her to be real when I was a child.

The bottom line is that we have learned to personify characteristics of nature which we then experience as thoughtforms. The angels we encounter during transition will be our loved ones and members of our collective. Our local creator personality, who is the life field for our collective, might be experienced as an Archangel or Deva. Very spiritually mature personalities remaining close as teachers might be sensed as ethereal beings. It all has a lot to do with how we have been taught.

Perception Adjustment

A *getting well* period is thought to be needed for the newly discarnate personality as they change their perception of what is real (at least *more real*). The transition process allows the personality to realize that old handicaps no longer apply. For instance, many of the reports we have heard include the expression of happiness in no longer needing a wheelchair or no longer being in pain.

We think this is also when we undergo a self-evaluation of our past lifetime. The self-evaluation is that dreaded judgment we are always warned about, but instead of our being judged by some authority on high, it is a personal process in which we sense how our actions affected others.

During the series of 4Cell EVP Demonstrations, (156) a question posed by Terry Dulin was *"What were some of your misconceptions about death and/or life on the other side?"* The response recorded by Vicki Talbott was her son Braden saying *"Regrets."* The message is one of the confirmations we have encountered indicating we undergo a life review. And, since it is from the perspective of those with whom we interacted, it has the potential to bring regrets for our actions during the physical lifetime. In the context of this 4Cell session, it was understood by Vicki that Braden was reluctant to answer because he felt it would pain her to know that she would have to face her life-deeds, good and not so good. This is apparently true for all of us.

Memory of our physical lifetime fades as we become accustomed to our new surroundings. As we have noted, our etheric communicators sometimes behave as if they are living in a world much as they experienced in the physical. This is true even to the extent that they might refer to a heavenly grandma's house at which members of the family in the etheric gather for a family dinner.

Note that Instrumental TransCommunication (ITC) messages are probably colored by the expectations of the practitioner. In fact, the message about grandma's house might begin as a thoughtform representing something like *"We are aware of each other and remain close in love."* It is the assumption of a person around the holidays that there would be a family gathering, and so it manifests in the ITC.

The messages we receive from loved ones are often collected close to the time of transition. It is not clear how soon we begin to move perceptually away from our old lifetime. We are told that our natural form is as a luminous being existing in a landscape easily influenced by our thoughts, and that ultimately, we are pure thought. I expect that we do not realize that aspect of our nature until we have gained great maturity, but it is likely that communication across the veil decreases as the person turns attention toward future opportunities to gain understanding.

As suggested by the Implicit Cosmology, the longer we are away from our physical form, the weaker the link to our physical form becomes in our worldview. The **Perceptual Agreement** Organizing Principle is defined as **Personality must be in perceptual agreement with the aspect of reality with which it will associate**. (27) If this principle accurately describes the influence of perception, it suggests that we cannot *go to* parts of reality that we cannot imagine. It also suggests that personalities who have not been directly associated with the physical for many cycles of their existence will likely have difficulty coming close to us. We will either experience them as a luminous being or by way of a representative. (Remember that *close* and *go to* are differences of perception in the etheric, not differences of distance.)

Purpose

My assumptions about why we experience a lifetime are explained in Essay 1: *Conditional Free Will* (Page 9), and especially in Essay 3: *Prime Imperative* (Page 49). The idea is that we exist to gain understanding about specific aspects of the nature of reality and our relationship with it. This purpose manifests as an urge to have life experiences that might provide opportunities to gain bits of that understanding. Reality is complex, and understanding is seldom absolute. Since our ability to perceive reality depends on our current understanding, clear perception requiring many opportunities to converge on understanding the actual nature of reality.

We do not gain this understanding by ourselves. Members of our collective share the same urge to gain understanding, but of slightly different aspect of reality. We are pretty sure that, during our transition, we in some way compare our current understanding with the intended. This is possibly done with the cooperation of our collective.

Next Venue for Learning

In the Implicit Cosmology, the urge to gain understanding is modeled as a basic characteristic of a life field inherited from Source. That means we appear to exist as Source's effort to understand itself. As we respond to the urge to understand our particular aspect of reality, we become perceptually more in agreement with Source. In that way, we are moving closer to Source, or are returning, as it is sometimes stated in other systems of thought.

In the Implicit Cosmology, the physical universe is modeled as an aspect of reality created by many personalities cooperating to hold the physical in their imagination as a collective thoughtform. In this model, each creator personality has issued many aspects of itself into the physical to experience and gain understanding. Together, we aspects populate the physical, and by interacting with each other and the physical, provide one another with unique learning opportunities.

While the interaction may seem severe such as an *"I will kill you this time and you kill me next"* kind of cooperation, it may also be an *"I will be your mother this time and you be mine next"* relationship. It depends on the kind of experience that will provide the kind of understanding we seek. If you think about it, a gardener and a nation's president can potentially gain the same understanding about the nature of kindness. Remember that what happens to us is not as important as how we react to what happens to us. It is our ability to find new understanding in experiences that is so important to our spiritual progression.

Self-Governing

An important point is that there are no enforcement officers on the other side making sure we do what we should. We are self-governing. By that, I mean that we are only able to experience what we can imagine and that is based on Worldview. That is why we tend to create what we already know. It is also what limits the aspects of reality with which we are able to interact.

Moving On

This essay provides a simplistic view of transition. Certainly, there are a lot of details that will make your personal experience different from everyone else's. However, this essay is generally accurate within the context of the Implicit Cosmology and reasonably reflects commonly reported near-death experiences.

The major differences between what you have been taught and what is explained here is the Implicit Cosmology. That, in turn, has been developed based on a foundation of current research and experience from various forms of transcommunication. If you do not accept the view posed in this essay, it might be good for you to review the evidence and be sure what you think is true is not just based on habit or faith. It is fine to develop your own cosmology, but if you do, be sure to explore the implications of your assumptions.

The final assumption is that at some point in the process of reviewing what you understand, the need for additional opportunities to gain further understanding will become apparent.

Working with other members of your collective, it is expected that you will select a new venue for learning. It may be in the physical, on earth or another planet, but it might also be in an entirely different, nonphysical aspect of reality.

The reason I have written this book, and *Your Immortal Self*, is that I know you have complete control over your destiny, if only you understood the principles. Mindfulness is your tool for assuring a productive and enjoyable transition.

We can believe what we wish but it is important that we understand the implications of what we believe. The more our worldview is in accord with the actual nature of reality, the more successful we will have been in this lifetime. What we do matters, here and hereafter.

References and Alternative Sources
Listed at the end of the book beginning on Page 357.

Intentional Blank Page

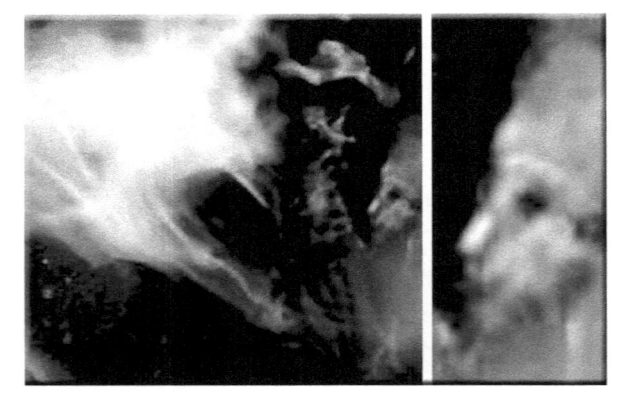

Moving water ITC Image collected by Tom and Lisa Butler.

This was collected by photographing light reflecting from moving water. You can see the original picture on the left and a slight enlargement of the face on the right. We captured two people in this session who appeared to be wearing a hat or possibly a crown.

Essay 17
The Hermes Concepts
2014

About This Essay

Consider the conversations you have with your friends and acquaintances about things paranormal. Would you say that at least some of your friends are well-informed about the more abstract concept such as the difference between a concept and a thoughtform?

My experience has been that few people have learned enough about these phenomena to independently form a point of view. Virtually everyone I know base their understanding on popular wisdom. When I try to understand their logical basis, it seems channeled material, preachers, television and friends are their source.

Okay, so that is our culture. Their beliefs exemplify thinking I consider a form of cultural contamination in which the ill-informed lead the even less informed. People are generally happy with their beliefs and I can think of their misconceptions as an entry-level understanding of a most complex and abstract subject; a good place from which to begin their journey of self-discovery.

As I remember, I billed the *Handbook of Metaphysics* (2) as a plain English introduction to the concepts intended to provide a framework of understanding. The idea was that, if my readers were better informed about the basics, they would be better prepared to understand new ideas. In a practical sense, that is the underlying intention of *Your Immortal Self* (1) and this book.

Here, I am speaking of modern people who have access to pretty good education and countless online sources of information. Do you think people living 6,000 years ago would have more objective understanding? Would I find a better conversation back then?

Think of the study of ancient text as a sort of metaphysical archaeology. The reason I am so fascinated with lessons attributed to Hermes and the *Katha Upanishad* is that they contain bits and pieces of information that are as abstract and profound as some of the material being discovered today.

Sure, I can use the ancient material to say that, even back then, others thought the same way. That is one of the things I do. But my real fascination, and the real message to you is the question about the source of that information. In this essay, and also Essay 18: *The Razor's Edge* (Page 309), are references to metaphysical concepts that are not widely understood today. Think of it! Ideas, such as the ancient teacher's version of the creative process appear to have fully matured before becoming part of the ancient teaching we see.

Remembering that our subject is all about the other side and communication across the veil, it seems reasonable to argue that our friends on the other side were as active back then are they are today. Perhaps even more so, judging by the sophistication of some of the concepts in the ancient text.

While we marvel at the fact that our ancestors knew some of these concepts, also wonder about how they learned. One of the proofs of the reality of the other side may be the evidence that there was transcommunication 6,000 years ago.

Introduction

Western metaphysical thought is greatly influenced by a wayshower known as Hermes, who is thought to have lived in Egypt some 6,000 years ago. The concepts attributed to him represent the foundation for most of Western religion and philosophy. However, intervening cultures have transformed Hermes the truth giver into more of an ideal which represents the greatest good associated with the desire to gain self-realization.

This essay is concerned with some of the metaphysical concepts thought to have been introduced by Hermes as they might be understood from the perspective of the contemporary paranormalist community. New understanding of how our mostly unconscious perceptual process produces conscious awareness is emerging in both mainstreams psychology and parapsychology. There is also a growing foundation of information indicating the nature of our immortality, our relationship with our physical body and the nature of the greater reality (etheric).

A basic assumption of this essay is that this new understanding provides a conceptual check on the ancient Hermetic teaching. This normalization of the ancient concepts with contemporary thought may help us better understand the original intent of teachings attributed to Hermes. (12)

> As you read this essay, keep in mind that my comments are based on a cosmology which is implied by my take on current understanding I have found amongst researchers in the paranormalist community.

Much of this cosmology has not been vetted, and so, it is important that you are mindful that the explanations here are conditional. It has yet to be seen if our greater community will embrace the cosmology.

There is a lot of *"It is not that way, it is this way,"* in this essay. The main reason is that emergent understanding is very different from what you might consider traditional metaphysics.

Who Was Hermes?

The Egyptians had a god named Thoth which was considered the mind of God as a teacher, source of writing, healing, art and music. (157) Many contemporary accounts describe Thoth and Hermes as the same person but it is more likely Thoth was an aspect of God, or ideal, while Hermes was probably an influential priest. As discussed below, based on the diagram typically associated with Thoth, the Thoth-Hermes character was a man who expressed the essence of Thoth.

Thoth
Image from: Biblioteca Pleyades Italia/Italy - EU

Of the many opinions about the reason an ibis is associated with Thoth, I like Edward Malkowski's interpretation because of the overall pragmatic view he takes with the history. (25) From his article:

> *For the ancient Egyptians, Ba animated a living person, whereas Ka was the energy emanating from that person. Although not an exact analogy, the Ka and the Ba are what traditional Western thought might refer as*

> spirit and soul. Another important aspect of Egyptian belief represented immortality, the ankh, depicted as the crested ibis.

and

> ... the meaning of a specific neter [Egyptian for god] was communicated in a visually symbolic manner. When a human was depicted with an animal head, this signified the principle as it occurs in man. If the whole animal was depicted it was a reference to a principle in general. Alternatively, a human head depicted on an animal represented that principle as it relates to the divine essence within mankind, not any person in particular, but the archetypal; as the immortal Ba is represented by a human-faced bird.

Amongst many titles, Thoth was considered the heart and tongue of the Sun God Ra and the means by which Ra's will was translated into speech. He also had the title of *"Three times great, great."* The Greeks thought Thoth and Hermes were the same, and thus gave Hermes-Thoth the title of Trismegistus (Greek for *Hermes the thrice-greatest*). The Romans referred to him as the god Mercury (*Mercurius ter Maximus* in Latin).

The Upanishads are a set of Sanskrit text thought to have been written 3,000-to-4,000 years ago. They are based on oral tradition about truths given to seekers by the gods. Some historians speculate that the oral tradition that preceded the Upanishads may have originated in Egypt some 2,000 years earlier, during the time of Hermes. As explained in Essay 18: *The Razor's Edge* (Page 309), it seems reasonable to speculate that the metaphysical concepts were spread by ancient traders across the Middle East and into the Indus Valley in what is Pakistan today, where the Upanishads are thought to have been written.

Some historians argue that Greek philosophers in the first millennium BC attributed their philosophical work to a famous person to increase the apparent importance of the work. If this is true, then it is safe to say that nearly all of the text attributed to Hermes was actually written by more contemporary philosophers.

A second confusing factor is that scholars of subsequent cultures have translated work attributed to Hermes from the perspective of their current beliefs. It is this babble of Hermetic attribution that one finds on the Internet today.

Thoth was considered the patron of scribes and the god of magic, healing and wisdom. It seems reasonable to argue that a man named Hermes, as a high priest of very early Egypt, was exalted as the spokesman for Thoth. In a practical view, there were probably many such priests over the centuries around the time associated with Hermes. For this reason, it may be most sensible to think of the *Hermetica* as the product of an early system of thought based on metaphysical concepts evolved from even earlier times.

Emerald Tablet

Because of what it includes, of all of the *Hermetica*, the *Emerald Tablet* is the one document I think may reasonably reflect the ancient teaching attributed to Hermes. It comes to us as it has been translated via a line of different cultures with different cultural references. Mindful of this, I have studied the text from the perspective of contemporary metaphysics. My thought is that, if I can make sense of the *Emerald Tablet* from a contemporary point of view, perhaps the ancient text will inherit a degree of new credibility.

There are as many different versions of the Emerald Tablet's origin as there are people writing on the subject. The most sensible consensus seems to be that the original version has been lost and the source for currently available translations is a Sumerian clay tablet containing Cuneiform script.

Some people claim the work was magically impressed onto a piece of emerald stone, but it is probable the name comes from the color of the original clay tablet. Early Egyptians used a form of glaze with which they covered the clay. It is known as blue-green Egyptian faience or Egyptian paste which is a sintered-quartz ceramic. As I understand the history, this technique was reserved for the most important artifacts.

Considering the content of the tablet, it was probably titled something like *The Truly Great Work*.

There are a number of different translations of the *Emerald Tablet*. I prefer the version translated from the Latin of Ficinus by Kircher, and into English by Dr. John Everard. It was first published in 1650. Even the translations attributed to Everard have slight differences. The one I use here is the first I encountered in 1990 during my early studies.

The translation is provided here in Italics. Each part is followed by how I understand the Tablet based on contemporary metaphysics.

The Emerald Tablet
by Hermes Mercurius Trismegistus

This is a lesson taught by a master to his initiates. It is concerned with the process of gaining progression (spiritual maturity).

1. *It is true and no lie, certain and to be depended upon, that which is above is as that which is below; and that which is below is as that which is above, for the performance of the one truly great work.*

This principle signifies that everything existing in the physical aspect of reality has its correspondence in the greater reality. Perhaps a clearer explanation of this principle is that everything in the physical (below) has been expressed from the etheric (above). In turn, that which has been expressed has an influence on the expresser (and other personalities).

A number of important lessons can come from this. There is direct correlation between effect (below or physical) and expression of intention (above or etheric). Be mindful of your intention (below to above) as it is can produce effects (above to below) which may or may not be what you envisioned.

The second lesson from this is what many people refer to as the Principle of Continuity. Reality exists as an unbroken thread of perception and expression from the intelligent core of Source to our conscious self and personal reality. As such, it is reasonable to extrapolate the nature of the greater reality from our local sense of reality (personal reality).

For quantum principles enthusiasts, the continuity of reality from our immortal self to Source is thought to be by way of a hierarchy of personalities. I refer to that as a nested hierarchy because there is a many-to-one relationship between a personality and aspects it has created in an effort to gain its intended understanding. This is also a *quantum-like* arrangement because it is stepwise, rather than one continuous flow of creation. However, it is also a fractal-like arrangement in that Source is the top fractal and each aspect of Source (you and me) is a sub-fractal of the Source fractal.

2. *And as all things are from only One Thing, by will of the one God, so all things have their origin in this one power, by adaptation to their individual purposes.*

Reality is the expression of Source (God) and is governed by organizing principles which emanate from the intention of Source. The combination of the expression of reality (the etheric) and the reason for that expression as organizing principles represent *The One Thing*.

We are being told that Source has created all things from Itself. The *One Power* is the Creative Process by which reality is adapted by individual aspects of Source to satisfy imagined purpose. The Creative Process is attention on an imagined outcome to produce an intended order. This process is limited by the ordering principles emanating from Source.

As such, *the One Thing* is reality and organizing principles as the expression of Source, and the *One Power* is the Creative Process by which the One Thing is adapted.

   3. *That only One Thing has the sun for its father, the moon for its mother.*

Sun is creative influence. Moon is receptive Self (immortal personality). The three aspects of creation are within each of us as attention (the Sun) on a visualized outcome (the Moon) with the intention (wind—Line 4) to make it so.

   4. *The wind carries it in its wings.*

The third part of the Creative Process (adaptation) is intention. It seems reasonable that the wind is the intention to make it so.

   5. *But its nurse is a spiritual Earth.*

Here, I think Earth is a reference to self (immortal self entangled with a human body in an avatar relationship; aka a person). We are the ones who are performing the Great Work, as we live in the physical (earth).

Thus far, Hermes has told us that we are empowered to gain progression by the very structure of reality, and through the organizing principles given to us by Source. The etheric is differentiated into new form through a person's influence based on the Creative Process.

   6. *That only One Thing is the true father of all things in the universe. Its power is integrating or perfecting after it has been united to a spiritualized Earth.*

Again, the only One Thing is Source's expression of organizing principles governing the behavior of reality, which undifferentiated, is Source's life field; the reality field. Reality can be modeled as life fields and the expression of life fields. Thus, the organizing principles are the Father of all things. This emanation from Source is perfected or changed as it is differentiated by Self.

We, as Self in the physical represents the Earth in this lesson. It is through increasing understanding that we perfect reality.

7. **Thou shalt separate the Earth from the fire, the subtle from the gross, by means of a gentle heat, and with great ingenuity.**

In most esoteric schools of thought today, *fire* is the intuitive aspect and earth is the empirical, objective aspect. To be consistent, *subtle* would be unconscious perception of reality as it is differentiated through the Creative Process. *Gross* would be reality as it is perceived as the product of the perceptual processes. *Heat* would be focused or directed intention and *ingenuity* is an excellent description of the kind of work required to learn how to think beyond cultural influences to follow the Mindful Way.

If this understanding is correct, Hermes is explaining the need to become aware of the difference between what is consciously perceived and the existence of a conceptual reality which underlies experience. This can be described as the difference between the etheric personality (immortal personality as *I am this*) and the person (conscious self as *I think I am this*).

8. **It ascends from Earth to heaven, and descends again to Earth. Thereby it receives the power of the superiors and the inferiors.**

"It" is based on the One Thing as reality and the one power as the Creative Process. But here, *"it"* is the product of the Creative Process which is correct understanding of reality. That is the ultimate objective of the Great Work.

Some people believe that this is a direct reference to the Kundalini and the seven Chakras. (The sixth and seventh Chakras are sometimes referred to as the Superior Chakras.) However, that is clearly a local system of thought.

The concept of earth as the physical person is consistently used in the *Emerald Tablet*. Also, the issue at hand is the Great Work. The Great Work is a way turned toward understanding (progression or spiritual maturity). To follow this path, we are told to change our awareness from body-centric to an immortal self-centric perspective.

Understanding is relative so that something understood tends to shine new light on that something, thus offering potential new understanding. Since a lifetime is a transient experience, acquisition of understanding appears to be for the benefit of the greater community or collective of personalities. Thus, we have access to understanding from the collective even as we contribute new understanding to the collective.

9. ***By this process thou wilt partake of life, love, and light, and the honors of the whole world; therefore, let all obscurity flee before thee.***

Hermes has identified our purpose in life and has described the process with which we can pursue that purpose. Now he is telling us that by living the life while consciously seeking understanding, we will align ourselves with the true nature of reality (Organizing Principles or Natural Law).

Near the end of this lesson (Line 13), Hermes identifies himself as an example of what living in accordance with the true nature of reality means. That is, teacher, expression of the principles in daily living and the potential effect of living in that way manifest as a (spiritually) successful person. In effect then, he is telling us that we too can be happy and respected citizens.

10. ***This is the strongest of all forces, overcoming every subtle and penetrating every solid thing.***

Hermes continues to refer to the etheric and principles from Source that permeates all of reality from the finest (undifferentiated etheric) to every aspect of the physical.

11. ***With this thou wilt be able to master all things and transmute all that is fine and all that is coarse.***

The Great Work described in the Cabala is the process of changing the young, immature Self into a Master of the principles governing the operation of reality. In the terminology of the Cabala, achieving God-Realization is described as a transmutation. While the process of transmuting the base metal of lead into a higher quality gold is a subject of earnest research in alchemy, it is often used as an analogy for the process of transmuting the ignorant seeker into a spiritually mature master. The process is achieved through adaptation of the organizing principles into all of the objects of reality.

12. ***So the world was created. Hence were all the wonderful adaptations of the One Thing manifested; but the arrangements that follow this great mystic path are hidden.***

Again, the One Thing is differentiation of reality by way of the Creative Process. The Creative Process is a person's intention acting on an imagined outcome to make it so. Differentiation is bound by organizing principles.

Hermes was explaining that all of reality was formed by way of the same principles he explained in the previous lines. An important concept here is

"For those who have eyes to see." The way described in this lesson is hidden to those who have not followed this path.

13. ***For this reason, I am called Hermes Trismegistus--one in essence but three in total aspect. In this Trinity are concealed the three parts of the wisdom of the whole world.***

See The Three Aspects of a Teacher, (Page 297).

14. **What I have to tell is now completed concerning the operation of the Sun.**

And so, Hermes has told his students the secret of creation, their purpose in this lifetime, and by doing so, has pointed them toward the Mindful Way of life.

Compare the advice of the *Emerald Tablet* with the Creative Process Discourse in Section I of *Your Immortal Self*. (1) An early version is at ethericstudies.org/creative-process/.

Paraphrasing the Emerald Tablet

Visualize Hermes giving his students a routine lesson.

Lesson Name: The Truly Great Work

a. I can tell you as your teacher that your thoughts and your deeds are directly related so that your thoughts affect your expression, and your perception of that expression affects your thoughts. (Line 1)

b. Reality is both singular as Source and complex as the expression of Source according to Its intention. This expression of intention represents ordering principles which govern the adaptation of reality to individual purpose. The world you live in is an aspect of the greater reality as it is expressed by way of the Creative Process. (Line 2)

c. The Creative Process requires the visualization of the imagined purpose with the intention to make it so. (Line 3 and 4)

d. You, the person as an etheric personality entangled with a physical body, are the creator in this lesson. (Line 5)

e. Thus, the creative influence produces all things in reality. The Creative Process finds expression through the informed intention of the person. (Line 6)

f. It is necessary to learn to distinguish between that which is part of actual reality as expression moderated by organizing principles and that which is perceived as real, but which is actually illusion. (Line 7)

g. Increased understanding of the actual nature of reality is contributed by the student to the collective of personalities in the greater reality, and thus merged, becomes available to the student as more profound understanding. (Line 8)

h. As such, you will find that understanding leads to clear sensing which enables a person to experience reality as it is, rather than as you have been taught. (Line 10 and 11)

i. Your increased understanding achieved through the Great Work may lead you to better living and increased contentment. (Line 9)

j. Thus, I have told you how the world has been created. But be mindful that these truths are not evident to those who have not stepped onto this way of learning. (Line 12)

k. As the teacher of this hidden way, I represent the three parts of a teacher. That is, I represent the understanding of the One Thing and The Great Work, I am an example of how you may integrate this understanding into daily life, and in me, you can see the possibilities of living this path. I am three times accomplished: as a teacher, role model and citizen. (Line 13)

l. And now you understand the Truly Great Work. (Line 14)

The Foundation Concepts Associated with Hermes

Your Immortal Self (1) includes a model of reality in which a number of organizing principles are used to define the fundamental nature of reality. In some systems of thought, these principles would be described as Natural Law; however, the ones listed in the book are rather different from what you may have learned. An older version of the Organizing Principles can be found on the Etheric Studies website. (27)

An important evolution in how we can study the early introduction of metaphysical concepts is the ability to say that, based on contemporary research, certain concepts have become reasonably well established. This is not to say that we are now 100% correct, only that we have entered into an age in which the concepts can be objectively and repeatedly examined.

> Perhaps the one reason I have persisted in writing about these concepts as long as I have is my sense of revelation contemporary transcommunication offers. Sometimes, I think of this time in our evolution as a people as the dawning of an age of understanding through mindfulness.

With this consideration, it is not too much of a stretch to argue that an ancient society might believe in one god but describe that one god in terms of how people relate to its characteristics. This is done today with Natural Law as people attempt to describe reality by way of its characteristics.

It is also noteworthy that articles about prehistoric Egypt are derived from Greek translations. I am not a language expert, but I know it can be shown that much of the metaphysical importance of the Aramaic language used by Jesus was lost by way of the Greek translations. (35) It is probable that similar loss in metaphysical meaning has occurred in the translation from Egyptian to Greek.

A useful technique for understanding other people's point of view is to normalize their concepts in more fundamental terms, and then to compare those with more familiar models. Doing that with what I can discover about the time of Hermes, and before the gods of ancient Egypt, it seems clear the Egyptians accepted the idea of one god with many aspects.

Support for the idea that the Egyptians had one god with many aspects is the early Egyptian name for god. The writers of the *Internet Sacred Text Archive* tell us:

> To the great and supreme power which made the earth, the heavens, the sea, the sky, men and women, animals, birds, and creeping things, all that is and all that shall be, the Egyptians gave the name Neter. (158)

According to Edward Malkowski: (159)

> From a modern Western perspective, their [the Egyptian's] religion has been billed as primitive and polytheistic, and appears as a mythological menagerie of gods. Nothing could be further from the truth. The source of this misunderstanding stems from the Egyptian word neter being translated into Greek as 'god,' which later took on the Westernised meaning of deity. The true meaning of neter was to describe an aspect of deity, not a deity to be worshipped. In essence, neters referred to principles of nature in a practical scientific way.

The Three Aspects of a Teacher

The *Emerald Tablet* gives us a good test of this perspective. One of the keys to who Hermes was is the assigned name, Hermes Trismegistus (Greek for Hermes the thrice-greatest). This title is found in the *Emerald Tablet* in which Hermes tells his students:

> 13. *For this reason, I am called Hermes Trismegistus—one in essence but three in total aspect. In this Trinity are concealed the three parts of the wisdom of the whole world.*

It is likely that there are multiple meanings in this line. The trinity is important throughout the Hermetic Wisdom, such as positive, negative, neutral, and body, mind, spirit. The concept of balance permeates the teaching. However, there is a more deeply hidden aspect of the trinity concept. Understanding of, and ability to properly manage the trinity of imagination, intention and attention is the foundation of *"the wisdom of the whole world."* This is the Creative Process.

The Greeks translated three-time great as Trismegistus. When considered with Line 12, the phrase, *"For this reason,"* makes it clear that the *"three parts of the wisdom of the world"* is a direct reference to the three aspects of all teachers. That is, a teacher represents the lesson to be taught (imagination), appears to the student as an example of what it is to understand the lesson (intention) and demonstrates the value of the lesson through application of the lesson in life (attention).

> 12. *So the world was created. Hence were all the wonderful adaptations of the One Thing manifested; but the arrangements that follow this great mystic path are hidden.*

These three aspects are also demonstrated by the way Jesus presented himself in the *Bible*: John 14.6: *Jesus saith unto him, I am the way, the truth, and the life: no man cometh unto the Father, but by me.* In this line, Jesus is showing himself to his followers as the three aspects of the teacher: follow me that I am the path; follow me as the Spirit of Truth; and, follow me as I have lived. See *Metaphysical View of John 14* essay at EthericStudies.org/metaphysical-view-john-14/. (160)

The *Hermetica* as Revealed Information?

It is a fair bet that scholars specializing in ancient civilizations are not likely to be students of metaphysics, and even less likely to be familiar with the concepts as they are informed by understanding gained via transcommunication.

An Internet search for information about Hermes, pre-history religions and migration of religious thought will produce a wide variety of scholarly and special interest commentary but little agreement. A good example of this disagreement is the origin of why Hermes is called Hermes Trismegistus. The first hint is that all of the English translations of the *Emerald Tablet* I have read use Trismegistus, which is a name given him by the Greeks much after his lifetime. I would expect the original text was more like *"... I am called Hermes Trice Great—one in essence but three in total aspect."* Did the Greeks understand the cultural significance of whatever Egyptian term represented Hermes in his time? According to Edward Malkowski (above), they probably did not in the same way they missed the meaning of neter.

Important Metaphysical Concepts Attributed to Hermes

Hermes is associated with astrology, alchemy and magic by way of the Hermetic philosophy which has come to be attributed to him. As noted above, part of the problem is that there is so much cultural contamination that attribution of any specific concept is intellectually risky. It is more probable that astrology, alchemy and magic are much later inventions based on then-current understanding of concepts attributed to Hermes.

The major concepts thought to have been introduced by Hermes include:

- **One God:** There is one God, which is comparable to Source or Infinite Intelligence.

- **Avatar Relationship:** The concept of man as a chariot with spirit as its driver comes to us from the Upanishads. (161) It is consistent with Line 7 and 8 of the *Emerald Tablet* in which a person, understanding and purpose are treated in terms of the two aspects of the gross and the subtle. See Line 1-III-3. in "The Story" (Page 312) in Essay 18: *The Razor's Edge* (Page 309).

- **Progression:** The purpose of gaining understanding about the self and reality. This is described in the *Emerald Tablet* as the Great Work.

- **Creative Process:** The *Emerald Tablet* is all about the major elements of the Creative Process.

- **Three Aspects of a Teacher:** This is actually a commandment of sorts. I have always maintained that *"Our lot is to learn, and having learned, our lot is to teach."* This would appear to be a consequence of progression as described in the status of Hermes as being thrice-great. As noted below, it is also mirrored in the Tarot. See Essay 3: *Prime Imperative* (Page 49).

- **Organizing Principles:** Hermes opens the *Emerald Tablet* with *"...that which is above is as that which is below; and that which is below is as that which is above...,"* which has become known as the Hermetic Law of Correspondence. The *One Thing* of Line 2 is the expression of Source. It is described in Line 6 as *"That only One Thing is the true father of all things in the universe."* Although not specifically stated, organizing principles are implicit in the concept of *"...by adaptation to their individual purposes"* as stated in the last part of Line 2.

It is important to note here that Natural Law, as it is taught today, is a much more contemporary invention. Cabala (often spelled *Kabbalah*, which is Hebrew meaning to receive or to accept) emerged out of Jewish mysticism around the 12th-century. It is a technical form of metaphysics that involves many interrelated, often secret concepts.

There are many organizing principles (Natural Law to some) described in the *Divine Pymander* which is traditionally attributed to the writing of Hermes Mercurius Trismegistus. (25) Considering this, it is interesting that virtually all of the Internet websites discussing Hermes and the principles only address the seven proposed in *The Kybalion*.

The Seven Organizing Principles of the Kybalion

Seven principles said to define the nature of reality have become the standard version of Natural Law. They probably became popular from the 1908 booklet titled *The Kybalion* by Three Initiates. (26) It should be noted that these are described from a physical or body-centric perspective. There are certainly more commonly cited principles attributed to ancient teaching from other sources, but it is only these seven I have found directly attributed to Hermetic teaching.

In keeping with my effort to normalize the *Hermetica* with more contemporary thought, I have added a suggested, more current understanding for each.

1. **The Principle of Mentalism.**

    The all is mind; The universe is mental.

This can be taken literally. In terms of the way we have been taught to distinguish mental from physical, the expression of reality from Source (God) is a thought. Emergent understanding from mainstream science indicates that we create our world as a thought exercise. This is supported in psi field studies. Reality is mental. We assign physical meaning to aspects of it, such as our body.

> Be careful not to be distracted by the cultural references for the thought concept. From the physical perspective, we as our body are solid and real while our thoughts are intangible and not real. But our physical perspective is only for this lifetime and restricted to our local culture. In fact, we are immortal, and our real perspective is that of your immortal self. From that more correct perspective, our body is a thought.

2. **The Principle of Correspondence.**

   As above, so below; as below, so above.

This principle signifies that everything existing in the physical aspect of reality has its correspondence in the greater reality. Perhaps a clearer explanation of this principle is that everything in the physical (below) has been expressed from the etheric (above). In turn, that which has been expressed has an influence on the expresser (and other personalities).

It is reasonable to think of reality as a collective thought, of which, our immortal aspect is one of the thinkers.

A number of important lessons can come from this. There is direct correlation between effect (below or physical) and expression of intention (above or etheric). We should be mindful of our intention (below to above) as it is can produce effects (above to below) which may or may not be what we envisioned.

A second lesson from this is what many people refer to as the Principle of Continuity. Reality exists as an unbroken thread of perception and expression from the intelligent core of Source to our conscious self and personal reality. As such, it is reasonable to extrapolate the nature of the greater reality from our local sense of reality (personal reality).

It is also reasonable to think this unbroken thread of perception and expression represents successive changes in understanding. In terms of understanding as the foundation of perception, reality is quantized. The continuity of reality is step-wise as one moves from relatively little perception of the actual nature of reality to more correct perception. But understanding is seen as relative, so that perception produces the potential for new understanding.

3. **The Principle of Vibration.**

    Nothing rests; everything moves; everything vibrates.

This principle is often understood as the closer one gets to God, the finer the vibration one must experience. However, our contemporary understanding is that vibration is a physical concept that has no evident, direct equivalent in the etheric. In fact, in the *Emerald Tablet,* Hermes uses *"subtle from the gross"* to indicate a difference between etheric and physical.

A more useful measure is relative understanding. In the Implicit Cosmology, (12) progression (perceptually toward Source) is achieved by gaining understanding. Understanding goes toward aligning Worldview with the actual nature of reality. It is Worldview that determines perception. As such, perceptual agreement becomes a determining factor for what a person can experience in reality. Stated as the Organizing Principle of **Perceptual Agreement**: **Personality must be in perceptual agreement with the aspect of reality with which it will associat**e.

4. **The Principle of Polarity.**

    Everything is dual; everything has poles; everything has its pair of opposites; like and unlike are the same; opposites are identical in nature, but different in degree; extremes meet; all truths are but half-truths; all paradoxes may be reconciled.

The Principle of Polarity is good advice and appears to be true in a practical sense as a physical principle, but it tends to lose meaning in the etheric. A life field is a singularity and its expression are not polar. Balance does exist as a beneficial behavior, but polar extremes and balance do not appear to be factors in the perception and expression functional areas of mind.

5. **The Principle of Rhythm.**

    Everything flows, out and in; everything has its tides; all things rise and fall; the pendulum-swing manifests in everything; the measure of the swing to the right is the measure of the swing to the left; rhythm compensates.

As with Principle 4, this is good advice for operating in the physical when rhythm is understood as regular, periodic cycles. The annual cycle of seasons, for instance, is a regular periodic cycle; however, the seasons are an artifact of what may not be a universally usual alignment of our planet's axis with its orbit around the sun.

There is no apparent support for the concept in terms of the etheric. The principle does appear to have meaning in the sense of a process. An example is the process of birth, youth, maturity, and death as we see in the cycle of the seasons.

The distinction is important. We see the process of gaining understanding typically experienced as a cycle, but not a regular periodic cycle. The initiation experience in ancient wisdom schools, for instance. The rhythm in question is a cycle of gaining information, living with that information as it becomes personal knowledge, and then a change in state of that knowledge to understanding. That cycle of learning is sometimes referred to as the *Dark Night of Soul* (162) when the cycle includes a period of mental anguish as the seeker integrates new understanding into Worldview.

The governing influence for the behavior of processes in the etheric appears to be the *Perceptual Agreement* Organizing Principle.

Some people relate the Principle of Rhythm to the Principle of Cycles, and that to the Principle of Reciprocity. My comments apply to all three perspectives, but also see my comments below for the Principle of Cause and Effect.

6. **The Principle of Cause and Effect.**

   Every Cause has its effect; every effect has its cause; everything happens according to Law; chance is but a name for Law not recognized; there are many planes of causation, but nothing escapes the Law.

This is a complex principle. It is a well-established physical principle, but its etheric counterpart is a little less definitive. As noted above, the concept of reciprocity is better related to cause and effect than to cycles or rhythm. The associated concept of reciprocity is mutual influence. While we think of force in the physical, we think in terms of influence in the etheric.

There does not appear to be a direct one-to-one exchange of influence, but rather the expression of influence which changes perception. That is, my thinking of you causes my perception of you to change in some way. There is a cause and effect, but it is not equal and opposite as we think of it in the physical.

7. **The Principle of Gender.**

Gender is in everything; everything has its masculine and feminine principles; Gender manifests on all planes.

Current versions of the *Hermetica* is full of reference to male and female aspects. But it does not fit well with my current understanding unless I rephrase the concept. So, instead of male and female, think in terms of two aspects of the Creative Process. (19) This might be better understood as the process of formation in which potential (female) is expressed through intention (male). This is consistent with Line 3 of the *Emerald Tablet*.

My Introduction to the Hermetic Concepts

Of course, I intended to be the first man on the moon, but it was in my teens that science clubs began to give way to curiosity about the nature of reality. Many of the science fiction books I had been reading had a strong metaphysical, magical sense that was complemented by such movies as *The Wizard of Oz* and my speculation about what it was really like *over the rainbow*.

The Rosicrucian's *"Thoughts have Wings"* (163) advertisements in the science magazines I read in my teens was sufficiently enticing for me to join and begin receiving their weekly discourses.

The Builders of the Adytum (BOTA) (6) gave me much-needed diversion from college study with a two-year course in the Tarot. They use the Tarot as a tool for teaching courses in the *Ageless Wisdom of Sacred Tarot* and *Cabala*. (164) Their deck is a version of the 1910 Rider-Waite deck modified by Paul Foster Case.

The term Cabala is derived from the Hebrew root *to receive, to accept* and represents a system of thought based on the earliest teaching attributed to Hermes. As previously noted, the teachings have been considerably altered in later cultures up to its more or less formal establishment in the Middle Ages, beginning probably with the seventh century.

The Tarot

One final point of reference to help you understand the importance of Hermes' contribution to (my) contemporary thought is the Tarot. While it is based on the *Hermetica*, it is so by way of centuries of reinterpretation and must be considered with discernment. Its value though, is that it embraces the spirit of the Great Work and much more that has been learned about human nature since the time of Hermes.

The Tarot is thought to be based on playing cards which have been adapted for use as a means of fortune telling. Occult versions of the cards began to show up in the 1300s. The Case deck I use consists of 78 cards with 22 Major Arcana and 56 Minor Arcana cards. Arcana from the Latin arcānum means secret, or as it is used in the Tarot, specialized knowledge which is unknown to or misunderstood by the average person.

While the Tarot is most commonly used for divination, in their occult use, the 22 Major Arcana, sometimes referred to as 22 Keys (keys to secret wisdom), represent the path to self-realization. Each card represents a step along the path beginning with *Key 0, The Fool*, which represents the person both at the beginning of the cycle of education as, ... well, as a fool, and at the end of the cycle following key 21 as the now enlightened, ... well, still a fool because the cycle is never-ending.

Important to my point about the evolution of the Hermetic concept is that the keys can be arranged in three rows of seven with Key 0 set above. The first row represents powers or potencies, those in the middle row represent laws or agencies and those in the bottom row represent conditions or effects. For instance, *Key 1, The Magician*, (self-conscious phase of mental activity, intention) represents the potential which works through the agency symbolized by *Key 8, Strength*, (authority over primal nature) to modify the conditions or effects typified by *Key 15, The Devil*, (erroneous belief in limitations). (165)

Tarot
Tarot Tableau representing the cards of the Major Arcana. This is the Paul Foster Case-Ann Davis deck. Images are from Builders of the Adytum bota.org.

Each row represents a progression from relatively little self-awareness toward greater understanding. Every element has been designed to have significance in that progression, even down to the colors. The male and female figures typically represent an aspect of a person (perception and expression, not sex) and water always represents the essence of mind.

As an example of the secret meanings encoded into the keys, the Fool's bag contains the same tools representing self-conscious phases of mental activity the Magician is working with on the table of Key 1. As the beginning of the cycle, the Fool is unaware of the tools, but then must have them out to see and feel in Key 1. However, after Key 21, the Fool represents the completed cycle and no longer needs to have the tools out to be able to work with them. He is master of them no matter where they are.

Note also the comparison between the three rows in the tableau and the three aspects of a teacher discussed earlier in this essay. The first row, *powers or potencies*, can be compared to fundamental organizing principles as the first aspect of the teacher. The second row, *laws or agencies*, can be compared to the application of the principles taught by the teacher as the second aspect. The third row, *conditions or effects of the principles as they are applied*, can be compared to the result of living in accordance with the principles of nature as the third aspect of the teacher.

> ***Know me as I express the principle,***
> ***as I live the principles***
> ***and as I benefited from living the principles.***

Resetting the Old Concepts

My first attempt to develop a metaphysical cosmology was the *Handbook of Metaphysics*. (2) It provides a good overview which can be useful to the reader as context for understanding new metaphysical ideas. I tried very hard to normalize the many different perspectives as a single view but found much disagreement amongst the various schools of thought. Even more problematic was the fact that so many systems of thought seemed to have originated from just a few original sources, themselves primitive from a contemporary perspective.

As I have discussed above, the problem of cultural contamination makes it necessary for us to find a new, contemporary anchor on which to reset the old teachings. I have a great deal of respect for the *Hermetica* and the Tarot, but it is now time to refresh our perspective. The new book, *Your Immortal Self,* (1) is what I believe to be a useful reset of the old concepts. It is also a leap forward from what we thought was true when I wrote the *Handbook*.

The take away I hope you will gain from this essay is that the ideas of our immortality and the need for our pursuit of understanding are fundamental lessons that have been given to us by our friends on the other side since the earliest days of civilization. These lessons have been refreshed many times by important wayshowers of all of the world's religions. There is a reason for this which we will all do well to head

Handbook of Metaphysics
A Plain English Discussion of
New Age Concepts

By
Tom Butler

References and Alternative Sources
Listed at the end of the book beginning on Page 357.

Intentional Blank Page

Video-Loop ITC Image collected by Tom and Lisa Butler.
    This example was used in an online visual perceptions study (atransc.org/visual-perception-study/). It is a head facing to your left. It is a profile from the neck up. The person may be a boy or boyish girl with short hair and appears to be wearing a dark shirt with a white collar like a sweater over a "T" shirt. The hair and face seem to be illuminated from your left and top. It had 81% correct recognition by online witnesses.

Essay 18

The Razor's Edge

2016

About This Essay

Lisa and I became involved in Eckankar (7) while we lived in Sacramento and before Lisa read Sarah Estep's book about Electronic Voice Phenomena, *Voices of Eternity*. (166) For me, it was the technical approach to metaphysics taught by Eckankar founder, Paul Twitchell that I found useful.

We moved away from Eckankar when we relocated to Kansas City, and our three years with them did not make me an authority on their beliefs. The reason I even mention the group is that Twitchell had a background in Eastern religions and brought many of their concepts into Eckankar. The foundation concept is that, as soul, we are co-workers with God, and that it is for us to understand Its nature as the *"Light and Sound of God."*

There are magical ideas in Eckankar. At least back then, I argued that you have not lived until you have sung the Hu with over 5,000 other Eckists at one of their conferences. Suddenly coming upon lucid awareness of an inner teacher during an experience referred to by Hindus and Eckists as the Darshan is reason enough for years of study and meditation.

> I have had experiences that seem at least similar to the Darshan. While they were a little unnerving at the time, I will say the effect has remained with me. There was no lightning strike of understanding. What came was the spontaneous shock of a rumbling sound and a subtle vibration throughout my body as if I was about to have an out-of-body experience. I was in a disassociated state each time.
>
> Rather than the desired enlightenment, from the experiences came a subtle certainty that I was on the right path, and that I really was a luminous being. I suppose it is that sort of realization I needed to enlighten my most stubborn perceptual processes.
>
> In a vivid dream during my Eckankar days, I remember crouching down near a little boy who was with his mother in a grocery store. He was standing with his hand on a shopping cart ... for security from an encroaching stranger, I suppose.
>
> I sang the Hu to him, and as I did, I was able to see the sound waves radiate from my mouth. As they did, they turned to shimmering light. It was from that dream that I realized we are all

wayshowers, like it or not, for the betterment or the detriment of our fellow seekers. The sound of our expression becomes the light of our fellow's understanding. This, in a similar fashion as the "Great Work" (Page 60) as described in Essay 3: *Prime Imperative* (Page 49).

I said the experiences spontaneously came to me, but it is important to note that, as with any system of study in which we become deeply involved, that very involvement has the potential over time to condition the mind to produce a desired response. The experiences were spontaneous in that I did not consciously trigger them, but they were predictable in that I set out to acquire that relevant understanding. You can do the same.

Another magical concept was the *Dark Night of Soul*. In that, it is thought to be common for a person to undergo a period of personal crisis just before passing through an initiation into greater understanding. None of these ideas are unique to Eckankar, but they fit in well with their lessons.

The Razor's Edge concept represents a way of thinking about what I refer to as the Mindful Way. There was even a movie made using it as the theme. An important point to remember about the Razor's Edge concept is that, like balancing a pencil by its pointy end, it is easier to fall off of that narrow way than to remain. Those who remain are special, indeed!

Introduction

The Razor's Edge refers to one of the ancient wisdoms concerned with the development of spiritual maturity. You may remember that there is a book titled *The Razor's Edge*, and later, at least two movies based on the book. (167) The story is about a person's journey to India to seek the transcendent meaning of life.

The phrase is found in the ancient Hindu Vedic Sanskrit text called the *Katha Upanishad*. The over 200 Upanishads describe the nature of ultimate reality and the path to gain spiritual maturity. They are thought to have been written 3,000-to-4,000 years ago and are based on oral tradition about truths given to seekers by the gods. There is also some evidence that Buddhism may have adopted concepts credited to the same origin.

Origin of the Upanishads
Upanishad is from Sanskrit, probably originally as *Upaniṣad*. *Katha* can be understood as story or legend and Upanishad refers to sitting down near something, presumably a teacher. The Upanishad originated from oral tradition that likely predates the usual 1,800-2,000 BCE stated for the origin of the written form.

Some historians speculate that the oral tradition may have originated in Egypt some 6,000 years ago, during the time of Hermes. It is thought that the metaphysical concepts were spread by ancient traders across the Middle East and into the Indus Valley in what is Pakistan today.

Existence of essentially the same important concepts in many different cultures provides hints about how they may have evolved. If a particular concept such as mind-body duality was taught in ancient Egypt, then it is not particularly unexpected to see it also taught in Hinduism and Christianity.

Katha Upanishad
The *Katha Upanishad* is told as a story about a boy, his father and the god of death. It is concerned with the nature of God as Source, individual seeking of greater understanding and the relationship between Self and the human body.

As with any text written in ancient times, *Katha Upanishad* has been translated by a number of modern scholars. This has resulted in many different versions. Because of this, it is necessary to focus on the underlying meaning, rather than taking them in a literal sense.

Invocation
It is common, when seeking personal understanding with the help of our unseen friends, to speak a prayer designed to align teacher, student and helper's intention.

According to Swami Krishnananda (74), every Upanishad lesson between a teacher and student begins and ends with an invocation of peace called the Shanti Mantra. As Swami Krishnananda's invocation is translated:

May we both be protected.
May both of us be taken care of properly.
May we study together.
May our teaching and learning be resplendent.
May there be no misunderstanding between us.

May there be no discord of any kind.
May there be peace, may there be peace, may there be peace.

According to Swami Krishnananda, "It means that there should be proper attunement of spirit between the Guru and disciple before they begin the study, for only then will the teaching be fruitful."

He also explains the three repetitions of "may there be peace," as "We have three kinds of troubles called tapatraya (internally, physical ones), externally from outside beings and from above given by the gods."

Middle East from Egypt to Indus Valley in Pakistan. (Google Maps)

The Story

As I understand the translations, *Katha Upanishad* begins with Gautama's sacrifice of all his worldly possessions to the gods, expecting good favor in return. Nachiketa notices that his father's sacrifice is insincere because the possessions were only those which were worn out. He asked his father, *"I too am yours, to which god will you offer me?"*

Gautama's terse response was, *"I give you to Yama* (the god of death).*"*

Nachiketa went to Yama's home but Yama was away and Nachiketa had to wait. Yama returned after three days and expressed regret that Nachiketa had to wait so long. He told Nachiketa, *"You have waited three days so ask three favors of me."*

Nachiketa asked for peace for his father and to learn the Sacred Fire Sacrifice. Yama agreed.

For his third request, Nachiketa asked to learn the mystery of what comes after death. Yama pleaded with him to ask something else. Nachiketa insisted, so Yama tested the boy by offering him wealth instead of the secret. The boy chose the path of spiritual understanding over the path of material possessions. This pleased Yama, and so he agreed to tell the boy about what came after death.

Yama explained that the key to understanding what comes after death is to understand that self, which is within each person, is inseparable from the Supreme Spirit, the vital force in the universe. In this regard, Part 1-II of *Katha Upanishad* is concerned with the need for a person to realize the importance of consciously deciding to seek understanding.

> Brahman is a Sanskrit word for Supreme Spirit or creative principle present in everything.

The part of the story concerned with The Razor's Edge is here and a brief discussion about the lesson is provided after. The material quoted here is from the Vedanta Spiritual Library: (22)

1-III-3. *Know the Self to be the master of the chariot, and the body to be the chariot. Know the intellect to be the charioteer, and the mind to be the reins.*

1-III-4. *The senses they speak of as the horses; the objects within their view, the way. When the Self is yoked with the mind and the senses, the wise call It the enjoyer.*

1-III-5. *But whoso is devoid of discrimination and is possessed of a mind ever uncollected - his senses are uncontrollable like the vicious horses of a driver.*

1-III-6. *But whoso is discriminative and possessed of a mind ever collected - his senses are controllable like the good horses of a driver.*

1-III-7. *But whoso is devoid of a discriminating intellect, possessed of an unrestrained mind [unmindful] and is ever impure, does not attain that goal, but goes to samsara.*

> Samsara is Hindu for the cycle of death and rebirth as life is bound to the material world.

1-III-8. *But whoso is possessed of a discriminating intellect and a restrained mind [mindful], and is ever pure, attains that goal from which he is not born again.*

1-III-9. *But the man who has a discriminating intellect as his driver, and a controlled-mind as the reins, reaches the end of the path - that supreme state of Vishnu.*

> Supreme state of Vishnu refers to self-realization or self-knowledge.

1-III-10. *The sensory objects are subtler than the senses, and subtler than the sensory objects is mind. But intellect is subtler than mind and subtler than intellect is Mahat (the Hiranyagarbha).*

> Mahat refers to the origin or Source of creation. *Hiranyagarbha* can be understood as the golden egg, where gold means objects of fulfilment and joy.

1-III-11. *The un-manifested (Avyakta) is subtler than Mahat and subtler than the un-manifested is Purusha. There is nothing subtler than Purusha. That is the end, that is the supreme goal.*

> The universe itself is a person, though without the limitations and prejudices of our human personality. This is what the science of Yoga calls the Purusha. The Purusha, meaning a person or conscious being, is a Sanskrit term for the Cosmic Being behind the universe, the spirit within all things. The entire universe is a manifestation of the Cosmic Person. This Cosmic Person endows every creature with personhood or a sense of self, not only humans but also animals and ultimately all of nature. (168)
>
> This agrees with the Implicit Cosmology.

1-III-12. *This Self hidden in all beings does not shine. But by seers of subtle and pointed intellect capable of perceiving subtle objects, It is seen.*

1-III-13. *Let the wise man merge speech in his mind, merge that (mind) into the intelligent self and the intelligent self into the Mahat. (Let him then) merge the Mahat into the peaceful Self.*

1-III-14. *Arise, awake, and learn by approaching the exalted ones, for that path is sharp as a razor's edge, impassable, and hard to go by, say the wise.*

1-III-15. *By knowing that which is soundless, touchless, formless, un-decaying, so also tasteless, eternal, odourless, beginningless, endless, subtler than Mahat and constant, man is liberated from the jaws of death.*

A Hindu Teacher's Translation

According to Gupta, (169) Yama's explanation is a succinct description of Hindu metaphysics, and focuses on the following points:

- *Yama said* [in **Line 1-II**] *there are the two life paths: pleasant/attractive and good/transcendental.*
    - *Pleasant/attractive, which is the path of material pleasures that tempts humans, leads to death.*

- o *Good/transcendental, which is the path of spiritual bliss, leads to immortality.*
- o *By a process of detached thinking, the clear minded choose the path of immortality and the muddle-headed fall for the path of pleasure and eventual pain and death.*
- *The sound Om! is the syllable of the Supreme Spirit*
- *The self is the same as the omnipresent Supreme Spirit. Smaller than the smallest and larger than the largest, the self is formless and all-pervading.*
- *The goal of the wise is to know this self.*
- *The self is like a rider; the horses are the senses, which self guides through the maze of desires.*
- *After death, it is the self that remains; the self is immortal.*
- *Mere reading of the scriptures or intellectual learning cannot lead to the realization of self.*
- *One must discriminate the self from the body, which is the seat of desire.*
- *Inability to realize Brahman results in one being enmeshed in the cycle of rebirths.*
- *Realization of the self leads to liberation from the cycle of life and death*

Universal Message

Preparing this essay has been an adventure for me. What I refer to as the Trans-Survival Hypothesis (10) and the resulting Implicit Cosmology (12) explain a model of reality which I believe represents current science and understanding gleaned from transcommunication. It is exciting to me that the *Katha Upanishad* very closely agrees with that cosmology.

While I had considered the Hermetic teaching and John 14 of the *Bible*, I had never looked at Hindu philosophy, mainly because of its convoluted terminology. After reading the *Katha Upanishad*, I took a close look at my prior understanding that religion was thought to have begun in the Hindus Valley as it evolved from Arian philosophy. Now I see that the philosophy contained in the Upanishads probably originated millennia earlier, in Egypt as the same source attributed to Hermes.

Here is a brief translation of the *Katha Upanishad* in terms of the Implicit Cosmology:

1-III-3 describes the avatar model used to describe the relationship between mind and body. In this model, mind is described in three parts: personality is the normally unconscious core intelligence which provides purpose; mind is the normally unconscious functional areas supporting perception and expression; and, Self is the conscious aspect of who we are.

In this model, the intellect is conscious self's expression of intention which can be thought of as one of two steering influences on normally unconscious mind. The second steering influence on mind is the urge or spiritual instinct to gain understanding expressed by personality. See Essay 3: *Prime Imperative* (Page 49).

What the world looks like to self is managed in mostly unconscious mind as Worldview, which is populated by cultural influences, prior understanding gain by the self/personality life field and human instincts. Worldview represents Self's personal reality: Self's perception of the world emerges into conscious awareness from the normally unconscious mind process which considers and translates environmental influences based on Worldview.

1-III-5 and **1-III-6** refer to the conscious influence of discerning intention on normally unconscious mind. This is one of the foundation concepts in the Mindful Way.

1-III-7 and **1-III-8** refer to the results of consciously applying accumulating understanding to discerning actions. This is another foundation concept in the Mindful Way.

1-III-9: Supreme State of Vishnu refers to self-realization or self-knowledge. This line tells us that mindfulness and conscious seeking may lead to the form of understanding we think of in contemporary terms as spiritual maturity.

1-III-10 and **1-III-11** describe the characteristics of Source. This is the *all is one* concept of a universal Infinite Intelligence, rather than a human-like god. Source manifests in the following ways:

1. In the avatar model, which is also described by the chariot analogy of the Upanishad, that which is sensed is transformed into subtle information able to be processed by the even more subtle mind.

2. The personality/mind/self complex or life field is subtle and not seen with the physical senses.
3. Personality, as the core intelligence of the life field, is that which is immortal, and which is subtler than mind. (Here I would argue that the life field integrates personality, mind and conscious self. In this way, personality is the source of purpose, conscious self is the experiencing aspect of personality and mind is judge as the functional aspect of experiencing.)
4. Mahat refers to the origin or source of creation, which is even more subtle than its aspects represented by the life field.
5. Source is seen as infinitely present (all is one, Mahat), infinite possibilities with the potential to manifest reality (Avyakta) and the intelligent core of reality (Purusha).

1-III-12 and **1-III-13** advise that the Self is in all things. An important cosmological point is that Source is the reality field and the life field of a person is an instance of the Source life field. As such, all things are either an aspect of Source or the expression of an aspect. The potential to create and Source, as the universal presence, is in each of us.

These lines tell us that those who have learned to sense the subtle characteristics of reality can sense the presence of self/Source in all things. We are advised to integrate this understanding into our worldview so as to align our perception and expression with the actual nature of Source.

1-III-14 and **1-III-15** Advise that we should find a teacher and consciously step onto the path toward greater understanding. The path is a narrow way in that it is far from the excesses seen on either extreme. It is also a difficult way because it is so easy for the seeker to turn toward misconception and illusion.

Finally, we are told that it is through correct understanding that we may step off of the wheel of reincarnation to move further toward universal understanding. (More likely, we will find ourselves with new and even more interesting challenges to understand.)

References and Alternative Sources
Listed at the end of the book beginning on Page 357.

Intentional Blank Page

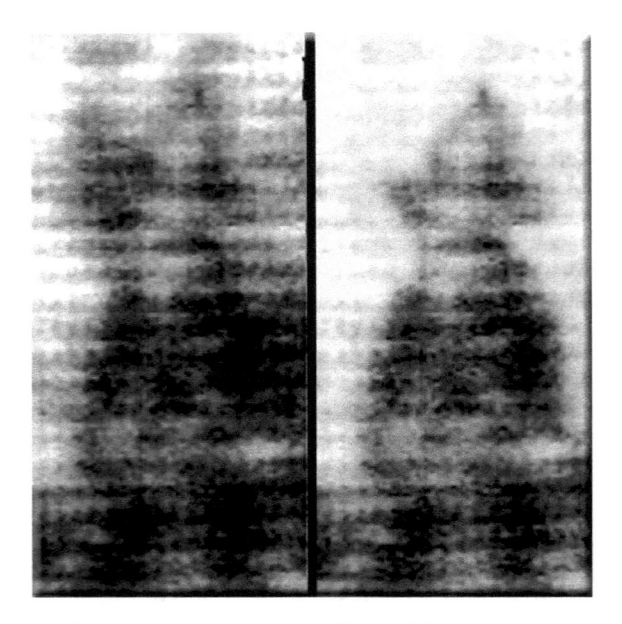

Video-Loop ITC Image collected by Tom and Lisa Butler.

The left side is the original. The background has been erased in the version on the right for clarity. We just don't know more than that this appears to be a man sitting on a bench.

Essay 19
Progression, Teaching and Community
2014

The Way of Progression
Through community comes knowledge
Through teaching comes understanding
It takes a collective

About This Essay

This essay is Discourse 12 of *Your Immortal Self*. (1) I included it here as a reminder about several of the reasons you exist.

The point is that you are part of a larger community. While we may not have a physical-world or societal reason to cooperate with our fellow persons, we do have an implied imperative to help others gain spiritual maturity. By doing so, we enable them to assist in our progression.

The Way of Progression is one of those thoughts that was stuck in my head one morning. I know it sounds kind of silly but take some time to consider the words. To teach, it is necessary to order our thoughts. In doing so, we gain a little understanding about that which we will teach. It does not work in front of a mirror. You need to interact with others to learn.

Spirituality

It is in our spirituality that we find our true meaning: who we really are and our purpose for existence. One of the best explanations for the meaning of spirituality I have read is offered by Deepak Chopra in his Huffington Post blog: (62)

> Spirituality is the experience of that domain of awareness where we experience our universality. This domain of awareness is a core consciousness that is beyond our mind, intellect and ego. In religious traditions, this core consciousness is referred to as the soul which is part of a collective soul or collective consciousness, which in turn is part of a more universal domain of consciousness referred to in religions as God.

Since we are an aspect of Source, in the Implicit Cosmology, *"our universality"* is the fact that our entanglement with Source makes us a citizen of the Source life field (reality, reality field). *"Soul"* is described as etheric personality which is who we really are, our immortal *I am this*. This is the *"core consciousness"* of Chopra's definition.

The *"collective Soul"* in Chopra's definition is referred to as the collective in the Implicit Cosmology, and it is modeled as the multitude of personalities related to this realm of reality. The *"universal domain of consciousness"* is modeled as the multitude of aspect which inherit their existence and purpose from Source. God then is Source, Infinite Intelligence, the One, or whatever name you use for the first cause of reality in which we are all one. This is the *"domain of awareness."*

Spirituality, then, is our sense of this connectedness from the perspective of our conscious self (*I think I am this*).

In the Implicit Cosmology, life is modeled as the building block of reality, and expressions of life are the venues for experiencing the nature of Source. We express the world around us, usually as a collective view because we share this venue with others. There are likely many realms similar to ours, each with many venues for learning. Together, they could be understood as *"universal consciousness."* This may be Jesus' *"many mansions"* from John 14:2.

Reality is modeled as a nested hierarchy of personalities so that the first round of aspectation is Source's effort to understand its nature. Our personality is an aspect of other personalities, probably many rounds removed from Source, which are further exploring Source's question: *who am I and what is my nature?*

> These assumptions are made to provide a beginning point for the model in much the same way the Big Bang is modeled as the beginning for the physical universe in physical sciences. The importance of the Implicit Cosmology is in the relationship of our *I am this* personality which is our unconscious core intelligence with our human avatar and our conscious self as *I think I am this*.

Spirituality, then, is our sense of this connectedness from the perspective of our conscious self (*I think I am this*).

Cooperation

Achieving understanding is not a solitary process. Just as our collective partners cooperate to help us while we are in this lifetime, so are we expected to help others who are in this lifetime with us. This does not mean we are our brother's keeper, but it does mean that we are unavoidably part of our brother's experiences. The part we play is our life experience, and so, we are unavoidably linked in the greater community and especially in our local community.

Friends as Teachers

With few exceptions, I have looked to a teacher to show me the way. I don't mean this in a mystical sense; it is just that I have managed to have someone I trust show me how to do extraordinary things before venturing out on my own. For instance, Lisa introduced me to Electronic Voice Phenomena; a mutual friend taught us how to mine for gold; a work friend introduced me to the Tarot. Another showed me how to navigate the Sacramental Delta waters at night. They did not just say, *"Here it is."* They took me through everything I needed to know to get started. It was always a natural sharing.

Selfless People

There is value in contemplating the meaning of the selfless concept. Dictionary.com defines **selfless** as *"Having little or no concern for oneself, especially with regard to fame, position, money, etc.; unselfish."* There are many ways in which selflessness is expressed: the family that takes in a foster child, a person who stops to help an animal, individuals who feed the poor and inspire selflessness in others, or the activists who dedicate themselves to a cause which probably should be important to everyone.

What causes people to be selfless? In some obscure way, there may be an evolutionary benefit. Evolutionists argue that there is a selfless gene which induces people to selfless acts. If they do not survive, their selflessness tends to benefit their family which would by inheritance also have the selfless gene. In that way, they argue, selflessness helps survival of the gene by helping the survival of family. (170).

While the selfless gene theory sounds reasonable, it is based on pretty tenuous evidence. When viewed from the perspective of human as avatar, the concept of cooperation amongst personalities seeking mutual progression is perhaps a more reasonable concept. The Prime Imperative would seem to result in an urge in every personality to help others, because by doing so, they are helping themselves and their collective. Ultimately, one could argue that selfless people are those who respond to the urge to answer Source's question. From a religious perspective, *"God desires that I do this."* See Essay 3: *Prime Imperative* (Page 49).

A Vision About Collectives

Just minutes after a meditation session at the Monroe Institute (171), I had the waking vision of the face of a clock. It was suspended in the air, face-up, but tilted toward me a little so that I could see that there were many black specks scurrying about on the white surface. My impression was that they were little stick people like those I would draw in a hurry. The space between three and four o'clock was an open hole and a few stick people had apparently fallen through.

The people on the face of the clock were somehow helping the people who had fallen through the hole. As the hour hand made a complete circuit, the ones in the hole came to the surface and a few of the others jumped into the hole.

My sense was that the stick people were all part of a collective of personalities, a soul group if you want, and they were doing all they could to help their fellows who had entered into a lifetime, symbolized by falling through the hole. I knew that they were helping one another progress by gaining in understanding and that none of them would be able to move on until all had made sufficient progress. The hour hand represented a lifetime.

Probably not all of them, but many of my helpers, friends, and guides, both on the other side and in the physical, are part of my collective. I am never very far from them, and just as I am a student, in turn, I am teacher, for we must all move as one.

First a Student, Then a Teacher

You have probably heard the phrase, *"Our lot is to learn and having learned, our lot is to teach."* The student becoming the teacher is the natural order of life. Civilization survives because lessons learned by elders are passed on to the children. We cannot be around other people, especially children, without influencing them in some way as accidental role models. However, in modern society, education is left to schools and hired teachers. There is a business of educating our young, leaving little more than role model duty to elders. Yet there are many examples in which individuals have taken the time to preserve knowledge for others.

The Internet has countless websites in which information for the sake of information is available at no charge. Looking past all of those commercial sites, one can find that there are many ordinary people who have spent time developing a personal website just to share what he or she has learned. An important point of this essay is that many of these people probably know they will not receive a thank-you. Yes, there is a human compulsion to leave a legacy for the next generation, but virtually all of these websites will vanish when the server fees are not paid. So discounting legacy, the most probable reason is that people build them out of the desire to help others.

Consider the model offered by the ancient wisdom schools. There is typically a path one must follow to gain masterhood. The first step on the path is for an individual to seek understanding by deliberately asking for a teacher. What follows is a series of lessons, tests and initiations as the seeker progresses from neophyte to initiate and finally to teacher and master.

The lessons provide the seeker with pertinent information. Tests are intended to determine if the seeker has assimilated the information in a way that can be expressed in novel ways. If so, then the seeker has demonstrated that the information has been assimilated as knowledge.

But here is the important part. During the initiation, the seeker is asked questions which may only indirectly relate to the lessons. If the seeker is able to provide reasonable answers that indicate understanding, then the seeker will be passed to the next level.

A selfless response to the urge to teach is difficult to ignore; however, the student is drawn to the lessons with usually only vague reasons for pursuing understanding. It is this teacher-seeker relationship I wish to address here.

The Nature of Understanding

There is a distinction between information, knowledge and understanding:

Information

Information can be compared to raw data. In the context of mind, it is sensory input from the environment brought by the physical senses and countless psi signals. As it is received, information is undifferentiated and is only sensible when organized into some context. In a practical sense, much of what you have read in this essay thus far is just information. Think of the bits and pieces of information as data points which must eventually be integrated into a sensible concept. Much of what people think they know is just information.

Knowledge

When information is considered, and the related concepts are understood, the resulting comprehension is considered knowledge. To be knowledgeable about something is to comprehend related concepts so that they can be reasonably well visualized. This is the key to the nature of knowledge. It must be comprehensible enough to be visualized as a concept. Here, *visualize* is used in the general sense of being imaged in some way in the mind. That might be sufficient recognition of an odor to name its source or formulation of many seemingly unrelated concepts into a form that can be described to others.

For instance, consider the old story of blind men examining an elephant: if each man represents a concept (leg, tail and such), then knowledge would be the ability of the men as a group to be able to describe an elephant. Knowledge is a combination of information, understanding concepts and comprehension of a global sense of meaning.

Understanding

Limits of Understanding

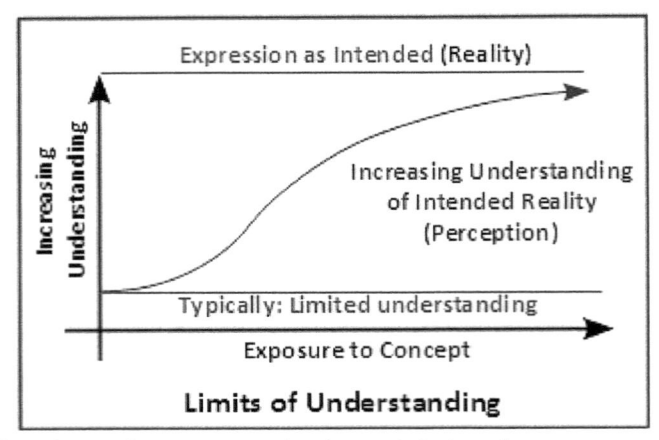

(Figure: graph with "Increasing Understanding" on y-axis and "Exposure to Concept" on x-axis, showing "Expression as Intended (Reality)" as upper bound, "Increasing Understanding of Intended Reality (Perception)" as an S-curve, and "Typically: Limited understanding" as lower line.)

In the Implicit Cosmology, perception is modeled as the outcome of a process that begins with encountering information that is external to the mind. We, as etheric mind and physical body, are immersed in an array of environmental signals. Some are physically sensed by the body, but most are psychically sensed. All of these signals are processed in our Attention Complex where they are first filtered according to whether or not we are interested. (38)

The degree to which understanding agrees with reality is also a function of our worldview. For instance, if we have been taught to be prejudiced about something, and we retain that prejudice, we will have that prejudice reinforced by the tone of the information coming to our conscious awareness. Unless we have become aware of the need to examine that perception, we will likely take what emerges from the unconscious perception process without question. Here, the exercises of mindfulness as a conscious effort to see things as they are intended become important in helping us see the world as it is, rather than as we are taught to think it is.

Understanding is not an absolute. As is shown in the *Limits of Understanding* Diagram (Page 324), correct perception (understanding) is typically limited upon first exposure to a concept. As a person seeks to better understand the concept, understanding approaches perception that agrees with reality. This is an obvious effect when, say, a new principle in science is introduced, but less obvious is the process of overcoming prejudices learned in youth and fears foisted upon the avatar relationship by human instincts.

As a person moves through a lifetime from youth to old age, things held to be true populate Worldview and tend to reinforce future perception. A conscious effort is usually required to break this cycle. Perhaps even more distressing, a person carries these *truths* into transition from this lifetime. The result can be a complex, reportedly emotionally painful period of *getting well* while Worldview is better aligned with reality.

The Process of Gaining Understanding

As a person gains in understanding via the accumulated lessons of many lifetimes (our own and/or those shared by other personalities in the collective), the person begins to understand the need for more mindfulness and the desirability of having a worldview that is in agreement with reality.

A magical thing happens when a person begins to consciously seek understanding. You may have heard the phrase: *"When the student is ready, the teacher will appear."* A similar saying was offered by psychologist Carl Jung: *"Synchronicity is an ever-present reality for those who have eyes to see."*

"For those who have eyes to see" is also a saying used in the Bible and secret wisdom schools. In fact, the whole idea of the occult or secret wisdom is not so much the secret but the need for the seeker to have sufficient understanding to recognize the lessons. An example is the concept of polarity: *Everything has an opposite.* The seeker is expected to realize that the middle way is the hidden knowledge to be derived from such concepts. In practice, this realization translates as perceptual understanding based on Worldview which is in alignment with reality.

At least one reason mindfulness works is that consciously seeking understanding effectively tells the unconscious part of mind to pay attention to external information that might help, rather than ignoring it as irrelevant. In many ways, the mostly unconscious mind is an obedient servant that shields us from a world that does not support what we think is true.

Questioning assumed truths moves the horizon of conscious self deeper into the unconscious process.

Worldview effectively prevents us from experiencing or believing anything that is not in agreement with our worldview. That is how it shapes our personal reality, even when it conflicts with actual reality. It is also how a person can obsessively believe something even though the facts seem overwhelmingly to the contrary.

Change must come incrementally. That is, if incoming information is clearly not acceptable to Worldview, it is ignored. However, information that is reasonably similar to what is in Worldview will sometimes be accepted via the *Maybe* outcome of the Perceptual Loop. The *Maybe* outcome is consciously experienced and incrementally changes Worldview. In that way, progression in understanding is a gradual thing. Engineers will recognize the curve in the *Limits of Understanding* Diagram (Page 324) as a naturally occurring rate of change found in many physical processes. Change is usually evolutionary and seldom catastrophic.

We Exist to Learn

Given the concepts described above, it is arguable that each of us is here to gain understanding. However, there is an old saying that *"When you're up to your neck in alligators, it's easy to forget you came to drain the swamp."* While we are in this lifetime and immersed in daily living, it is easy to forget why we came. In a practical sense, there are few signs to remind us of our Prime Imperative. (See Essay 3: *Prime Imperative* (Page 49).

Progression may seem to be a solitary path. Perhaps that is why they call such paths *personal improvement.* In fact, the old saying that we are all connected is more truth than platitude, and the path we may seem to be walking alone is more a march of the multitude of personalities who await our progression while doing what they can to assist us.

We who are seeking understanding are a community, a collective of like-minded people that exists in the physical as personality entangled avatars and in the etheric as immortal beings.

It is worth repeating the opening understanding:

The Way of Progression
Through community comes knowledge
Through teaching comes understanding
It takes a collective

The Prime Imperative as a Spiritual Obligation

While we all share an urge to gain understanding, in a very real sense, we have a sort of spiritual obligation to respond to that urge. While it may seem to be a sacrifice to teach or help others learn, every such effort is rewarded with new understanding and resulting progression that likely would not be possible otherwise.

It is much too complex of a concept to say that each of us needs to serve in some way in the physical. As I have noted above, there are many ways we teach. Perhaps the majority of us are too early on the path to be more than inspiration to others. However, if we have reached a point in our evolution at which we are accustomed to self-evaluation, it is time to begin finding ways to serve by teaching.

In a very real sense, progression can go only so far without teaching and that requires a community.

The Transition Experience

As I understand it, all of us experience a short period of disorientation as we begin the process of transition out of this lifetime. The first milestone of transition appears to be the often-reported period of *getting well*. That is thought to be followed by a period of self-assessment and finally the completion of transition as we enter into a new venue for learning. The new venue could be in the physical, but it could as well be in some other aspect of reality.

Personal reality is a product of Worldview, and we take Worldview with us into transition. That is probably why, during the initial shock of transition, it seems to be relatively easy for a personality to remain perceptually close to familiar surroundings. It is during this time that most communication with people in the physical seems to occur. A common report is the departing visit to say goodbye to a good friend.

The *getting well* period is needed to allow the personality time to realize that old handicaps no longer apply. This may also be when we undergo a self-evaluation of our just finished lifetime. This is that dreaded judgment we are warned about. However, instead of our being judged by some authority on high, it is a personal process in which we sense how our actions affected others from the perspective of those whom we affected. Perhaps it is from this self-evaluation that we determine our future lessons.

The period of adjustment apparently begins as the sense that we are in a world very much as we experienced in the physical; again, clinging to the familiar. However, if we expect a heaven from the religious point of view, we will likely find ourselves in that heaven. As we understand it, over time, we find that these are only constructs of our worldview, which itself is slowly realigned to better agree with the greater reality.

As we understand the process, we are born into a venue for learning at the end of this transition (a new lifetime), with a point of view, a degree of maturity in our understanding of the nature of reality gained from past experiences and an urge to learn specific lessons.

We are not all-wise when not in a lifetime. Instead, we have an imperative to learn and a degree of perceptual maturity which helps to form our point of view. It is this maturity and resulting point of view which determine what aspect of reality we can be in agreement with. In effect, the greater our understanding, the wider range of venues and opportunities for learning we have available to our personality.

> To emphasize this point, the relationship between personality, Attention Complex and conscious self remains during transition. The only real difference is that the human influence is no longer as strong. Part of transition may well be the process of sorting through Worldview to flush out that human influence. Meanwhile, to put it in the vernacular, *if a person was an idiot in life, he or she is likely going to be an idiot in transition.*

An important point is that there are no enforcement officers making sure we do what we should. We are self-governing. Our understanding manifests as perception and that determines what aspects of reality we can visit (Principle of Agreement).

It appears that transcommunication is possible during transition, but once a personality transitions into a new venue of learning, it appears that all communication into the physical stops unless it is assisted by other, more mature personalities.

See Essay 16: *What is it Like on the Other Side* (Page 267).

Self-Realization

This essay can be summarized as *Understanding Self-Realization*. The reason to study metaphysics is not just to map reality. It involves a process of discovery intended to help us learn how to gain personal understanding so as to fulfill our purpose. It is okay to ignore the details if you find the guidance.

Here, I do not mean purpose in a mystical sense. An acorn fulfills its purpose by growing into an oak tree. If the acorn tries to be a willow, it will likely end up being a pretty sad-looking oak tree.

Metaphysics tells us that our purpose is to align our personal reality with the actual nature of reality. If we ignore that Prime Imperative, we will likely end up being less than we could be. There is no need for me to explain this point. As you live your life as a person, you must surely see the benefits of living in accordance with organizing principles and the resulting discomfort if you do not.

Fulfilling your purpose may or may not be the same as what society considers being successful. Mainstream society is dominated by people trying to be willows. Only a few are sufficiently self-aware to recognize there is a purpose that underlies the dynamics of our society. It is for you to live in this society as a useful citizen without allowing its less aware point of view contaminate your worldview.

Remember that it is not what happens to you. It is how you react to what happens to you.

References and Alternative Sources

Listed at the end of the book beginning on Page 357.

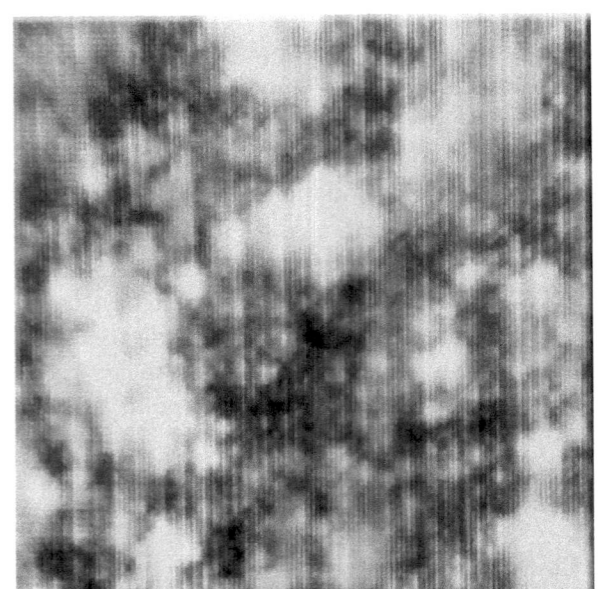

Intentional Blank Page

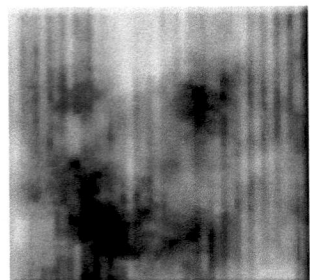

Video-Loop ITC Image collected by Tom and Lisa Butler.

Like the example on page 325, this example was used in an online visual perceptions study (atransc.org/visual-perception-study/). On the left is the original video frame, rotated 90 degrees. It had considerable, mostly random patches of color. You should be able to see the dog on the right Just above and a little to the right of center.

This example, right, is the head of a dog facing toward you and to your left. You can see his eyes and snout. A little of the neck is visible and just a hint of ears. The animal appears to be very alert and appears to have short hair.

Along with two color version, this example had 81% correct recognition.

Essay 20
Law of Silence
2017

About This Essay

As seekers, we live in a rather different world than our friends and family who are not on the Mindful Way. Paying more attention to our thoughts produces greater sensitivity to community dynamics and a different way of thinking that is often misunderstood by our friends. Knowing the teacher must be asked, it can be painful to remain quiet when we know people around us are expressing belief in baseless cultural wisdom. A practical consequence of increased spiritual maturity is often uncomfortable awareness of the actual nature of our world. Yet, we must remain a part of the world in order to continue our development; perhaps to be of service along the way.

You may notice that I address many of these issues in my writing. While some are the focus of an essay, most are embedded in those essays as part of my explanation of the main point. My objective is to explain every aspect of the metaphysics, as I understand the concepts in the context of the Implicit Cosmology. (12) If I were to design a college course on the subject, tests would be in the form of asking you to describe your understanding of these *incidental* concepts.

The Law of Silence is one such *incidental* concept. I wrote this essay while I was actively supporting *The Otherside Press Magazine*. (172) The magazine was a startup publication of the American Society for Standards in Mediumship & Psychical Investigation (ASSMPI). (173) Most of the essays I wrote for the magazine were slanted toward the need for and benefits of a cooperative community. Personal progression is facilitated by participation in one.

Law of Silence was written to guide people toward effective communication. Understanding the dynamics of communication is fundamental to participation in a cooperative community. I believe the reason so many people tend to be lurkers, rather than active participants is that they have spoken when it was best to remain quiet and were somehow discouraged from further attempts to communicate. Of course, I am not a psychologist, so this is only based on my observations and personal experience.

Many argue that the Bible includes reference to the Law of Silence, but from my reading, it is used in the Bible in the form of "Thou shalt not speak..." in the context of not speaking untruths or out of turn. For instance, in **1 Timothy 2:11** states: *"Let the woman learn in silence with all subjection."* And, in **1 Timothy 2:12** *"But I suffer not a woman to teach, nor to usurp authority over the man, but to be in silence."*

My first real encountered with the Law of Silence came from the teaching of Paul Twitchell, founder of Eckankar. (25) As a technical metaphysician, Twitchell's lectures were able to help me make sense of this rather abstract concept.

It took me many years and a lot of weird conversations to realize I see the world rather differently than others. For instance, I remember being excited about how I could actually see the pecking order described by Ardrey in his book, *African Genesis,* (41) as it was demonstrated when a flock of birds made claim to a fence. Try as I did, none of the people I told about it showed any interest.

I saw the same sort of dynamic when a recently divorced friend tried to explain his grief to others. He inevitably went away with a sense of emptiness, as his friends failed to respond as he expected. I spent a lot of time with him as we talked it out. In the end, the message was that he needed to look elsewhere for sympathetic ears.

It was about then that the book, *I'm OK - You're OK* (69) came out. That was one of the more useful self-help books of the time, and it gave me a way of telling others that there was a logical flow of communication.

The Law of Silence is a tool which can help you be a more effective communicator in any venue. It is a more effective tool when considered in the context of your immortal self.

Introduction

The **Law of Silence** can be defined as **Sharing important information with someone who does not respond as well as expected can dissipate the sense of the importance of the information.** This does not amount to a law so much as it is a cautionary wisdom, but it is a foundation concept for the secrecy often required by ancient wisdom schools and advice to be at least considered by everyone.

Secrecy has been part of the ancient wisdom schools since the early days of civilization. There were a number of reasons for secrecy that involved the nature of the schools and local society. Of those, the wisdom of *keeping our own counsel* remains important today. However, paradoxically, personal progression is greatly facilitated by mastering meaningful communication of ideas. So, which is it? Should we not share our thoughts about these concepts or should we openly talk about them? The answer is that sometimes it is best to remain silent, but sometimes it is more important to speak out. Understanding three aspects of communication will help to resolve this paradox.

Suspended Judgment

The Principle of **Suspended Judgment** is described in the Implicit Cosmology as **Increased understanding may come from unexpected or unwanted outcomes of expression.** (27) As we mature into adults, Worldview, what we unconsciously think is true about the world, is mostly populated with cultural beliefs. These beliefs are too often local prejudices and baseless lore that, if acted on, may lead us into a less than beneficial way of living. From this perspective, the purpose of life can be described as a quest to align Worldview with the actual nature of reality. That is, to replace beliefs with *correct* understanding.

A good way to take control of our unconscious mental processes is the Mindful Way. (48) By consciously making the practice of suspended judgment a habitual part of our relationship with our environment, it is possible to train our unconscious perceptual process to avoid premature decisions and thereby gain understanding that might not come if we simply accepted our first impressions.

Suspended judgment involves avoiding good-bad, agree-disagree, believe-disbelieve decisions about experiences so as to allow time, sometimes months or years, for more information to develop. It is important to be responsible for our actions, including seeking to accomplish specific tasks. The idea is that realizing something other than an intended result does not mean failure. Instead, the unexpected result offers an opportunity to reevaluate the assumptions on which the actions were based.

Suspended Judgment can be considered a more contemporary expression of the Law of Detachment. Deepak Chopra provides a good description of the **Law of Detachment**: (174)

> In detachment lies the wisdom of uncertainty, in the wisdom of uncertainty lies the freedom from our past, from the known, which is the prison of past conditioning. And in our willingness to step into the

unknown, the field of all possibilities, we surrender ourselves to the creative mind that orchestrates the dance of the universe.

The idea of the Law of Detachment is that being set on a particular result of some effort will blind us from seeing the potential benefits of alternative results. There is a problem of literal interpretation, though. Virtually everything in society tells us to become engaged, making the idea of being detached from the outcome of our actions seem like we do not care. We can resolve this in our mind by making a distinction between detachment and objective.

If we set out to achieve a final objective, our attachment might be in achieving that objective but not how we achieve it. For instance, if our objective is to have affordable transportation, programming for a sports car might miss the point. When a really nice sedan becomes available, the Law of Detachment tells us that we should take it. Alternatively, it might be wiser to decide not to buy a car if we live and work in a large city. Deciding that would meet our objective.

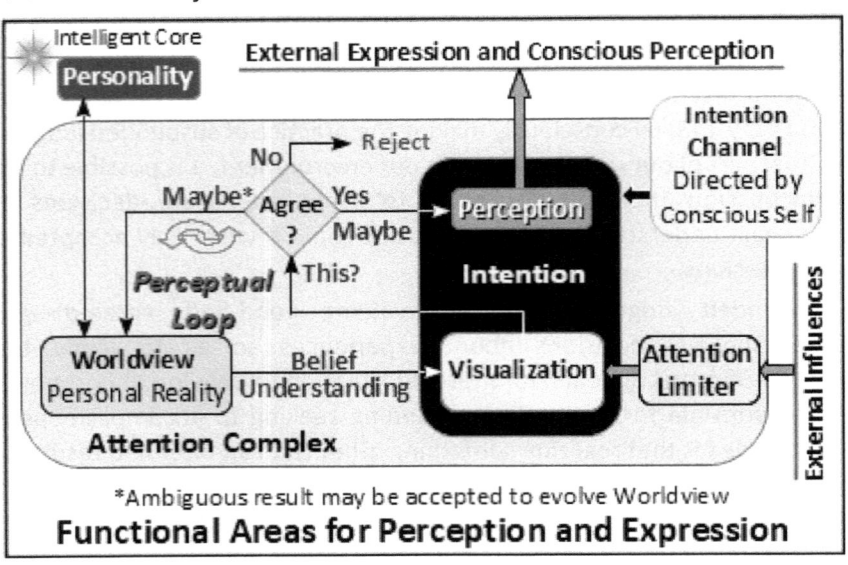

Functional Areas for Perception and Expression

*Ambiguous result may be accepted to evolve Worldview

The principle of suspended judgment supports a more general perspective. Our unconscious mental processes are hard-wired to make decisions. That means there is an *accept, reject* or narrowly defined *conditional accept* decision for everything we do, sense or visualize. Consciously deciding to replace a *reject* outcome with a wait and see response teaches the unconscious mind to be more lenient in the *conditional accept* decision. This has the potential to make the content of Worldview more accessible to evolution toward greater understanding.

The second benefit of suspended judgment can be found in how the mostly unconscious mind decides to ignore some environmental information based on our prior interests. In fact, there must be countless signals coming to our mind, yet only a few are presented to our conscious awareness. Our expression of intention helps to direct this filtering process, but mind learns very slowly, so it takes time. By habitually maintaining the attitude of suspended judgment, we create more ambiguity in our interests, thus making the door for what gets through the filter open a little wider. The effect is that we become more aware of many events in our environment we would otherwise ignore. Paranormal experiences fall into that category because we have been taught from birth that they do not exist.

Perceptual Agreement

The Principle of **Perceptual Agreement** is described as **Personality must be in perceptual agreement with the aspect of reality with which it will associate.** (18) This principle provides a mechanism by which order can be imposed on a system without many complex rules limiting behavior. Rather than some ethereal being telling us what we can and cannot do, or where we can go in *heaven,* this principle assures that we will gravitate toward aspects of reality which we are best able to understand. The practical consequence is that freedom of access to the greater reality is increased by gaining understanding which is in accordance with the actual nature of reality.

The self-limiting aspect based on understanding appears to be made practical by the *Maybe* state of the Perceptual Loop shown in the *Functional Areas of Perception Diagram* (Page 334). New understanding occurs when the new input is different but reasonably close to prior understanding. When it is, the new input is allowed into Worldview, **albeit after it is visualized by our Perceptual Loop based on prior understanding held in Worldview**. In this model, in response to the inherited urge to gain understanding, personality turns toward that which is familiar but with a preference for new experiences. Put a different way, without the intervention of conscious intention to habitually see actual reality, the perceptual processes tend to turn us toward the familiar.

There is a requirement in Rupert Sheldrake's Hypothesis of Formative Causation–(15) for Nature's Habit to be evolved under the influence of creative solutions to environmental problems. The *Maybe* solution in the Perceptual Loop provides this mechanism.

For our discussion, a consequence of this principle is that communication between two people is best when they are in perceptual agreement. That is, when their worldview leads them to see the world more or less the same way. This agreement need not be complete, but communication improves as the differences decrease. When describing an important, personal experience to others, it is more likely they will understand if they have a similar worldview or perception of reality.

For Those Who Have Eyes to See

As it happened, most of our mental processes are unconscious. (67) (16) (68) As the arbiter of our mostly unconscious perceptual process, (18) our worldview determines what we become consciously aware of and the attitude we will have toward that emerging awareness. You have probably heard the phrase, *"For those who have eyes to see."* It can be found in the Bible and in modern philosophy. For instance, Carl Jung told us that *"Synchronicity is an ever-present reality for those who have eyes to see."*

The phrase, *eyes to see,* is another way of saying *perceptual agreement*. In its simplest form, if we are not familiar with a concept, then we may not understand references to it. This becomes important to the Law of Silence when abstract concepts or emotion-laden information is involved. If our audience does not have relevant information in Worldview because of lack of relevant experience or due to inattention, then communication is impaired, and the person will likely not respond as expected. This disconnect in understanding is the source of such unfortunate responses as inattentional blindness or incredulity blindness.

It Takes a Community

Some of you will be familiar with the **Golden Rule**: *Do unto others as you would have them do unto you.* This sage wisdom is in the literature of just about every religion and is one of the first lessons most of us learned as a child. I am in my seventies now, and for the life of me, I have never been able to see the practical logic of the saying. Sure, I understand the idea that being kind to others will encourage them to be kind to me, but it doesn't really work that way.

I needn't go into the dynamics of prejudice and cultural fallacy. The point is pretty obvious. The urge to assure survival of our gene pool is so deeply embedded in our thinking that every action is based, first on a spontaneous instinctual response and then on what we have been taught is real. For all of us, our every action is determined by our worldview. My action can only slightly modify another person's worldview.

The cure is mindfulness, (48) but one must come to that without being forced. Habitually examining our every action is the only way I know to override the influence of instincts and cultural training. If one steps onto that path of mindful reeducation, in time, this examination becomes automatic to produce a more understanding reaction to the world. Under that condition, the Golden Rule makes sense.

Even though the literal interpretation of the Golden Rule seems to be only wishful thinking, the underlying esoteric lesson it conveys is one of the most important wisdoms paving our way to spiritual maturity. Three of the organizing principles in the Implicit Cosmology, (12) *Cooperative Communities* (Page 337), *Rapport* (Page 338) and *Perceptual Agreement* (Page 335) support this conjecture:

Cooperative Communities

The **Cooperative Community** Principle is defined as **Personalities are attracted to communities of like-minded people cooperating to facilitate progression.** (27) In effect, we inherit a sort of Prime Imperative from our local source to gain understanding about the nature of reality. Understanding is relative but converges on a sense of the actual nature of reality as a person progresses. In the spiritual sense of *"what is our purpose,"* progression is accomplished by learning through experience. See Essay 3: *Prime Imperative* (Page 49).

Note the distinction between collectives and communities. A **Collective** is defined as **Personalities related by a shared source tend to cooperate to favor the progression of mutually entangled individuals.** Members of a collective are bound together by their common source. There is evidence that many personalities cooperate in the etheric as part of a collective to fulfill the intention of their common source.

In the physical, we are attracted to communities of like-minded people for mutual cooperation toward gaining understanding. Cooperative communities likely support people from different collectives with rather different inherited objectives. The intention to gain understanding is a common factor. Cooperation is facilitated by rapport which is established between personalities as they interact in the community.

As an old adage goes, *"Our lot is to learn, and having learned, our lot is to teach."* Each member of a cooperative community fills both the role of seeker and teacher simply by interacting with other members of the community.

Rapport

Rapport can be defined as ***Personalities are interconnected by links of cooperation (influence) forming a matrix of relationships (cooperating community).*** One person's awareness of another personality produces a link of influence between the two life fields. The nature and intensity of this link is a function of one or both personality's visualization, attention and intention. The nature of this link of rapport also depends on the clarity (intensity) of awareness and the reason for the awareness. These links are dynamic and are thought to facilitate cooperation.

Perceptual Agreement is a moderating factor for rapport. Presumably, the more two people agree on the nature of reality, the stronger the link of rapport. This is something that can wax and wane from moment-to-moment, as one or the other party is more or less at ease. This idea is well-characterized by the book, *I'm OK - You're OK.* (69)

The links of rapport, coupled with the Principle of Perceptual Agreement, would serve to integrate reality toward a single thought. Using the chorus model, that would look like a super chorus of individual singers singing the same song. This super-unity, I think, is the objective of our existence. It also explains why Hans Bender, (36) speaking through the FEG medium, was so determined about how our actions in the physical are disrupting the greater community in the etheric.

The Three Aspects of a Teacher

The teacher-student relationship is inherent in the Cooperative Community Principle. It seems reasonable to imagine that a long past wayshower might have told his students something to the effect of: *"As you seek so do you teach, as you participate in the great work."* In keeping with the habit of secrecy, such a profundity would have been deliberately made obscure. Thus, most of these lessons had a public version. The seeker-teacher relationship is all about cooperation, and that might have been popularized as the Golden Rule.

The *Emerald Tablet* attributed to Hermes (28) is an example of hidden knowledge that might have preceded the Golden Rule. Of the fourteen lines, Line 12 and 13 are arguably teaching the importance of cooperation.

> 12. *So the world was created. Hence were all the wonderful adaptations of the one thing manifested; but the arrangements that follow this great mystic path are hidden.*

The One Thing is the organizing principles acting on the etheric, which is Source as the reality field. Hermes is explaining that everything in reality was formed by way of the same principles he has explained in the previous lines. An important concept here is *"For those who have eyes to see."* The way described in this lesson is hidden to those who have not followed this path.

> 13. *For this reason I am called Hermes Trismegistus--one in essence but three in total aspect. In this Trinity are concealed the three parts of the wisdom of the whole world.*

It is likely that there are multiple meanings in this line. The trinity concept is important throughout the Hermetic Wisdom. For instance, positive, negative and neutral, and body, mind and spirit. However, there is a more deeply hidden aspect of the trinity concept. Understanding of, and ability to properly manage the trinity of imagination, attention and intention is the foundation of *"the wisdom of the whole world."* See the *Functional Areas of Perception* Diagram (Page 334).

Trismegistus was the honorary name given to a person who had achieved mastery over himself and the principles, and who has integrated them into a way of life to live a good life. When considered with Line 12, the phrase, *"For this reason,"* makes it clear that the *"three parts of the wisdom of the world"* is a direct reference to the three aspects of all teachers. That is, a teacher represents the lesson to be taught, appears to the student as an example of what it is to understand the lesson and demonstrates the value of the lesson through application of the lesson in life.

These three aspects are demonstrated by the way Jesus presented himself in the Bible: John 14.6: *"Jesus saith unto him, I am the way, the truth, and the life: no man cometh unto the Father, but by me."* In this line, Jesus has told his disciples that he personifies the path that they must follow to return to God. He is showing himself to his followers as the three aspects of the teacher: follow me that I am the path; follow me as the Spirit of Truth; and, follow me as I have lived.

With these considerations, I believe the Golden Rule can be understood as a much-simplified expression of the concepts related to community, cooperation and rapport: **Teach others as they teach you**. These principles and the three aspects of a teacher are fundamental enablers for our personal progression. The challenge is to balance the wisdom of silence with the advice that we should be wayshowers for others.

Practicing the Law of Silence

Keep your own counsel is excellent advice, especially for people who are new to a path of learning; however, it is also excellent advice to find others with whom to safely discuss important information. The task is in knowing when to share and when to remain silent.

There are obvious cues for when to refrain from sharing important information. For instance, a conservative Christian is predictably offended by metaphysical subjects that might seem to conflict with popular understanding of biblical teaching. A discussion of the sanctity of all life would not be a good subject in a gun shop or at a Second Amendment meeting unless you are looking for a fight.

Sharing news of family problems with someone who has no useful references to understand your stress will likely not produce a supportive response. In the same way, sharing your excitement about having learned a very conceptual lesson about life, with someone who is not yet ready for self-improvement, is likely to drain away your enthusiasm as you try in vain to explain your excitement. The key to when to share is in understanding perceptual agreement. Make sure your listener is at least reasonably close to you in Worldview.

At the same time, be mindful that you are always a teacher. We seldom experience grand awakenings. Instead, we learn in small increments. Be patient with the uninitiated people around you. If you think you understand a little something about the world that might be of help to your friend's progression, it is important that you share what you know. However, do so in increments that will be understandable.

The rest of that story is that you should not be attached to the outcome of the telling. People seldom react as we expect, so be open to whatever your friend's reaction might be. Because people learn in small increments, they are more likely to understand in the future if you expose them to the concept now. If you are not prepared to accept a *"So what"* response, keep your counsel.

Community

The most important thing you can do is to find a community of like-minded people and cultivate a culture of sharing and mutual support. It is by cooperating in a community that you will fulfill your urge to gain understanding.

> It should be pretty clear by now that I have an extensive background in what I like to refer to as Etheric Studies. I do not consider myself an expert; however, unlike most people around me, I have thought longer and harder about many of the concepts related to progression than others in the community. While I try to remain open to new ideas and unexpected improvements in my understanding, like everyone else, my worldview has considerable momentum. It is difficult for me to see past that which I think I have come to understand.
>
> This is important to you in that encountering a very knowledgeable person in a presumably cooperative community comes with special cautions. Learn to speak with others as if you have a question you wish them to answer. Don't assume they have superior knowledge. Just be aware that your next important lesson might come from a stranger's offhand comment or even an apparently wrong comment about the nature of reality.
>
> Contemplate a lot on what it means to have an open mind. Everything has a message for you, should you know how to look. Especially know that people like me have become so accustomed to teaching that it is difficult for us to stop and listen. Oh, and looking back, I realize that sometimes I say the strangest things. One can only hope that there is an unexpected message in what I say for someone.

References and Alternative Sources

Listed at the end of the book beginning on Page 357.

Intentional Blank Page

Video-Loop ITC Image collected by Tom and Lisa Butler.

The faces we find in visual ITC may be from discarnate people, but they may also be from the memory of people still in the physical. We do not have sufficient controlled study to be sure. Our inclination is to think it may be a combination of the two, in which we in the physical facilitate message from those who are on the other side.

In this case, I think we captured my father's face. We had asked him to come through in that session, but we also had the photograph on my desk. Notice the same profile. It is common to see faces in ITC that have the bottom part obscured.

Essay 21
Informed Regret
2017

About This Essay

I close this book with a warning of sorts. You may have noticed that some of our most brilliant people turn their attention to philosophical musing in their later years. It is probably true that we all would, given the right situations in life.

Perhaps it is necessary for us to have lived a lifetime in order to become sufficiently self-aware to realize that we might have been much happier had we taken a different, perhaps more spiritual path in life. It is a truism that all of the financial gain in the world will not keep us from transition or that moment in which we must realize the implications of our actions toward others.

We think it is also a truism that what we do now matters here and hereafter. It is not that being good lets us into heaven. It is that the more aware we are of the actual nature of reality, the more aware we can expect to be during our transition. By extension, that means the more we are able to influence our next learning experiences.

I have to admit that this is all speculation. However, having a pragmatic temperament, I would not say these things if I did not see reasonable support for their validity.

So, this essay is a reminder that it is time to begin following the Mindful Way. Lucidity is not something we decide to have. It is developed over many years as a lifestyle. Do yourself and the rest of us a most important service by turning your intention toward understanding. You will be glad you did.

Abstract

In the song, *If I Were Brave*, singer Jana Stanfield (175) answers the question, *"What would I do today, if I were brave?"* with, *"If I were brave, I'd walk the razor's edge."* *If I were brave* is a reference to learning to follow our spiritual instincts. The *Razor's edge* represents the path we follow while obeying those instincts.

The Implicit Cosmology (12) posits that we inherit a spiritual urge to gain understanding about the nature of reality. (23) (See: Essay 3: *Prime Imperative* (Page 49)) This urge is as compelling as our human's survival instincts, but as a practical matter, it is the human instincts which drive much of our thinking. In practice, we must come to the realization that there is a need to consciously seek progression, as our instincts urge. Even though we are often reminded of the need to follow our spiritual instincts, most of us who do, come to this realization late in life. This essay focuses on the concept of informed regret. The original title was *"I Could Have Had a V8."* The intention of this essay is to help you decide before there is need for regret.

Introduction

Remember the advertisement: *"I Could Have Had a V8"*? The idea is that you already knew about the good-for-you wonders of the V8® brand drink, but without thinking, you had something else to drink like a less nutritious soda pop. Consider the sense of regret the phrase is intended to evoke.

Focus on the feeling that comes from realizing you have made the wrong decision. Perhaps you just realized you should have bought the blue car, rather than the red one you were driving when you got a speeding ticket. Perhaps you realized that you just said something mean to a friend that can never be taken back.

Try to enter into that consciousness for a moment. It is a sense of lost opportunity. A foolish, perhaps unthinking decision that cannot be undone. Yet, there is the promise to yourself that, next time, you will remember to have a V8!

This is about the emergent feeling that can be found at the edge of your conscious awareness. In contemplation, look between your unconscious and conscious awareness; the region in which your lucidity is developing. That is where, in your mind, those urges to act first emerge.

Here, I wish to evoke that sense of lost opportunity in your spiritual life. Assume for a moment that you have lived your life thus far without stepping onto a path of deliberate spiritual progression. Sure, you have taken a few courses and read a lot of books about things spiritual, but if you have not consciously decided to follow a particular way of learning, perhaps you are really just dabbling.

Everything changes when (if) that moment comes in which you realize that you really are an immortal being. As the implications of that realization sink in, it must become evident that seeking is for your immortal self, and not just to improve this lifetime. The ultimate meaning must come that what you do now matters the rest of your eternity.

As it Happened for Me

Other than the promise of Santa Clause and those tattletale elves my mother claimed were always watching me, my first real brush with a formal system of thought came when my mother made me attend Sunday School. The church's preacher lived in the house next to ours. My brush with faith abruptly ended when I saw his wife run out of the house, only to have him grab her and literally kick her back in.

Faith quickly gave way to science when I talked my kindergarten teacher into reading books to me on astronomy. It seemed important to know about astronomy if I was to be the first person on the moon. It took me decades to realize that, in many disciplines, science is just another form of faith. Over time, my sense of world order changed to be the convergence of scientific theory with engineering principles. Science became learned speculation for me while engineering became clever application of science.

In 1987, Lisa introduced me to Electronic Voice Phenomena (EVP). Then in 2000, we assumed leadership of the Association TransCommunication (ATransC) and I became deeply involved with the study of the survival hypothesis.

While I mingled my engineering career with the study of metaphysics, my focus was always on the practical aspects of human potential. Think of it as personal improvement, which is a body-centric pursuit. Even though I studied telepathy, out-of-body travel and such, it never really occurred to me that I was an immortal being only temporarily enjoying the human condition. I was told that very thing by many teachers, but I was told otherwise in hundreds of different ways. Information about my immortality had little effect on my sense of world order.

We published 57 quarterly ATransC NewsJournals through the Spring of 2014. As we developed articles for them, the realization dawned on me that EVP are not just interesting phenomena like ghosts and Astral projection. They represent a kind of revelation about our immortality.

I think Essay 2: *The Mindful Way* (Page 37) that I wrote for that last NewsJournal was my *"I could have had a V8"* moment. I was 71 by then and becoming aware of my physical mortality. Rather than looking for ways to increase my potential as a human, I should have been seeking ways to improve my lucidity by consciously purging lazy beliefs. I had been told, but the warnings were abstractions that had little effect on my sense of order until I realized that the voices in EVP could only have been initiated by immortal personalities.

> The original title of *The Mindful Way* essay in the last ATransC NewsJournal was actually *Mindfulness*. I changed the title as it became evident that my writing was about a system of thought.

Humankind Has Been Told Many Times

Here, I want to give you a few references which will help you understand the difference between dabbling along the seeker's path and purposefully traveling the Mindful Way.

The *Katha Upanishad* is a 4,000 years old ancient Hindu Vedic Sanskrit text which tells how the God of Death explained the nature of life to a seeker. Swami Krishnananda tells us that the God of Death explained that there are two paths in life from which we must choose: *The Good and the Pleasant.* (74) below is, first, Panoli's Sanskrit-to-English translation of lines 1-11-1 and 1-11-2 of the *Katha Upanishad*, followed by the Swami's explanation:

> **The Panoli Translation:** Different is (that which is) preferable; and different, indeed, is the pleasurable. These two, serving different purposes, blind man. Good accrues to him who, of these two, chooses the preferable. He who chooses the pleasurable falls from the goal. **1-11-1 Katha Upanishad** (22)
>
>> **Swami's paraphrase:** "There are two things in this world, and people pursue either this or that. These two may be regarded as the path of the pleasant, and the path of the good. Most people choose the former, and not the good. The pleasant is pleasing, but passing, and ends in pain. It is different from the good. But while the good need not necessarily be pleasant, the pleasant is not good."
>>
>>> **Swami's explanation:** Both come to a person, and we are free to choose. But we choose the tinsel because it glitters. An experience seems to be pleasant because of the reaction of our nerves. A condition that is brought about as a result of a reaction is passing, and not being. Lack of discrimination is the reason for choosing pleasure; confusion of mind causes

a wrong choice. When you grope in darkness, you fall into the pit, but you know it only after the fall. Similarly, the senseworld is darkness, and sense-objects come to ruin you, but the misguided mind cannot understand this. "Good comes to a person who chooses the good. But he who chooses the pleasant falls short of his aim."

The Panoli Translation: The preferable and the pleasurable approach man. The intelligent one examines both and separates them. Yea, the intelligent one prefers the preferable to the pleasurable, (whereas) the ignorant one selects the pleasurable for the sake of yoga (attainment of that which is not already possessed) and kshema (the preservation of that which is already in possession). **1-11-2 Katha Upanishad** (22)

Swami's paraphrase: "The dull-witted person chooses the pleasant: he wants to pass the day somehow. He does not know where or how the good is. The hero who is endowed with the power of discrimination, chooses the ultimate good."

Swami's explanation: When the pleasant and good come to us, they come together, in a mixed form, so that you cannot understand them. The best example for this is the world itself: you can use it as a passage to eternity, or for your pleasure. Yama (God of death) tested Nachiketas (The Seeker in this Upanishad) in the same way as this world tests us. Temptations come every day, in everything we see. We are caught in them because we are unable to distinguish between right and wrong. We do not know what will happen tomorrow. But our ignorance is so dark that we expect more pleasure, forgetting that death may come any moment. Death is the best teacher; there is not a better one: Understanding dawns by meditation on death. Suppose death comes to you in five minutes. Suppose you know it. What will you do? Will you act as you act now? You will act differently. It is true that we may die any moment. Yet, we do not think of it. Who prevents us from choosing the good? It is lack of understanding, which hides the defective side and shows only the pleasant aspects.

The Razor's Edge

The second stanza of *If I Were Brave* by Jana Stanfield, (175) has the words:

If I were brave, I'd walk the razor's edge,
where fools and dreamers dare to tread.
And never lose faith, even when losing my way.
What step would I take ... today, if I were brave?
What would I do today, if I were brave?

The *razor's edge* is a phrase also found in the *Katha Upanishad*.

Arise, awake, and learn by approaching the exalted ones, for that path is sharp as a razor's edge, impassable, and hard to go by, say the wise. **1-III-14 Katha Upanishad** (22)

As I learned from Paul Twitchell's lectures, (155) spiritual seeking requires courage. The best illustration I can give is the first time a person experiences what Twitchell referred to as Darshan, a Sanskrit term (darśana) related to coming in sight of a deity or a holy person. Darshan sometimes produces a similarly traumatic sense as that when a person undergoes a difficult exit during what is sometimes referred to as an out-of-body or Astral projection experience. Some people simply move into that dissociated awareness, but for some, it is like pulling a cork from a wine bottle.

It can be very fearful when a person suddenly enters into such a dissociated state for the first time. An important function of a teacher is to prepare the seeker for this, because if the experiencer's mostly unconscious mind goes into the *"Stranger! Danger!"* mode, it may be difficult for the student to be open to future experiences. It is possible that a fearful response can cut of further development for this lifetime.

Another example of the dangers of seeking is how difficult it is to remain on the path while managing our human's instinctual responses to life's situations. It is said that, even an advanced personality that has chosen to enter into a lifetime for some purpose, is apt to slide into ignorance of the Mindful Way because of the appealing nature of the physical world.

The Mindful Way is traveled moment-by-moment. Every action, every decision has the potential of moving us further from understanding. As we progress, our judgment becomes more in tune with the Mindful Way, but it takes time and there is always the unexpected challenge which might distract us from the way.

What Would You Do?

Taking the question of *"What would I do today, if I were brave?"* in the context of the first stanza of the song, the answer is that, if I were brave, I would be a seeker. Here, I define a **seeker** as **one who realizes the need to consciously seek spiritual maturity.** In the context of the *Katha Upanishad*, this means to wisely select the spiritual way, rather than seeking fame and wealth. In the context of our current reality, it means to learn to habitually live this lifetime while turning toward opportunities to gain understanding, seeking understanding in all experiences while allowing that understanding to influence our choices.

In ancient times, taking the Mindful Way may have demanded making a choice between seeking material success and spiritual progression. Our consciousness has changed since then, and we seem more able to find progression without rejecting a prosperous way of life. In our time, this means living the life in a mindful way ... wherever life takes us.

While I might judge other people's actions from a rational living or good citizenship perspective, I am not prepared to judge those actions from the perspective of spiritual seeking. It is clear that we sometimes need difficult experiences to gain important understanding. As Twitchell often noted, a wealthy businessman might well be a spiritual master, as might be the janitor serving his building.

The measure of the good choice rather than the choice of pleasure, as discussed in the *Katha Upanishad*, is not material versus spiritual. It is probably best described as an ethical one. See Essay 5: *Ethics as a Personal Code for Mindfulness* (Page 75). From my experience, by seeking to follow a personal code of ethics, I find myself also turning toward a more spiritual perspective in my daily living. Mine begins with Seth's *"Do not violate,"* which is elaborated on with such ethical principles as respect and judgment. Those are expressed as such ideas as *"just because I can, doesn't mean I should"* and *"I will not impose my will on others."*

Develop such a code for yourself and then apply it to every aspect of whatever you do in life, be it living in a cave or seeking success in the corporate world. This is at least one version of that V8 drink you can have, should you decide to consciously seek spiritual maturity.

My Learned Point of View

Ethical Understanding	Ethical Principles	Ethical Expressions
Do not violate *Seth*	• Respect • Kindness • Do no harm • Justice • Fairness • Suspended judgment • Courage • Discernment	• Just because I can, doesn't mean I should • Mindfulness is a way of life • Citizenship means cooperation • How will my actions affect me and others? • Is it a belief or a supportable understanding? • I will not impose my will on others • Lessons come from new experiences • Contemplation not meditation

Possible Mindfulness Personal Code of Ethics

Your mindful way will not look like mine. You live in a different personal reality and have had different experiences which are leading to future experiences which are different from my future experiences. Even so, I can act as a role model for you in the spirit of *teach me as I teach you*. Here is an overview of the way I have come to deal with my world. You can think of this as what to do or what not to do, depending on how you view my example.

I should preface this by saying I began dabbling with seeking in my teens but did not begin seriously considering the reality of my immortality until I was in my early forties. By now, I do have a good sense of the wisdom of lucidity but have not learned to integrate it so that I can consider myself anything like an adept. I am a role model because of my book learning more than because of my wisdom and spiritual maturity.

A Pragmatic Point of View

Part of writing *Your Immortal Self*, (1) was finding a way to explain the implications of survival. The most important consequence of survival is that our mind is not in our body. We are an etheric life form which has joined in an entangled relationship with a human body to experience the physical aspect of reality. This has been a gradual realization for me. It has taken many years to even partially integrate the implications into my view of reality.

To become entangled with our human avatar, we are born into this lifetime. Because we share Worldview with our human, our first few years are spent populating Worldview with learned things from our family community and the media ... and learning to cope with our human's instincts. The result is that we naturally develop a body-centric perspective of reality. We think we are our body, and for most people, we think our body's urges are our urges.

For many of us, our senior years brings the realization that there is more to us than our body. Even fewer of us consciously become a seeker of greater understanding about our etheric nature. When this happened, the process of changing point of view from body-centric (we are our body) to an immortal self-centric perspective can begin.

The first lesson of the immortal self-centric perspective is that our conscious self is our experiencer and our mostly unconscious mind is our judge. We always experience reality from the perspective of our experiencer as a video camera-like perspective. This is true when we are awake and when we are disassociated as in meditation, dream or delirium.

I say that our mind is mostly unconscious mind because, as judge, it includes our perception and expression functions. The judge uses Worldview as a standard to decide if our conscious self will experience incoming information. (16) It also colors how we experience that information by changing it into a familiar form. The result is that we, as conscious self, tend to experience reality as we think it is supposed to be, based on experience ... what we have been taught.

Worldview only changes in small increments. The one influence we have on it is our conscious intention to understand the implications of what we think is true. Part of learning to experience reality as it is, rather than as we have been taught, is the long process of examining everything we think. Is it true? Does it make sense? Do we understand the implications of what we think is true? Do we agree with those implications?

When we believe something, it is important that we also believe the implications. What are the consequences of our beliefs; of our actions?

If we are persistent in questioning our beliefs, in time, we develop a clear sense of reality that is more in agreement with its actual nature. The more this is true, the greater our lucidity. That is how we begin to see further into the mysteries of reality. We do not become enlightened beings by wishing it so. We do so by doing the work. We accept responsibility for what we believe to be true and work very hard to align our beliefs with the actual nature of reality. That is the way of the seeker.

My Advice

- **Live your life by engaging with it, rather than avoiding potentially stressful experiences just because they might get your blood pressure up.**

  We are here to gain understanding about the nature of reality through experience. Some of the most important understanding comes from the most challenging experiences. Think as an immortal self, not as a person living in fear of death. Being killed is not the worst thing that can happen to us.

- **At least intend to get a good night's sleep without cluttering it with a lucid dreaming** *to do list.*

  For me, lucid dreaming is a lot of work for little gain. I can make better progress simply by paying attention to life and habitually contemplating daily experiences. Any understanding you think you gain from lucid dreaming is apt to be just your worldview offering bits of memory to answer your requests ... telling you a likely story. There is no way of knowing if your conclusions are the right ones. For me, dream interpretation is a game of *stump the chump.*

- **Never forget the influence of cultural contamination on your experiences.**

  Experience through your senses and not through those of your teacher. Observe the influence of cultural contamination on the thinking of your friends and opinion setters.

- **Be pragmatic. Belief is the safe harbor for those who wish to abdicate their self-determination.**

  It is not necessary to be sure about the choices you make. There may be many ways to the same end. All you can do is pick the way which seems best and try it out. Living is all about testing what you think is right. That means a good life is one in which a person seeks understanding without preconditions of right and wrong, correct or incorrect. It is not a *the last one with all of the toys wins* kind of life.

- **Accept that *stuff happens*.**

    If, in retrospect, you made the wrong choice, admit it, clean up the mess and look for the lessons as you move on. Bad mistakes are the stuff of great life lessons.

- **The implication of pragmatism is that it is good to unburden ourselves of those *keep in case they are useful* fragments of our life.**

    Our consciousness changes as we gain in understanding. Relics tend to hold us back. Pragmatism means making useful decisions and moving on.

- **You are a creator. Take responsibility for your creations.**

    It is true that we live in a shared venue for learning, but we are ultimately the creator of our reality by the way we react to our circumstances as victim or student. It is not about what happens to us, but how we react to what happens to us. Resist being attached to a particular outcome of your creative efforts. Do this by focusing on the basic intent rather than a particular solution.

- **Meditation is important, but as a tool, not as a haven.**

    Undertake a course of action in which you decide to develop a channel to a deep meditative state, whatever that means for you. The Monroe Institute Hemi Sync is a good approach for that. (3) The more familiar that deep state of mental relaxation is to your mind, the more useful it will be as a tool.

- **We are all deep-trance mediums.**

    By that, I mean in those moments in which you pause to think of something, you slip into that deep trance state ... and then so quickly back that you do not notice you were disassociated. The more familiar the path to that deep state, the better access to your collective, friends and teachers on the other side. The alpha frequencies of mind are your working meditative level of awareness. Seek to make that part of your waking awareness.

- **Learn to think from the perspective of an immortal being temporarily entangled with your human during this lifetime.**

    This is important! Think about what it means.

- **We experience the world based on information that has been translated by our worldview.**

This is not a New Age idea. It is solid science that will likely serve your seeking better than any way of meditation. Also, remember that your worldview includes your human's instincts. Strong emotions such as obsession, fear, anger and envy are your human's reaction. As you learn to understand this, you will find ways to enter into one of these emotional states as needed, but then as quickly return to balance with relatively little discontent. Be mindful of the potentially negative influence of cultural contamination.

- **The first secret wisdom of the ancient way is balance.**
- **The second is that the teacher is also the student.**

The hidden way is the narrow way which requires suspended judgment to tame the inner judge. Judgment is easy and swift, but once a decision is made, it requires much effort to undo.

- **There is no magic, only the influence of organizing principles.**

Learn to recognize the fundamental principles governing the operation of reality and to use them in your life. (27) The ability to spout baseless belief appears as wisdom to the uninformed. Increasing lucidity means increasingly seeing actual reality. Hyperlucidity is thinking we see actual reality, when in fact, we only see what we have been taught to see.

The reason I warn away from gaining wisdom through lucid dreaming, meditation and the help of mediums is that the information must come first to our mostly unconscious mental perception which wants to color it in familiar terms. Thus, the narrow way is followed by testing everything, thereby teaching our perceptual processes to be less judgmental. Learning to test perception quickly leads us to more pragmatic sources of understanding.

- **Your mediums, teachers and friends probably do not realize this.**

Learn to understand the concept of coloring in mediumship.

- **Learn to practice suspended judgment.**

It is important to learn from those who have come before. But, while using that information to further your understanding, resist taking it as truth or rejecting it as false. Civilization is based on the teacher-student relationship. Assuming you know more than your teacher denies you of opportunities to learn. As we are told in the *Katha Upanishad*, find trusted teachers.

Of course, there is always more, but these are some of the lessons I have learned. This does not mean I have learned to follow them, but part of my learning to integrate these lessons is teaching them, so I thank you for being my student and look forward to being yours.

References and Alternative Sources

Listed at the end of the book beginning on Page 357.

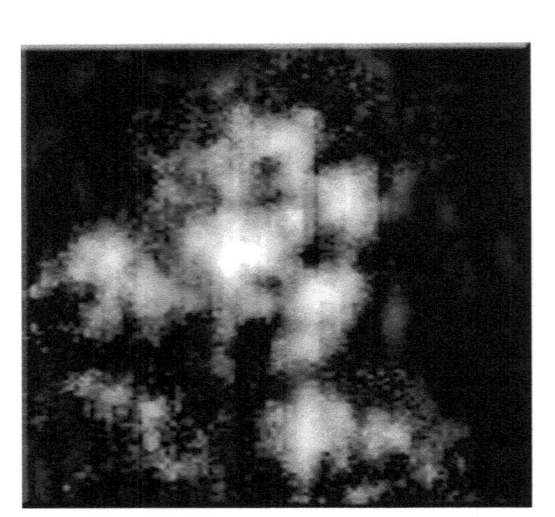

Intentional Blank Page

Video-Loop ITC Image collected by Tom and Lisa Butler.

This is a difficult example to make out, but I think it is important. There is a man and a woman in apparently intimate profile. The man's mouth is at the very center of the picture. He is facing toward your right.

The woman's face is behind his so that you see the left half. Her mouth is about at the man's chin. Her head is turned slightly up as if snuggling his cheek.

As with most of our work, close examination will reveal other face-like features in the picture.

We do not wish to make too much of this. Our impression is of a very loving demonstration between two people. We do not know if they are on this side of the veil.

References and Alternative Sources

1. Butler, Tom. *Your Immortal Self, Exploring the Mindful Way.* AA-EVP Publishing, 2016. ISBN 978-0-9727493-8-1. ethericstudies.org/books-tom-butler/
2. Butler, Tom. *Handbook of Metaphysics: A plain English Discussion of New Age Concepts.* The Christopher Publishing House, Hanover, 1994. ISBN: 0815804857. ethericstudies.org/books-tom-butler/#Handbook_of_Metaphysics
3. Butler, Tom. "The Monroe Way." *Association TransCommunication.* 2008. atransc.org/monroe-way/.
4. *The Silva Method.* silvamethod.com.
5. Delphi University. delphiu.com.
6. Builders of the Adytum. bota.org.
7. *Eckankar.* eckankar.org.
8. National Spiritualist Association of Churches. nsac.org.
9. Butler, Tom. "Glossary of Terms." *Etheric Studies.* 2014. ethericstudies.org/glossary-of-terms/.
10. Butler, Tom. "Trans-Survival Hypothesis." *Etheric Studies.* 2015. ethericstudies.org/trans-survival-hypothesis/.
11. Doyle, Bob. Determinism. *The Information Philosopher.* 1968. informationphilosopher.com/freedom/determinism.html.
12. Butler, Tom. "Implicit Cosmology." *Etheric Studies.* 2015. ethericstudies.org/implicit-cosmology/.
13. Butler, Tom. "Life Field." *Etheric Studies.* 2014. ethericstudies.org/life-field/.
14. Butler, Tom. "Etheric Fields". *Etheric Studies.* 2014. ethericstudies.org/etheric-fields/.
15. Sheldrake, Rupert PhD. "Morphic Resonance and Morphic Fields." *Rupert Sheldrake.* sheldrake.org/research/morphic-resonance/introduction?.
16. Carpenter, James. "First Sight: A Model and A Theory of Psi." *James Carpenter.* 2014. drjimcarpenter.com/about/documents/FirstSightformindfield.pdf.
17. Butler, Tom. "Perceptual Agreement." *Etheric Studies.* 2015. ethericstudies.org/perceptual-agreement/.
18. Butler, Tom. "Perception." *Etheric Studies.* 2015. ethericstudies.org/perception/.

19. Butler, Tom. "The Creative Process." *Etheric Studies.* 2014. ethericstudies.org/creative-process/.
20. Butler, Tom. "The Cosmology of Imaginary Space." *Etheric Studies.* 2014. ethericstudies.org/cosmology-imaginary-space/.
21. Kinser, Patricia Anne. "Brain Structures and their Functions." *Serendip Studios.* 2000. serendip.brynmawr.edu/bb/kinser/Structure1.html.
22. Panoli, Vidyavachaspati V. Translator. "Katha Upanishad." *Vedanta Spiritual Library.* celextel.org/upanishads/krishna_yajur_veda/katha.html.
23. Butler, Tom. "Prime Imperative." *Etheric Studies.* 2017. ethericstudies.org/prime-imperative/.
24. "About Spiritualism." *Etheric Studies.* 2018. ethericstudies.org/about-spiritualism/.
25. Trismegistus, Hermes Mercurius. "The Divine Pymander of Hermes Mercurius Trismegistus." *Internet Sacred Text Archive.* sacred-texts.com/eso/pym/index.htm.
26. Initiates, Three. "The Kybalion." *Marja de Vries.* marjadevries.nl/universelewetten/kybalion.pdf. Also see kybalion.org/.
27. Butler, Tom. "Organizing Principles." *Etheric Studies.* 2015. ethericstudies.org/organizing-principles/.
28. Butler, Tom. "The Hermes Concepts." *Etheric Studies.* 2016. ethericstudies.org/hermes-concepts/.
29. Butler, Tom. "How We Think." *Etheric Studies.* 2014. ethericstudies.org/how-we-think/.
30. "Highlights of Tarot." *B.O.T.A.* bota.org/botaineurope/en/tarot/.
31. Ropeik, David. "The Greatest Threat of All: Human Instincts Overwhelm Reason." *Psychology Today.* 2015. psychologytoday.com/blog/how-risky-is-it-really/201501/the-greatest-threat-all-human-instincts-overwhelm-reason.
32. Butler, Tom and Lisa. "ATransC 33-1 NewsJournal." *Association TransCommunication.* 2017. atransc.org/wp-content/uploads/2016/10/33-1-Spring-2014-ATransC-NewsJournal.pdf.
33. Cunningham, Paul F. "The Content-Source Problem in Modern Mediumship Research." *Rivier University, Department of Psychology.* The Journal of Parapsychology, 76(2), 295-319., 2012. rivier.edu/faculty/pcunningham/Publications/CunninghamJP_Fall-2012-Vol-76-(2)-295-319.pdf.
34. Waite, Arthur Edward. The Pictorial Key to the Tarot. 1911.
35. Grimes, Roberta A. "How Gospel Analysis Can Be Combined with Afterlife Evidence and Traditional Science to Help Us Better Understand

Consciousness." *Proceedings, The Academy for Spiritual And Consciousness Studies, Inc.* 2013.

36. Butler, Tom and Lisa. "Hans Bender's Message at Reno Séances." *Association TransCommunication*. 2013. atransc.org/hans-benders-message/.
37. Jane Roberts Learning Center. sethlearningcenter.org/.
38. Carpenter, James C, Ph.D. *First Sight: ESP and Parapsychology in Everyday Life.* Rowman & Littlefield. 2012. ISBN 978-1-4422-1392-0 (ebook). firstsightbook.com.
39. Butler, Tom. "EVP Online Listening Trials." *Association TransCommunication*. 2008. atransc.org/evp-online-listening-trials/.
40. Gullà, Daniele. "Computer–Based Analysis of Supposed Paranormal Voice: The Question of Anomalies Detected and Speaker Identification." *Association TransCommunication*. 2004. atransc.org/gulla-voice-analysis/.
41. Ardrey, Robert. *African Genesis: A Personal Investigation into the Animal Origins and Nature of Man.* New York: Atheneum, 1961.
42. Draper, Grenville. "A Brief Guide to Darwin's Theory of Natural Selection (Evolution)." *Department of Earth Sciences.* 2004. www2.fiu.edu/~draper/Darwin.pdf.
43. Russell, Ronald. "*Far Journeys* by Robert Monroe: An Excerpt from the new book." *Intuitive Connections Network.* 1992. intuitive-connections.net/2007/book-monroe.htm.
44. Myers, Steve. "Myers Briggs Personality Types." *Team Technology.* teamtechnology.co.uk/tt/t-articl/mb-simpl.htm.
45. Merrill, David W. and Reid, Roger H. *Personal Styles and Effective Performance.* Chilton Book Company, 1981.
46. Gurstelle EB, de Oliveira JL. "Daytime parahypnagogia: a state of consciousness that occurs when we almost fall asleep." *US National Library of Medicine, National Institute of Health.* 2004. ncbi.nlm.nih.gov/pubmed/14962619.
47. Roberts, Jane. *The Nature of Personal Reality.* San Rafael, Novato: Amber-Allen Publishing and New World Library. 1974. ISBN 1-878424-06-8.
48. Butler, Tom. "The Mindful Way." *Etheric Studies.* 2014. ethericstudies.org/mindfulness/.
49. Gert, Bernard and Gert, Joshua. "The Definition of Morality." *The Stanford Encyclopedia of Philosophy.* 2016. plato.stanford.edu/archives/fall2017/entries/morality-definition/.
50. Kraut, Richard. "Aristotle's Ethics." *The Stanford Encyclopedia of Philosophy.* 2001, revised 2014. plato.stanford.edu/archives/fall2017/entries/aristotle-ethics/.

51. "The Belmont Report: Office of the Secretary, Ethical Principles and Guidelines for the Protection of Human Subjects of Research." *The National Commission for the Protection of Human Subjects of Biomedical and Behavioral Research.* 1979. hhs.gov/ohrp/humansubjects/guidance/belmont.html.
52. Parapsychological Association. parapsych.org/home.aspx.
53. Society for Psychical Research. spr.ac.uk/main/.
54. *The Rhine.* rhine.org/.
55. Academy for Spiritual and Consciousness Studies, Inc. ascsi.org/.
56. Sudduth, Michael. "Super-Psi and the Survivalist Interpretation of Mediumship." *Cup of Nirvana.* 2009. michaelsudduth.com/wp-content/uploads/2016/01/SurvivalMediumship.pdf.
57. Simmonds-Moore, Christine. "What is Exceptional Psychology?" *Journal of Parapsychology* (#76 supplement, 54-57). 2012.
58. APStaff. "What is Anomalistic Psychology?" *Goldsmiths, University of London.* 2015. gold.ac.uk/apru/what/. (Visited 4-6-2018)
59. Butler, Tom. "About Etheric Studies." *Etheric Studies.* 2007. ethericstudies.org/about-etheric-studies/.
60. Stemman, Roy. "Skepticism: The New Religion." *Etheric Studies.* Originally from Paranormal Review, 2010. ethericstudies.org/skepticism-new-religion/.
61. Mayer, Gerhard. "What About parapsychology and Anomalistics? Results of a WGFP and GFA Member Survey." *Jouranl of the Society for Psychical Research*, Vol. 81.4, 2017.
62. Chopra, Deepak. "Only Spirituality Can Solve The Problems Of The World." *Huffpost Healthy Living.* 2010. huffingtonpost.com/deepak-chopra/only-spirituality-can-sol_b_474221.html.
63. Storm, Lance. "The Sheep - Goat Effect." *Psi Encyclopedia (SPR).* 2016. psi-encyclopedia.spr.ac.uk/articles/sheep-goat-effect.
64. Leary, Mark. "A Research Study into the Interpretation of EVP - Three parts." *Association TransCommunication.* 2013. atransc.org/radiosweep-study2/.
65. Mügge, Kai. "Society for Research in Rapport and Telekinesis (SORRAT)." *Felix Experimental Group.* 2008. felixcircle.blogspot.com/2008/09/sorrat.html.
66. Bargh, John A. "Our Unconscious Mind." *PScience Associates.* 2015. (Published: Scientific American, Vol. Volume 310.) pscience.com/wp-content/uploads/2013/12/UNCONSCIOUS-unconscious-mind-shapes-our-day-to-day-interactions-Bargh-SciAm-2013.pdf

67. Max-Planck-Gesellschaft. "Decision-making May Be Surprisingly Unconscious Activity." *Science Daily*. 2008. sciencedaily.com/releases/2008/04/080414145705.htm.
68. Bargh, John A. and Morsella, Ezequiel. "The Unconscious Mind." *US National Library of Medicine, National Institutes of Health*. 2008. ncbi.nlm.nih.gov/pmc/articles/PMC2440575/.
69. Harris, Thomas A. M.D. *I'm OK – You're OK* (the book.) *M.D.* 1969. ISBN-13: 978-0060724276. drthomasharris.com/.
70. Butler, Tom. "Debunking Survival Under Cover of False Academic Authority." *Etheric Studies*. 2014. ethericstudies.org/scientist-attack-medium/.
71. Strieber, Whitley. *Communion: A True Story by Whitley Strieber*. Avon, 1987. ASIN: B01NH0M6TG.
72. Miller, Glenn M.D. "Voyager Syndrome." *Glenn Miller M.D.* glennmillermd.com/voyager-syndrome.
73. Nashawaty, Chris. "The Jerusalem Syndrome: Why Some Religious Tourists Believe They Are the Messiah." *Wired Magazine*. 2012. wired.com/2012/02/ff_jerusalemsyndrome.
74. Krishnananda, Swami. "Commentary on the Katha Upanishad." *The Divine Life Society Sivananda Ashram, Rishikesh, India*. swami-krishnananda.org/katha1_0.html.
75. Butler, Tom. "EVP Online Phantom Voices." *Association TransCommunication*. 2012. atransc.org/phantom-voices/.
76. Butler, Tom. "Radio-Sweep: A Case Study." *Association TransCommunication*. 2009. atransc.org/radiosweep-study1/.
77. Butler, Tom. "EVPmaker with Allophones: Where are We Now?" *Association TransCommunication*. 2011. atransc.org/evpmaker-study-where-are-we-now/.
78. Heinen, Cindy. "Information Gathering Using EVPmaker With Allophone: A Yearlong Trial." *Association TransCommunication*. 2010. atransc.org/information-gathering-using-evpmaker/.
79. Butler, Tom and Butler, Lisa. *There is No Death and There are No Dead*. Reno, AA-EVP Publishing. 2003. ISBN: 0-9727493-0-6. atransc.org/books-atransc/
80. Butler, Tom. "Personality-Centric Perspective." *Etheric Studies*. 2014. ethericstudies.org/personality-centric-perspective/.
81. Butler, Tom. "Irrational Nature of Gun Ownership." *Etheric Studies*. 2017. ethericstudies.org/irrational-nature-of-gun-ownership/.
82. Butler, Tom. "Ethics as a Personal Code for Mindfulness." *Etheric Studies*. 2016. ethericstudies.org/code-of-ethics/.

83. Berg, Sarah. "Jack's Masks - Lord of the Flies." *eNotes.* scienceleadership.org/blog/jack-s_masks_-_lord_of_the_flies.
84. "Biography of Wilhelm Reich." *The Wilhelm Reich Infant Trust.* 2011. wilhelmreichtrust.org/biography.html.
85. Science and Engineering Indicator 2006, "Chapter 7: Science and Technology: Public Attitudes and Understanding." *National Science Foundation.* 2006. wayback.archive-it.org/5902/20150818094952/http://www.nsf.gov/statistics/seind06/c7/c7s2.htm.
86. Dace, Ted. "The Anti-Sheldrake Phenomenon, Attacking Morphic Resonance." *Rupert Sheldrake.* 2010. sheldrake.org/reactions/the-anti-sheldrake-phenomenon.
87. Butler, Tom. "Point of View." *Etheric Studies.* 2015. ethericstudies.org/point-of-view/.
88. Koukl, Greg. "Sagan and Scientism." *Stand to Reason.* April 22, 2013. str.org/articles/sagan-and-scientism#.WTNbdWjyuUl.
89. Kelley, Charles R. Ph.D. "What is Orgone Energy?" *Kelley-Radix.* 1999. kelley-radix.org/downloads/what_is_orgone_energy.pdf.
90. Charman, Robert A. "An Unusual Form of Radiation has a Reproducible Effect in the Laboratory." *Association TransCommunication.* 2010. atransc.org/unusual-energy/.
91. "Rupert Sheldrake." *Wikipedia.* en.wikipedia.org/wiki/Rupert_Sheldrake.
92. *International Skeptics Forum.* internationalskeptics.com/forums/forumindex.php.
93. *Skeptical about Skeptics.* skepticalaboutskeptics.org/.
94. "Wikipedia:Requests for arbitration/Paranormal." *Wikipedia.* 2007. en.wikipedia.org/wiki/Wikipedia:Requests_for_arbitration/Paranormal.
95. "Wikipedia:Requests for arbitration/Pseudoscience." *Wikipedia.* en.wikipedia.org/wiki/Wikipedia:Requests_for_arbitration/Pseudoscience.
96. Lower, Stephen. "Pseudoscience." *Chem1 Tutorial.* 2008. chem1.com/acad/sci/pseudosci.html.
97. Shermer, Michael. Why People Believe Weird Things: Pseudoscience, Superstition, and Other Confusions of Our Time. W. H. Freeman and Company, New York. 1997. michaelshermer.com/weird-things/.
98. "User Talk:Tom Butler." *Wikipedia.* 2014. en.wikipedia.org/wiki/User_talk:Tom_Butler#Arbitration_Enforcement.
99. "Electronic voice phenomenon." *Wikipedia.* en.wikipedia.org/wiki/Electronic_voice_phenomenon.
100. Radin, Dean. "Experiments Testing Models of Mind-Matter Interaction." *Dean Radin.* 2006. deanradin.com/FOC2014/Radin2006MarkovRNG.pdf.

101. "Welcome to the CRV-REG Study." Sponsored by The International Remote Viewing Association. crvreg.org/.
102. Butler, Tom. "Critiquing ITC Articles written by Imants Barušs." *Etheric Studies.* 2010. ethericstudies.org/failure-to-replicate-itc/.
103. Desai, Rajiv. "Imitation Science." *Dr Rajiv Desai: An Educational Blog.* 2013. drrajivdesaimd.com/2013/12/01/imitation-science/.
104. Evrard, Renaud, Glazier, Jacob W. "Beyond the Ideological Divide in Near-Death Studies: A Tertium Quid Approach." *Journal of Exceptional Experiences and Psychology,* Winter, 2016, Vol. 4. ISSN 2327-428X. exceptionalpsychology.com.
105. Butler, Tom. "A Model for EVP." *Association TransCommunication.* 2017. atransc.org/model-for-evp/.
106. Carroll, Robert T. "Pathological Science." *The Skeptic's Dictionary.* 2015. skepdic.com/pathosc.html.
107. Alcock, James. "Electronic Voice Phenomena: Voices of the Dead?" *The Committee for Skeptical Inquiry.* 2004. csicop.org/specialarticles/show/electronic_voice_phenomena_voices_of_the_dead.
108. Banks, Joe. "Rorschach Audio: Ghost Voices and Perceptual Creativity." *Paranormal.* 2001. peeranormal.com/wp-content/uploads/2016/08/Banks-2001-Rorschach-Audio-Ghost-Voices-and-Perceptual-Creativity.pdf.
109. "Psi Encyclopedia." *Society for Psychical Research.* spr.ac.uk/publications/psi-encyclopedia.
110. "Wikipedia:Fringe theories." *Wikipedia.* en.wikipedia.org/wiki/Wikipedia:Fringe_theories.
111. Sanger, Larry. "Why Wikipedia Must Jettison Its Anti-Elitism." *Larry Sanger.* 2004. larrysanger.org/2004/12/why-wikipedia-must-jettison-its-anti-elitism/.
112. "CZ:Paranormal Subgroup." *Citizendium.* en.citizendium.org/wiki/CZ:Paranormal_Subgroup.
113. "Wikipedia:WikiProject Skepticism." *Wikipedia.* en.wikipedia.org/wiki/Wikipedia:Wikiproject_Rational_Skepticism.
114. "JREF appears to be dormant - TB." *James Randi Educational Foundation.* randi.org/.
115. Carroll, Bob. *The Skeptic's Dictionary.* skepdic.com/contents.html.
116. "Wikipedia:Fringe theories/Noticeboard." *Wikipedia.* en.wikipedia.org/wiki/Wikipedia:Fringe_theories/Noticeboard.
117. "Pseudoscience." *Wikipedia.* en.wikipedia.org/wiki/Pseudoscience.

118. Cardoso, Anabela. "ITC Journal." *Instrumental Transcommunication.* itcjournal.org/.
119. MacRae, Alexander. "Report of an Anomalous Speech Products Experiment Inside a Double Screened Room." *Journal of the Society for Psychical Research.* 2003. spr.ac.uk.
120. Clark, Keith. *iDigitalMedium.* idigitalmedium.com/.
121. Rinaldi, Sonia. Institute for Advanced Research in Transcommunication Instrumental. ipati.org/index_en.html.
122. "Attribution-ShareAlike 3.0 Unported (CC BY-SA 3.0)." *Creative Commons.* creativecommons.org/licenses/by-sa/3.0/.
123. "Talk:Rupert Sheldrake/Archive 19." *Wikipedia.* 2014. en.wikipedia.org/wiki/Talk:Rupert_Sheldrake/Archive_19.
124. Sheldrake, Rupert. "Wikipedia Under Threat." *Rupert Sheldrake.* sheldrake.org/about-rupert-sheldrake/blog/wikipedia-under-threat.
125. "Category:Pseudoscience." *Wikipedia.* en.wikipedia.org/wiki/Category:Pseudoscience.
126. "User:9SGjOSfyHJaQVsEmy9NS." *Wikipedia.* Revision as of 12:53, 17 January 2009. en.wikipedia.org/w/index.php?title=User:9SGjOSfyHJaQVsEmy9NS&diff=264662941&oldid=264658824.
127. "Wikipedia:Requests for arbitration/Martinphi-ScienceApologist." *Wikipedia.* 2007. en.wikipedia.org/wiki/Wikipedia:Requests_for_arbitration/Martinphi-ScienceApologist.
128. Braude, Stephen. "Investigations of the Felix Experimental Group: 2010–2013. 2014." academia.edu/7593753/Investigations_of_the_Felix_Experimental_Group_2010-2013.
129. Nahm, Michael. "The Development and Phenomena of a Circle for Physical Mediumship." 2014. anomalistik.de/images/pdf/sdm/sdm-2014-08-nahm.pdf.
130. Alvarado, Carlos S. "Questions about the Physical Phenomena of the Felix Circle." *Parapsychology.* 2014. carlossalvarado.wordpress.com/tag/felix-circle-felix-experimental-circle-kai-muegge-physical-mediumship-stephen-braude-michael-nahm/.
121. *Paranormal Review.* Society for Psychical Research. spr.ac.uk
132. Mulacz, Peter. "Fall of the House of Felix?" Spring 2015. http://doczz.com.br/doc/1053347/fall-of-the-house-of-felix%3F. Published on *Paranormal Review.* Society for Psychical Research. spr.ac.uk
133. Society for Scientific Exploration. scientificexploration.org/.

134. Lizbeth A. Adams, Ph.D. CIP. with contributions by Timothy Callahan, Ph.D. "Research Ethics." *University of Washington School of Medicine.* 2014. depts.washington.edu/bioethx/topics/resrch.html.
135. Broude, Stephen. "Felix Experimental Group." *Psi Encyclopedia.* 2016. psi-encyclopedia.spr.ac.uk/articles/felix-experimental-group.
136. "Libel and Slander." *The Free Dictionary by Farlex.* legal-dictionary.thefreedictionary.com/Libel+and+Slander.
137. Braude, Stephen. "Provincialism in the Life Sciences: A Review of Rupert Sheldrake's A New Science of Life." *academia.edu.* 1983. academia.edu/2643570/_Radical_Provincialism_in_the_Life_Sciences_A_R eview_of_Rupert_Sheldrakes_A_New_Science_of_Life.
138. Zahradnik, Walter von Lucadou & Frauke. "Predictions of The Model of Pragmatic Information About RSPK." *Parapsych.org.* 2004. archived.parapsych.org/papers/09.pdf.
139. *Forever Family Foundation.* www.foreverfamilyfoundation.org/.
140. Butler, Tom and Lisa. "A Visit to the Felix Experimental Group." *Association TransCommunication.* 2010 ATransC NewsJournal, Vols. 29-3, Page 4. atransc.org/wp-content/uploads/2016/10/29-3-Fall-2010-ATransC-NewsJournal.pdf
141. Noory, George with Braude, Stephen E. "Student Loan Crisis Parapsychology." *Coast-to-Coast AM with George Noory.* 2016. coasttocoastam.com/show/2016/02/22.
142. Braude, Stephen. "Editorial." *Journal of Scientific Exploration.* Vol. 30, No. 1, pp. 5–9, 2016. scientificexploration.org/docs/30/jse_30_1_Editorial.pdf.
143. Nahm, Michael. "Further comments about Kai Mügge's alleged mediumship and recent developments." *Journal of Scientific Exploration.* Vol. 30, No. 1, pp. 56–62, 2016. scientificexploration.org/docs/30/jse_30_1_Nahm.pdf.
144. Ventola, Annalisa. "A Brain Response to a Future Event?" *Public Parapsychology.* 2007. publicparapsychology.blogspot.com/2007/11/brain-response-to-future-event.html.
145. Butler, Tom. "Wikipedia Arbitration." *Etheric Studies.* 2016. ethericstudies.org/wikipedia-arbitration/.
146. "Ethical and Professional Standards for Parapsychologists: Aspirational Guidelines." *Paraphychological Association.* 2005. parapsych.org/section/42/ethical_and_professional_standards.aspx.
147. Butler, Tom. "Practitioner Advocacy Panel." *Etheric Studies.* 2016. ethericstudies.org/practitioner-advocacy-panel/.

148. Lea, Robert. "Science is not the enemy: A Response to 'Limits of science, trust and responsibility.'" *Skeptic's Boot.* 14 March 2017. skepticsboot.blogspot.co.uk/2017/03/science-is-not-enemy-response-to-limits.html.
149. Butler, Tom. "Source." *Etheric Studies.* 2015. ethericstudies.org/source/.
150. Singer, Emily. "A Comeback for Lamarckian Evolution?" *MIT Technology Review.* 2009. technologyreview.com/news/411880/a-comeback-for-lamarckian-evolution/.
151. Appleyard, Bryan. "Rupert Sheldrake's Alternative Science." *Bryan Appleyard.com.* 2012. bryanappleyard.com/rupert-sheldrake-alternative-science/.
152. Joshi, Sheila. "James Carpenter's First Sight model and neurological damage-induced psi openings." *Blog: Neuroscience and Psi.* August 11, 2012. neuroscienceandpsi.blogspot.com/2012/08/james-carpenters-first-sight-model-and.html.
153. "Esalen Survival After Bodily Death Conference Archive." *SurvivalAfterDeath.Info.* survivalafterdeath.info/articles/esalen/index.html.
154. Butler, Tom. *Two Worlds, One Heart.* Edmonton, Commonwealth Publications, Canada. 1995. ISBN: 1-55197-053-8. ethericstudies.org/books-tom-butler/
155. Twitchell, Paul. *The Flute of God.* First. Eckankar, 1969. p. 94. ISBN is for second edition. ISBN-13: 978-1570430329.
156. Butler, Tom. "4Cell EVP Demonstration." *Association TransCommunication.* 2005. atransc.org/4cell-evp-demonstration/.
157. Crystal, Ellie. "Thoth." *Crystal Links.* crystalinks.com/thoth.html.
158. "The Egyptians' Ideas of God." *Internet Sacred Text Archive.* translated by E. A. Wallis Budge in 1895. sacred-texts.com/egy/ebod/.
159. Malkowski, Edward. "Before The Pharaohs: The Evidence for Advanced Civilisation in Egypt's Mysterious Prehistory." *New Dawn.* 2013. newdawnmagazine.com/articles/before-the-pharaohs-the-evidence-for-advanced-civilisation-in-egypts-mysterious-prehistory.
160. Butler, Tom. "Metaphysical View of John 14." *Etheric Studies.* 2015. ethericstudies.org/metaphysical-view-john-14/.
161. Butler, Tom. "The Razor's Edge." *Etheric Studies.* 2016. ethericstudies.org/razors-edge/.
162. Butler, Tom. "Winter Solstice as a Parable for Progression." *Etheric Studies.* 2015. ethericstudies.org/winter-solstice-and-progression/.
163. The Rosicrucian Order. AMORC. rosicrucian.org.

164. Pick, Bernhard. "Chapter I. Name and Origin of The Cabala," "The Cabala: Its Influence on Judaism and Christianity." *Internet Sacred Texts Archive.* 1913. sacred-texts.com/jud/cab/cab03.htm.
165. Case, Paul Foster. *The Tarot: A key to the Wisdom of the Ages.* Richmond : Macoy Publishing Company, 1947.
166. Estep, Sarah. *Voices of Eternity.* New York : Fawcett Gold Medal Book, Ballantine Books, 1988. ISBN 0-449-13424-5. atransc.org/resources/books.htm.
167. Maugham, W. Somerset. *The Razor's Edge.* Vintage, Reprint: 2003, 1944. ISBN-13: 978-1400034208.
168. Frawley, David. "Yoga, the Purusha and the Cosmic Being" (from the book: Yoga and the Sacred Fire). *American Institute of Vedic Studies.* 2012. vedanet.com/the-purusha-principle-of-yoga/.
169. Gupta, K. Soul. "The Eternal Being." *Spirituality and Us.* 2014. spiritualityandus.org/2014/09/14/soul-the-eternal-being/.
170. Judson, Olivia. "The Selfless Gene." *Manchester.edu.* 2007. users.manchester.edu/Facstaff/SSNaragon/Online/texts/201/Judson-SelflessGene.pdf.
171. Monroe, Robert. "The Monroe Institute." *The Monroe Institute.* monroeinstitute.org/.
172. *The Otherside Press.* theothersidepress.com/.
173. American Society of Standards in Mediumship and Psychical Investigation. assmpi.org/.
174. Chopra, Deepak. *The Seven Spiritual Laws of Success: A practical guide to the fulfillment of your dreams.* Amber-Allen Publishing & New World Library, 1994. chopra.com/the-seven-spiritual-laws-of-success. ISBN 1-878424-11-4.
175. Stanfield, Jana. "If I Were Brave." 1914. janastanfield.com/2014/09/if-i-were-brave/.
176. Coleman, Patrick John. "A Shaman's Guide To The Dark Night Of The Soul." *Medium.* 2014. medium.com/concrete-shamanism/a-shamans-guide-to-the-dark-night-of-the-soul-e9e699a6a787.

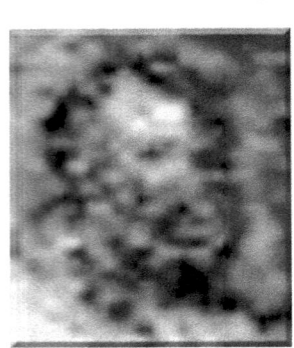

Intentional Blank Page

Video-Loop ITC Image collected by Tom and Lisa Butler.

This is the image I use for my avatar on the Internet. You should be able to see a man's head facing toward your right shoulder. The skin around his right temple and eye is brightly lit. I picked him because he has a full beard and presents as a scholarly person. He is wearing a ruff (pleated neck band) which indicates he lived in the sixteenth century.

Using a picture of myself seems to move the focus from what I do to how I look. While I think it is important that we always use our real name, it is also important to me that my writing take center stage.

Well, … it is also an interesting test. I have literally never received a comment or qurestion about my avatar. Do people only see a blob of light and dark and feel sorry for me? Do they realize they are possibly looking at a paranormally produced image of a long-dead person? Seems not.

Index

1 Timothy 2:11 332
1 Timothy 2:12 332
1932 Nobel Prize in Chemistry 190
4Cell EVP Demonstrations 280
A Priori 167
Academia.edu 209, 242
Academic-Layperson Partition 76, 134, 141, 191, 207, 216, 224, 245, 251
Academy for Spiritual and Consciousness Studies 84
Acupuncture 167
African Genesis 50, 332
Ageless Wisdom of Sacred Tarot 304
Alchemy 298
Alexander MacRae 198
Alien Abduction 105
American Association of Electronic Voice Phenomena 84
American Society for Standards in Mediumship & Psychical Investigation 331
Ancient Wisdom Schools 5, 256, 332
Angels 264, 278
Ankh 288
Anne Wipf iv
Anomalistic Psychology 3, 86, 133, 233, 246, 227, 233, 251
Anticipation Corollary (First Sight) 59
Apple Man 23
Aramaic 41
Aramaic Language 296
Ardrey 332
Arian Philosophy 315
Association of Electronic Voice Phenomena 143
Association TransCommunication 75, 142, 168, 198, 345
Astral Projection 345
Astrology 69, 298
Astronomy 345
ATransC Idea Exchange 153, 246
ATransC NewsJournal 198
Attention Complex 72

Avatar 14, 38, 229, 235, 257, 267, 272, 279, 316, 321, 351
Avyakta 314, 317
Axis Mundi 273
Ba - Spirit 287
Belmont Report 81, 213, 250
Bernard Gert 78
Bernie Sanders 68
Bible 279, 298, 315, 325, 332, 336, 339
Bidirectionality Corollary (First Sight). 183
Big Bang 320
Black Box Analysis 13, 43, 256
Blind Men and an Elephant 134, 148, 251
Body Mind 72, 279
Body-Centric 235, 244, 257, 293, 345, 351
Book Burning Mentality 163
Braden 280
Brahma 279
Brahman 313
Brain Stem 26
Brian Cox 251
Buck Fever 233
Buddha 38
Buddhism 37, 132, 310
Builders of the Adytum 4, 33, 304
Cabala 121, 293, 299, 304
Captain Kirk 50
Carl Jung 109, 325, 336
Carl Sagan 199
Carlos Alvarado 209
Case (Tarot) 33
Censorship 169
Chakras 292
Chariot (Self) 37, 298, 313, 316
Charlatans 167
Charles R. Kelley 169
Chiropractic 166
Chorus 53, 338
Christianity 311, 340
Christopher Publishing House 271
Circular Referencing 168
Citizen Scientists 84, 134, 238, 241

Citizendium 196, 201
Class A EVP 184, 188
Coast-To-Coast AM 220
Cognitive Psychology 239
Collective 147, 293, 320, 337
Coloring 105, 269
Communion
   A True Story by Whitley Strieber 106
Complementary Health 166
Complementary Medicine 170
Continental Drift 187
Cooperative Communities 60, 227, 337
Creative Process 13, 109, 256, 291, 295
Cuneiform Script 289
Cycle of Life and Death 315
Dark Night of Soul 302, 310
Darshan 309, 348
Darwinian 261
David Merrill 69
David Ropeik 34
Declaration of Principles 29, 58, 77
Deepak Chopra 92, 93, 319, 333
Defamation 214
Delphi University 4
Delusion 136
Detachment 121
Determinism 10
Deva 112, 264, 278
Devil 279
Dictionary.com 321
Direct Voice 239
Divine Pymander 29, 299
Dualist 3, 88, 244, 229, 251
Earth Plane 63
Eckankar 4, 309, 332
Edward Malkowski 287
Egypt 41, 60, 223, 286, 296, 311, 315
Egyptian Faience 289
Electromagnetism 235
Electronic Voice Phenomena ... 5, 49, 178, 225, 309, 345
Elementals 279
Elves 345
Emerald Tablet 41, 60, 109, 260, 289, 339
Emergent Science 187
Esoteric Schools 205

Ethereal Beings 265
Etheric Studies 88, 253, 341
Ethics 80, 270
Ethics Panel 129
Evil 157
EVPmaker 138
Exceptional Experience Psychology 87, 133, 183, 228, 246
Facebook 68
Far Journeys 63
Federal Government 182, 243
FEG Medium 209, 214, 221, 241, 338
Felix Experimental Group 208, 209, 221
Ficinus 289
File Drawer Effect 181, 183, 239
First Sight Theory 5, 14, 42, 57, 59, 92, 101, 105, 119, 123
Food and Drug Administration 169
Forebrain 26
Forever Family Foundation 217
Foundation Instinct 50
Founder's Syndrome 83
Fractal 23, 260
Friedrich Jürgenson 193
Fringe 195, 243
Frontier Science 187
Frontier Subject 195
Gautama 312
Gerhard Mayer 90
Gertrude Schmeidler 92, 155
Gestalt 15, 17, 19
Ghost Box 139
Ghost Hunting 84, 133, 170, 192, 226
Ghosts 345
Glossary of Terms 8, 248
God of Death 346
God's will 10
God-Realization 293
Golden Rule 336, 338
Great Work 60, 121, 295, 304
Greek 288, 296
Greg Koukl 166
Guides 265
Gupta 314
Halloween cobweb 215
Handbook of Metaphysics 75, 271, 285, 306

Hans Bender 41, 219, 338
Hebrew 299, 304
Hemi Sync 353
Hermes. 5, 29, 33, 41, 109, 223, 260, 285, 286, 298, 311, 315, 339
Hermetic 58
Hermetic Law of Correspondence 299
Hermetic Principles 31
Hermetic Teaching .. 41, 60, 256, 286, 315
Hermetic Wisdom 9, 297, 339
Hermetic Wisdom Schools 205
Hermetica 289, 303
Hillary Clinton 68
Hindbrain 68
Hindu Philosophy 315
Hindu Vedic Sanskrit 310, 346
Hinduism 37, 56, 66, 132, 279, 311
Hindus Valley 315
Hiranyagarbha 314
Hobbyists 134
Holography 235
Horses (Chariot) 37, 315, 246, 313
Hubble Telescope 185
Huffington Post 92, 319
Hyperlucidity 32, 136, 145, 232, 354
Hypothesis of Formative Causation 13, 18, 24, 65, 69, 105, 116, 154, 260, 335
I'm OK - You're OK 102, 332, 338
Ibis 287
Ideoplastic 115, 272
If I Were Brave 343, 347
Immutable Laws 31
Implicit Cosmology 136, 254, 259, 315, 331
Inadvertency and Frustration Corollary (First Sight) 123
Inattentional Blindness 336
Incarnation 66
Incredulity Blindness 93, 155, 336
India 310
Indus Valley 288, 311
Initiations 205
Instrumental TransCommunication 40, 86, 120, 132, 183, 198, 203, 220, 225, 238, 254, 280, 363
Intended Order 9, 21, 31, 41, 109, 138, 147, 186, 285

Intention Channel 107
International Skeptics Forum 171
Internet Sacred Text Archive 296
Irving Langmuir 190
James Carpenter ... 5, 15, 42, 92, 101, 119, 123
James Randi 197
James Randi Foundation 88
Jana Stanfield 343, 347
Jane Roberts 41, 47, 75, 80, 278
Jerusalem Syndrome 125
Jesus 41, 122, 279, 296, 298, 320, 339
Jewish Mysticism 299
John 14 .... 298, 315, 320, 339, 366
John Everard 289
Joshua Gert 78
Journal of Scientific Exploration 181, 209, 221
Journal of the Society for Psychical Research 90
Judge (Inner) 47, 126, 156, 260, 269, 317, 351
Judgment 280
Junk Science 190
Ka - Soul 287
Kabbalah 299
Kai Mügge 41, 221
Kal Niranjan 279
Karma 131
Katha Upanishad 27, 37, 132, 285, 298, 310
Keith Clark 198
Key 0, The Fool (Tarot) 305
Key 1, The Magician (Tarot) 306
Key 15, The Devil (Tarot) 306
Key 8, Strength (Tarot) 306
Kircher 289
Kundalini 292
Lamarckian 261
Large Hadron Collider 185
Larry Sanger 196
Law of Detachment 333
Law of Silence 205, 331, 332
Liminality Corollary (First Sight) 123
Liminal 123
Lisa .... vi, 4, 42, 140, 142, 217, 231, 309, 321, 345

Lord of the Flies 153, 201
Lucidity.. v, 6, 21, 32, 39, 57, 73, 102, 109, 123, 145, 232, 267, 343, 354
Ludwigs2 167
Luke 41
Magic 298
Magical Thinking 125
Magnetic Field Detector 240
Mahat 314, 317
Major Arcana 5, 304
Mandelbrot Set 23
Mark Leary 93
Max Vorworm 272
Mediumistic Practitioners 134
Mercury 288
Metaphysical Archaeology 285
Michael Nahm 209, 221
Middle East 288, 311
Minor Arcana 304
Monroe Institute 4, 322, 353
Morality 80
Morphic Field 260
Morphogenesis 10, 24, 56, 65
Motto 165
Mount Olympus 256
Myers Memorial Medal 210, 212
Nachiketa 312
National Spiritualist Association of Churches 28, 57, 77, 137
Native American Shaman 153
Natural Law .. 27, 29, 58, 75, 78, 122, 255, 256, 263, 265, 296
Naturalist 241
Nature Spirit 112, 264, 278
Nature's Habit 10, 24, 26, 56, 144, 154, 335
Nature-Nurture 10
Near-Death Experience 84, 212, 240, 267, 67, 282
Neter 288
New Age .. 5, 11, 28, 35, 50, 131, 180, 206, 279, 354
NewsJournals 185, 345
Nexus Personality 52, 59
Nonlocality 13, 186
Normalist 3, 87, 227, 238, 251
Normalize 122, 138, 296, 300
Ohm's Law 22
Polytheistic 297
Om 315
Omnipotence 52
Omnipresence 52
Omniscience 52
One Power 109, 291
One Thing 109, 223, 260, 295, 339
Opportunistic EVP 138
Organizing Principles 28, 30, 58, 256, 263, 296
Orgone Energy 169
Original Research 195
Ouija Board 240
Out-of-Body Experience 267
Pakistan 288, 311
Panoli 346
Paranormal Arbitration 175
Paranormal Review 210
Paranormalists 83, 132
Parapsychological Association 83, 211
Paris Syndrome 125
Pathological Science 190
Paul Foster Case 304
Paul Twitchell 279, 309, 332, 348
Peer Review 135, 248
Perceptual Agreement 31, 110, 116, 147, 155, 264, 281, 301, 335
Perceptual Loop 72, 100, 336
Personal Code of Ethics 81, 144, 250, 270, 349
Personal Styles 69, 97, 154
Personalness Corollary (First Sight) 15
Pet theories 137
Peter Mulacz 209
Phantom Lights 239
Phantom Voices Syndrome 159
Physical Energy 235
Physical Universe Hypothesis 86, 227
Pinnacle Instinct 50
Planchette 240
Pledge 143
Poisoned Atmosphere in Wikipedia ... 197
Practitioner Advocacy Panel 249
Presentiment 234
Prime Imperative .. 62, 132, 256, 259, 321, 327, 337

Index

Principle of Continuity 290, 301
Principle of Perceptual Agreement 14
Principle of Suspended Judgment 333
Profiling 270
Pseudo-Researchers 217
Pseudoscience 161, 168, 242
Pseudo-Scientists 217
Pseudoskeptic 165
Psi Encyclopedia 194, 212, 221, 224
Psi Field 86, 179
Psi Functioning 179
Psi+ Normalist 87, 228, 238, 244
Psychic 86
Psychokinesis 86
Purusha 314, 317
Quackery 171
Quacks 166
Quantum Entanglement 186
Quantum Mechanics 235
Quantum Mysticism 186
Quantum Principles 290
Radio-Sweep 138, 159, 181
Random Event Generator 179
Razor's edge 343
Research Ethics 213
Retro Familiar 32
Rhine Research Center 83
Rider-Waite (Tarot) 304
Robert A. Charman 170
Robert Ardrey 50
Robert Carroll 197
Robert Monroe 63
Roger Reid 69
Rosicrucian 303
Roy Stemman 88, 168
Rupert Sheldrake 5, 10, 24, 102, 116, 144, 171, 175, 177, 202, 215, 243, 260, 335
Sacramental Delta 321
Sacred Fire Sacrifice 312
Sagan 166
Samsara 313
Sanskrit 288, 311, 346
Santa Clause 345
Sarah Estep 143, 309
Scientific Punditry 168
Scientism 89, 166, 172, 175, 177

Second Amendment 340
Selective Reporting 239
Selfless Gene 321
Seth 41, 47, 75, 80, 277, 278
Seth Material 49, 75
Shanti Mantra 311
Sheep-Goat Effect 42, 92, 155
Shiva 279
Silver Bullet Syndrome 138
Skeptic's Boot 251
Skeptical About Skeptics 173
Skepticism
    The New Religion 88
Social Engineering 270
Social Media 152
Society for Psychical Research 83, 194, 210, 221, 224
Society for Scientific Exploration 209, 211
Sonia Rinaldi 198
Spirit Box 139
Spiritual Instincts 11, 26, 46, 67, 114, 131, 343
Spiritualism 4, 28, 88
Spiritualismlink.com 216
Spiritualist 50, 58, 105, 137, 255
Spiritualist Society 126, 219
Star Trek 50
Status Quo 166
Stendhal Syndrome 125
Stephen Braude 209, 221
Steve Crow 126
Stochastic Amplification 186, 235, 240
Storytelling 106
Subject Matter Expert 196
Sumerian 289
Sun God Ra 288
Sunday School 345
Super-Psi Hypothesis 86, 87, 228
Supreme Spirit 313
Supreme State of Vishnu 316
Survival Hypothesis 86, 87, 229
Swami Krishnananda 311, 346
Switching Corollary (First Sight) 101
Tapatraya 312
Tarot 5, 33, 299, 304
Temperament 155

Terminal Lucidity 73
Territorial Imperative 50
Terry Dulin 280
The Kybalion 29, 299
The One Thing 41, 60, 109, 223, 260, 291, 299, 339
The Otherside Press Magazine 331
The Prime Imperative 258
The Razor's Edge 310
The Silva Method 4
The Tower (Tarot) 33
The Wizard of Oz 303
Theory of Evolution 54
There is No Death and There are No Dead 142
Thomas Harris 102
Thoth 287
Time 234
Tourists AA 63
Transform EVP 138
Transmigration of Souls 25, 56
Transmutation 293
Trans-Survival Hypothesis 315
Trapdoor Defense 250
Traveling Perspective 14
Trice-telling 153
Trimurti 279
Trinity Concept 339
Trismegistus 288, 339
Trojan Horse 163, 173, 224, 246
Troll 178
Two Worlds, One Heart 275
Type A (Kinds of Phenomena) 239
Type B (Kinds of Phenomena) 240
U.S. Department of Health and Human Services 81, 213
United States 34
US Government 161, 168, 178, 192
User:Second Quantization 177
V8 Drink 344
Vedanta Spiritual Library 313
Veil of forgetfulness 267
Vetting 247
Vicki Talbott 280
Villager-Explorer Effect 155, 156
Vishnu 279, 313
Washington School of Medicine 213
Weighting and Signing Corollary (First Sight) 15
Whacko 166
Whitely Strieber 106
Wikipedia 5, 102, 152, 163, 167, 170, 175, 188, 192, 217, 221
Wikiproject
    Rational Skepticism 197
Wilhelm Reich 166, 169, 182, 243
Wilhelm Reich Museum 169, 243
Witness Panel 194
Woo-Woo 166
Yama 312
Yoga 38, 314
Your Immortal Self 253
Zen Buddhist 156, 2066
Zodiac 70